Fuels and Petrochemicals Division

Core Programming Topic at the 2011 AIChE Spring Meeting and 7th Global Congress on Process Safety

Chicago, Illinois, USA
13-17 March 2011

ISBN: 978-1-61782-905-5

Printed from e-media with permission by:

Curran Associates, Inc.
57 Morehouse Lane
Red Hook, NY 12571

Some format issues inherent in the e-media version may also appear in this print version.

Printed by Curran Associates, Inc. (2011)

For permission requests, please contact AIChE
at the address below.

AIChE
3 Park Avenue
New York, NY 10016-5991

Phone: (203) 702-7660
Fax: (203) 775-5177

www.aiche.org

Additional copies of this publication are available from:

Curran Associates, Inc.
57 Morehouse Lane
Red Hook, NY 12571 USA
Phone: 845-758-0400
Fax: 845-758-2634
Email: curran@proceedings.com
Web: www.proceedings.com

TABLE OF CONTENTS

(10a) Welcoming Session Sponsored By Fuels & Petrochemicals Division

This welcoming session sponsored by Fuels & Petrochemicals Division (F&PD) is a new initiative organized by the AIChE's national and F&PD technical programming team as a preamble to the division's sponsored technical sessions. This will be opened by a short introduction by the chair of the division. Besides recognizing leadership in his introductory short presentation, the chair will highlight the key elements of value additions the division brings to its members. The presentation will be concluded with a special request to join the division where professional opportunities are limited only by members' imagination, to positively impact academia and industry and thereby society at large. For details of the division and its many facets of activities and functions
please follow the link www.aiche-fpd.org.

(10b) What Are We Going to Do with All This Natural Gas? - Ramifications of the Dramatic Expansion of Natural Gas Available to the Market Due to North American Shale Gas Production and LNG Liquefaction

Sam Thomas, Chart Industries, Inc., Garfield Heights, OH

The success in Shale and other unconventional formations has dramatically expanded the N. American Natural Gas reserves and promises to also do so in many other parts of the world. New LNG liquefaction capacity in the Middle East and Asia along with additional large scale capacity from Australia and Papua New Guinea coming on-line in the next 5 years opens the opportunity for a global trade in Natural Gas stabilizing historically volatile pricing. How will this important energy and petrochemical feedstock be utilized? The success in improving techniques to allow unconventional gas reserves to be proven and significantly increase the available reserves has not as yet been fully appreciated by markets and policy makers. This presentation reviews the ramifications of natural gas becoming a larger part of the global energy supply for which there are numerous competing interests. Natural Gas holds the ability to significantly reduce carbon emissions by substitution for coal in power generation, it can be used as a transportation fuel either as LNG, CNG or GTL clean diesel increasing energy independence of some of the largest economies in the world simultaneously reducing carbon emissions. It could enable the USA to maintain its position as a leading petrochemical producer if used for ethylene feedstock. How will these competing interests play out and how will governments respond?

See more of this Session: Fuels & Petrochemicals Division Welcome Session and Energy Leadership Presentation
See more of this Group/Topical: Topical 7: 14th Topical on Refinery Processing

(14a) The Energy Challenge: Providing Sustainable Supplies to Meet Growing Demand

F. Emil Jacobs, ExxonMobil Research and Engineering Company, Annandale, NJ

As economies grow and living standards improve, the world faces a dual energy challenge: meeting the growing demand for energy while mitigating emissions and their impact on the environment. ExxonMobil believes that addressing this dual challenge requires an integrated set of solutions which includes increasing the supply of all commercially viable energy sources, reducing emissions, and improving energy efficiency in all sectors of the economy. Technology will be a key enabler of continued progress in all of these areas.

Many experts agree that the immense scale of global energy demand necessitates a continuing role for oil and gas as a substantial portion of the energy supply mix for at least the next several decades. These experts also forecast that the overall energy mix will shift toward a relatively higher level of sustainability through a combination of increased efficiency across the economy, growth in lower-carbon sources of energy, and development and deployment of technologies that help mitigate emissions.

ExxonMobil has developed a number of technologies that contribute to increased supplies of energy, reduced emissions, and higher efficiency in its operations and in consumer use of its products. The presentation will provide an overview of ongoing research and development activity in a broad range of technology areas including algae-based biofuels. It will also highlight some of company's current technologies including lithium-ion battery separator films, lightweight plastics that improve automotive efficiency, and processes that could advance carbon capture and storage.

See more of this Session: Spring Meeting Keynote Address - Sponsored by the AIChE Fuels & Petrochemicals Division
See more of this Group/Topical: Liaison Functions

(19a) Acid Gas Removal Unit: The Challenge of Deep COS Removal

Kumar Sugavanam[1], Volker Giesen[2] and Torsten Katz[2], (1)Gas Treatment Process Technology, BASF Corporation, Houston, TX, (2)BASF SE, Germany

The design of the Acid Gas Removal Unit (AGRU) is dependent on parameters such as feed gas conditions and the desired specifications to be achieved. Whereas the removal of CO_2 and H_2S were key design parameters in the past, increasingly sour gases also now contain considerable amounts of carbonyl sulfide (COS).

COS is one of the potential sulfur containing components in natural gas processing and as such has to be removed for safety and environmental reasons. Due to the chemical and physical properties of the molecule, the removal can be challenging especially when a deep sulfur removal specification is being targeted.

Although various options for removal can be employed, BASF's aMDEA™ technology can offer a significantly deeper COS removal rate for specific applications such as LNG. BASF as one of the leading suppliers of acid gas removal technologies will present on the benefits and the advantages of deep COS removal with aMDEA™ along with relevant case studies to demonstrate the same.

See more of this Session: LNG Feed Gas Treatment
See more of this Group/Topical: Topical 6: 11th Topical Conference on Gas Utilization

(19b) Development of Solvent-Free Process for FLNG Pretreatment Application

Shain Doong, Saadet Ulas Acikgoz, Thomas A. Jasonowicz, Bhargav Sharma and Lubo Zhou, UOP LLC, a Honeywell Company, Des Plaines, IL

Offshore LNG production on a floating platform (FLNG) offers a cost effective way to develop natural gas fields remote from land and pipeline infrastructure. There are over 800 such "stranded" gas fields worldwide with reserve in the range of 0.5~2 trillion cubic feet, which are ideal for FLNG projects. In recent years, interest in developing FLNG solutions to monetize off-shore stranded gas has rapidly grown and FLNG is now considered at the forefront of offshore industry focus. The liquefaction and related offshore technologies have also advanced to a point that FLNG projects have become technically viable.

UOP has several technologies that are well suited for acid gas removal or pretreatment of FLNG. A new FLNG pretreatment system has recently been developed by UOP. The new process uses UOP's Separex™ membrane for bulk removal of CO_2 (as well as water and H_2S), followed by UOP's MolSiv™ adsorption unit for final polishing to reach the product specification of less than 50 ppm CO_2. This new process not only eliminates the amine solvent unit, but it also employs commercially proven technologies with off-shore experience. It is anticipated that this process will be more robust than other solutions with respect to the sea motion and more flexible to the variation of the feed gas supply. The benefits of this new process will be discussed in this paper.

See more of this Session: LNG Feed Gas Treatment
See more of this Group/Topical: Topical 6: 11th Topical Conference on Gas Utilization

(19c) Successful Start-up of a DGA System

Mahdi Nouri, Bechtel Oil, Gas & Chemicals Inc., Houston, TX

Aminoethoxyethanol (AEE, DGA®) is one of the common solvents in a gas processing plant for capture of CO2 to avoid carbon dioxide freeze-up at cryogenic temperature levels. The principal considerations affecting the performance of the plant when starting-up the gas treating unit are the product specifications, amine concentration, lean and rich amine loading (acid gas mole/amine mole), the number of absorption and regeneration stages, and equipment sizes.

The purpose of this paper is to review and illustrate the effect on plant performance of implementing key operating parameters, tuning controllers, and training of operating personnel for meeting the treated gas specifications, minimizing start-up time, number of show-downs, and capacity enhancement. This paper will address stabled operation in a gas treating unit using DGA aqueous solution by presenting the operating data and amine quality during start up period.

See more of this Session: LNG Feed Gas Treatment
See more of this Group/Topical: Topical 6: 11th Topical Conference on Gas Utilization

CO_2 Tolerant N_2 Rejection from Natural Gas

Heinz C. Bauer (speaker), heinz.bauer@linde-le.com
Claudia M. Gollwitzer, claudia.gollwitzer@linde-le.com
Martin Gwinner, martin.gwinner@linde-le.com
Linde Engineering, 82049 Pullach, Dr-Carl-von-Linde Str 6-14, Germany

AIChE Spring Meeting, March 2011
11th Topical Conference on Natural Gas Utilization
Chicago, IL, March 13-17, 2011

Abstract

Associated gas from the oil industry has been flared for decades. The prevailing attitude towards CO_2 emissions and other pollutants requires smart approaches to processing low BTU gases in a cost effective and environmentally responsible manner.

The injection of nitrogen into oil reservoirs for Enhanced Oil Recovery (EOR) represents a new challenge for cryogenic nitrogen rejection technology. Over the duration of the production, the nitrogen concentration in the associated gas will steadily increase over time from a few percent up to 30 vol % or more of the entire gas flow.

A new approach for CO_2 tolerant cryogenic nitrogen rejection from sales gas with varying nitrogen content applicable especially downstream NGL extraction will be discussed.

Introduction

Natural gas streams, or fractions thereof, typically have to comply with specifications, if they are to be used for standard applications. A summary of typical pipeline gas specification are shown in Table 1.

		France	Italy	UK	Canada	USA	Japan
H_2S	mg/Nm³	7	6.6	5	6	6	1 to 5
total S	mg/Nm³	75	150	50	240	18	8 to 30
CO_2	vol %	3	3	2	2	3	n/a
O_2	vol %	n/a	0.6	n/a	0.4	n/a	n/a
H_2O dew point	deg C	n/a	-5 at 70 bar	-10 at any pressure	4 lbs/MMSCF	4 lbs/MMSCF	-10 at 80 bar
HC dew point	deg C	n/a	0 at 1 to 70 bar	-2 at 1 to 70 bar	-10 up to 55 bar	-10 at op. pressure	-1 at 1 to 80 bar
gross calorific value	BTU/SCF	990-1,160	885-1,145	>1,065	>995	>1,065	>1,090

Table 1 Typical Sales Gas Specifications[1]

In many countries or pipeline networks N_2 is not limited as a pure component. Rather, the nitrogen concentration is addressed via the gross calorific value or is lumped together with CO_2. US pipeline operators limit[2] the total concentration of inerts (N_2 and CO_2) to max 4 vol % with an upper threshold of max 3 vol % N_2. Pure methane has a gross calorific value of 1,009 BTU/SCF. Therefore, it is not possible to significantly dilute lean natural gases with N_2 before reaching the lower level for the heating value.

Nitrogen-rich gas, which does not meet sales gas specifications without processing can be used either locally for power generation or heating purposes. While boilers are very forgiving for low BTU gases with a high N_2 content, only a few low efficiency gas turbines can be operated stably beyond 30 vol % N_2 in the fuel gas. Gases with N_2 concentrations in excess of 30 vol % N_2 can be expected from the end-flash systems of LNG plants when the natural gas feed contains more than 3 to 4 vol % N_2.

The production of gas from reservoirs containing excessive N_2 may be deliberately postponed, if production is not economically feasible. However, if a N_2 rich gas results as associated gas or end-flash gas with other valuable products like crude oil or LNG, there is a strong economic incentive to upgrade these low BTU gases either to sales gas or to gas turbine fuel. Permanent flaring of such undesired by-products has been banned and shall be avoided in the future.

Process Alternatives

Excess N_2 is removed in Nitrogen Rejection Units (NRU). The rejected N_2 typically has a concentration of 1,000 vppm CH_4. As this stream is vented to the atmosphere local authorities may allow a maximum CH_4 concentration of 10,000 vppm or even as little as 100 vppm. Bulk removal of N_2 with high recovery rates and stringent emission limits tends to be complex and expensive. Thus, different techniques have evolved for various plant sizes.

- PSA systems (e.g. Molecular Gate® by BASF)
- Membranes (e.g. MTR[3])
- Lean oil absorption (e.g. AET[4])
- Cryogenic processes

Non-cryogenic processes, such as the first three entries, are typically limited to applications of less than 50 MMSCFD feed gas. Non-cryogenic treatment of greater amounts of feed gas requires higher investment and operating cost (due to recycle compression). Cryogenic treatment of greater amounts of feed gas is proportionally lower.

Pre-treatment for Cryogenic Processing

Cryogenic nitrogen rejection requires the removal of water upstream of the low temperature section to avoid corrosion and freezing problems. Molecular sieve bed driers are readily available for this application in all required sizes and with favorable economics.

Another prominent feed gas contaminant is CO_2, which may be present in well streams in concentrations of less than one percent up to more than 50 vol%[5], in pipeline gases after NGL extraction with a limited CO_2 content and in end-flash gases of LNG plants with very low CO_2 concentrations of a few vppm only. Rejection of CO_2 with amines is a well established technology, but still requires substantial capital for investment and operation. Thus, there is a significant commercial incentive to avoid or at least shrink the effort for CO_2 removal.

Lean pipeline gas frequently has been processed in cryogenic plants to recover valuable C_{2plus} components. In these cases the actual CO_2 concentration is even below the pipeline gas specification as per Table 1 to avoid freezing during NGL extraction, which typically takes place at a temperature level of about -120 deg C. Figure 1 illustrates the corresponding tolerable CO_2 concentration range, which is about 0.01 mol/mol (equivalent to 1 mol%).

In a nutshell, pipeline gas downstream NGL extraction has a fairly constant and reliable CO_2 concentration and is a candidate for innovative nitrogen rejection systems, which are tolerant with respect to moderate CO_2 content in the feed stock.

Figure 1 Solubility of CO_2 in cryogenic liquids

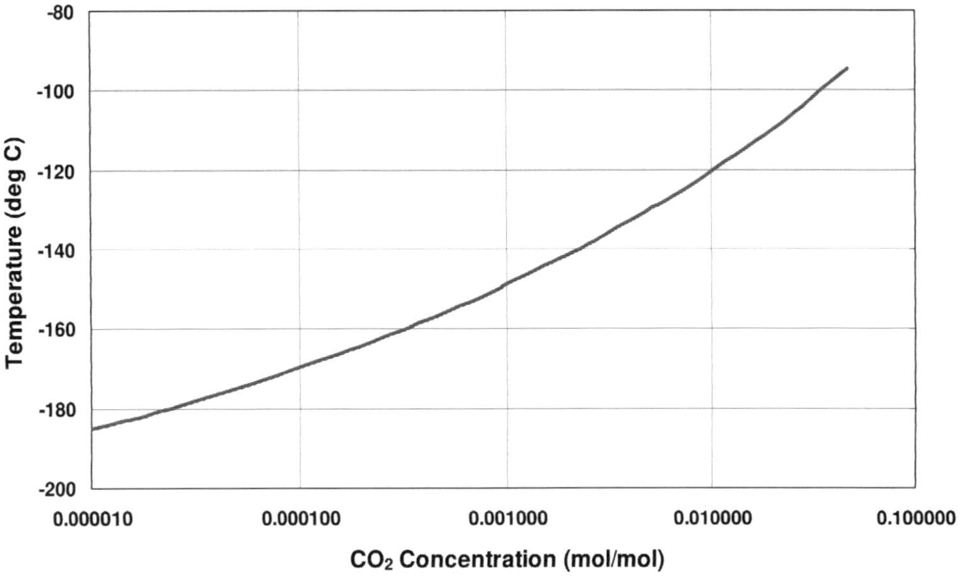

Competing Processing Schemes

Cryogenic N_2 rejection from pipeline gas requires at least one rectification column, which separates light components including N_2 as overhead stream from the hydrocarbon rich bottoms. More complex double column configurations with or without enrichment column will not be discussed, as they are not very suitable for small N_2 feed gas concentrations.

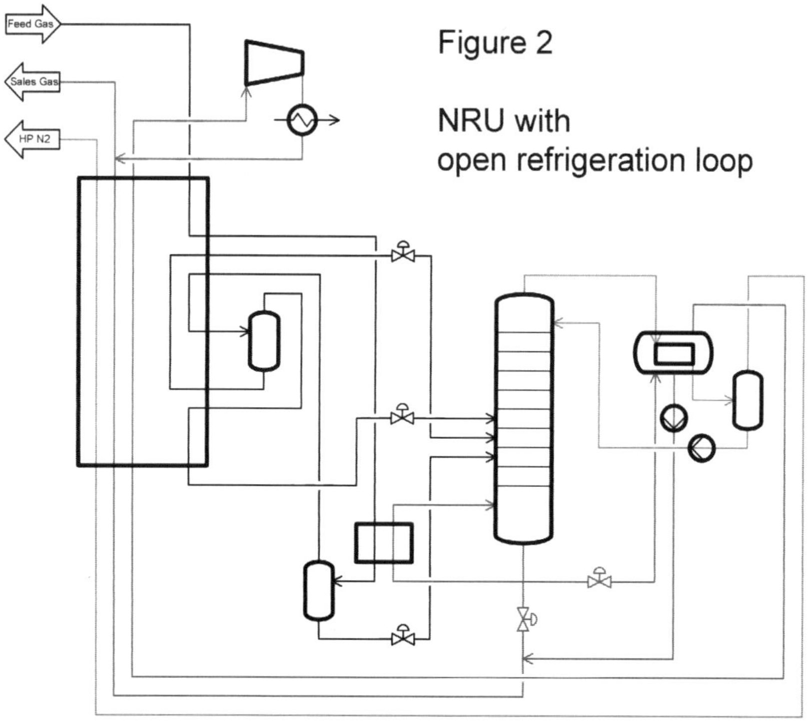

Figure 2

NRU with open refrigeration loop

If the N_2 rich fraction can be used under elevated pressure, e.g. to maintain the reservoir pressure for EOR, it can be given to battery limits at about 25 bar without compression. A simple process sketch for such a concept is shown in Figure 2, where part of the column bottoms are used under low pressure as coolant for the reflux condenser. This arrangement is simple and efficient, as the heat duty of the condenser can be satisfied easily with this once-through operation. However, the bottoms need to be very lean in CO_2, as the condenser is operated in a temperature range of -150 to -160 deg C. Further, the content of C_{2plus} must not exceed a few percent to avoid vacuum suction conditions in the refrigerant compressor.

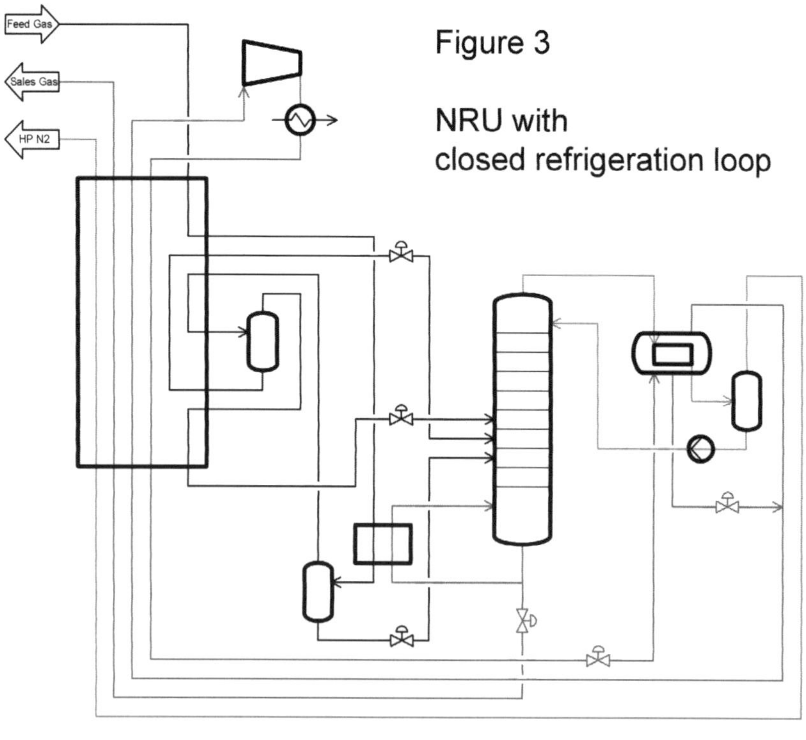

Figure 3

NRU with closed refrigeration loop

These limitations for the refrigerant composition can be avoided, if a closed refrigeration loop is used for the reflux condenser (see Figure 3). The perfect refrigerant would be high purity methane with all its shortcomings concerning availability and storage during idle time of the plant.

Overall, the ideal refrigerant should be readily available in the plant and should be sufficiently free of unwanted components like CO_2 and C_{2plus} components. Figure 4 illustrates a novel process configuration, where the coolant for the reflux condenser is taken from a side stream of the column, no longer from the bottom stream. To avoid contamination of the side (product) stream by feed streams a partition is installed inside the column, which splits the column in the middle section into two rather independent segments. While both segments are coupled hydraulically, the composition on the divided trays can be kept different. Especially, heavier components including CO_2 and C_{2plus} from the feed lines are pushed to the bottom section without contaminating the product side of the partition. This rectification concept has been commercialized by BASF in 1985[6] and has proven to be reliable and efficient since then. More details about dividing wall columns can be found e.g. under the footnote[7].

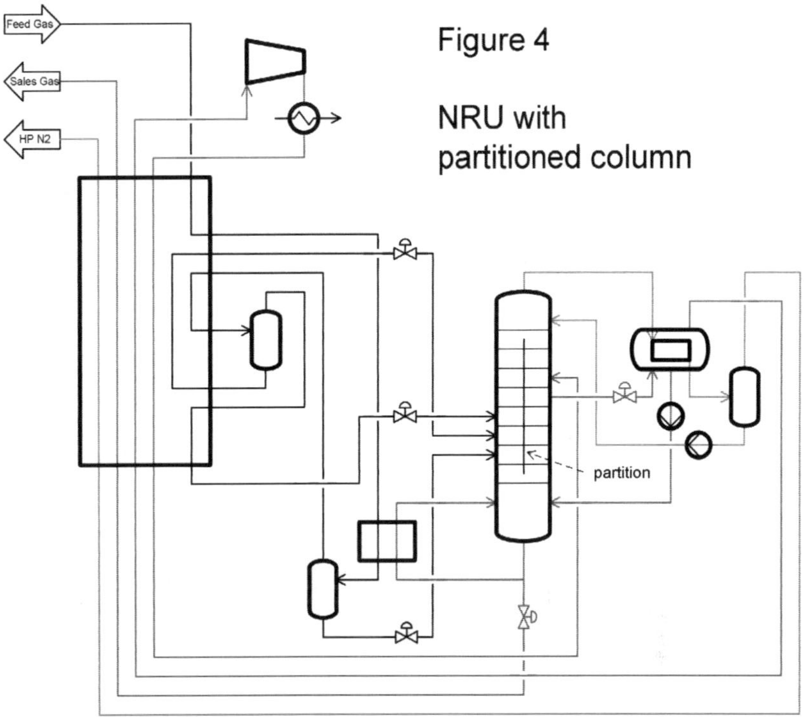

Figure 4

NRU with partitioned column

Quantitative Evaluation of the CO_2 Tolerance

A preliminary assessment of the characteristics of the process as per Figure 4 encouraged our team to start a quantitative evaluation of its performance under different operating conditions. In the context of her thesis one of us (CMG) carried out many computer runs with a powerful process simulator to identify the admissible CO_2 concentration without solid precipitation in any part of the process under variation of

- the feed gas composition,
- the feed gas pressure, and
- the N_2 concentration in the feed gas.

Figure 5 shows the main results of this extensive work. In the lower part of Figure 5 red symbols mark the highest CO_2 concentration, under which a conventional plant as per Figure 2 can be operated. The different symbols refer to different feed gas compositions and pressures. All entries are in the range of 10 to 100 vppm and indicate clearly that a full fledged amine system for the entire feed stream is required to avoid operational problems in the cold part of the process due to freezing CO_2.

Figure 5 CO_2 Concentration Thresholds for Efficient Operation

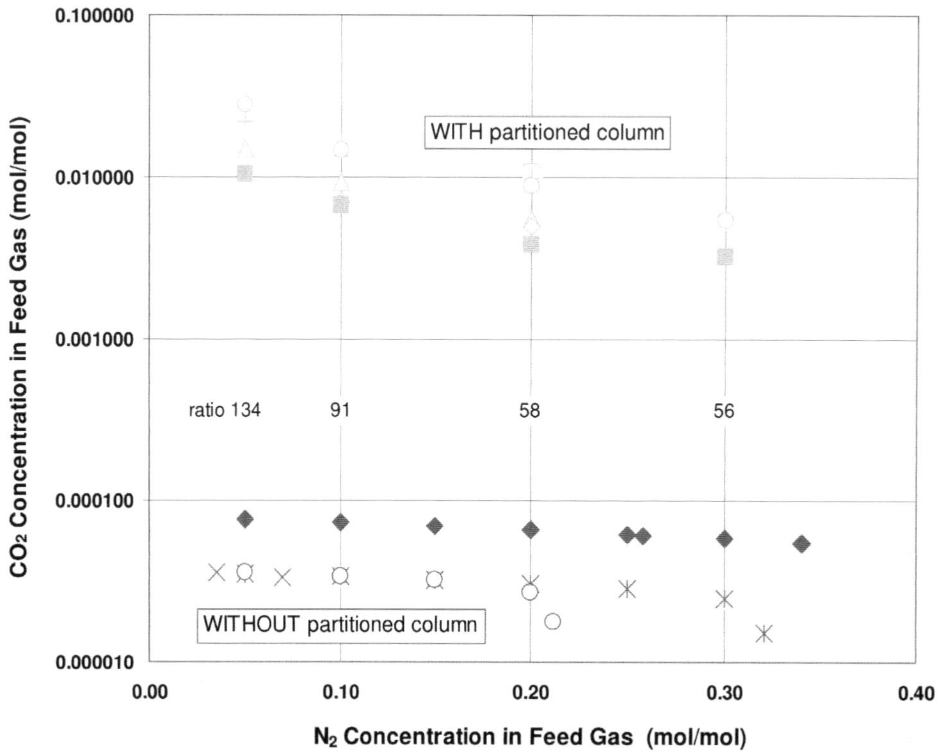

The investigations of the process according to Figure 4 unveiled a significantly different behavior under variable conditions.

For a given set of parameters the power consumption for recycle and sales gas compression (sales gas compressor not shown in Figures 2 to 4) back to feed gas pressure was constant for small CO_2 concentration and increased, when the CO_2 content in the feed stream was raised. This behavior is shown for selected parameters in Figure 6. All simulation results could be regressed fairly well by the simple correlation

Shaft Power (CO_2 conc=x) equals $Ax^2 + Bx$ + Shaft Power (CO_2 conc=0) (1)

starting from a CO_2 concentration equivalent to x=-B/A, which is one of the roots of equation (1). The thermodynamic interpretation of these findings is such that the process according to Figure 4 can be operated without risk of solid CO_2 precipitation up to a CO_2 concentration of -B/A. From then on some adjustments of the process parameters like feed temperature to the column etc need to be made to keep the process operable. These adjustments, however, increase the energy consumption up to a certain value, from where on steady operation is no longer feasible.

The numerical analysis of the primary results provided reliable figures for the parameter -B/A (intersection of the horizontal and the parabolic curves in Figure 6, marked with a red circle), which is understood as concentration threshold for efficient operation of the process with a partitioned N_2/CH_4 fractionator. The values for these thresholds are shown in the upper part of Figure 5 with green symbols. The tolerable CO_2 concentration of the novel process is clearly superior to that one of the conventional process. A value for the CO_2 concentration ratio between the two processes is given, which is a worst case evaluation as the most favorable operating conditions for the conventional process are compared to the most challenging operating conditions for the novel process.

Figure 6 Influence of CO_2 Concentration on Power Consumption

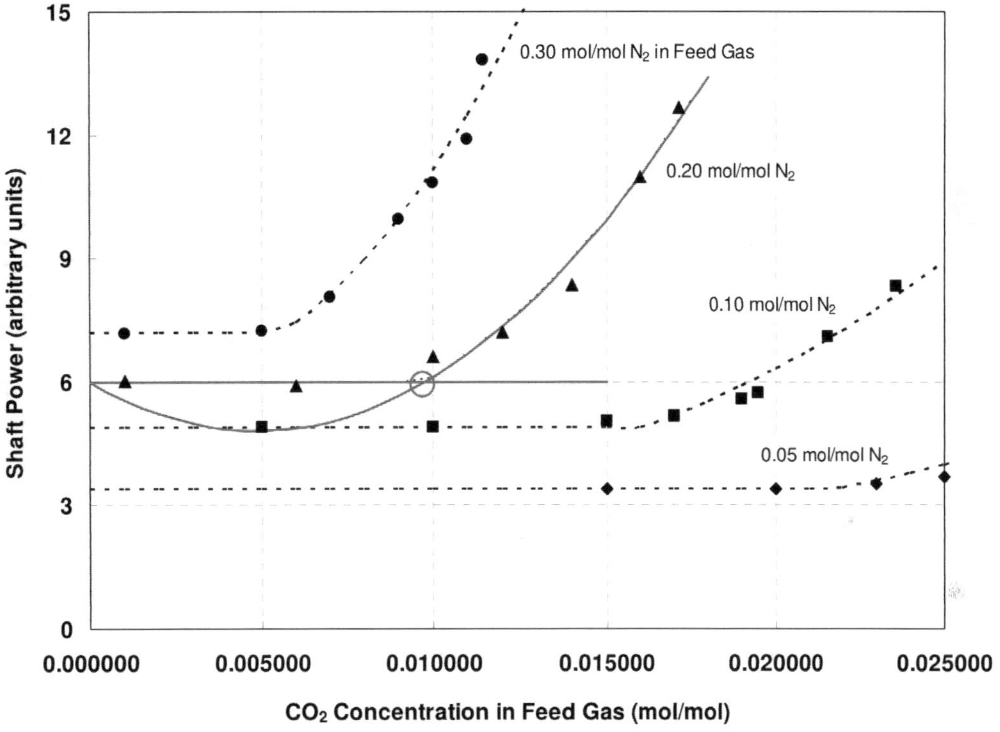

Conclusions

It has been shown that a process with a dividing wall in the main N_2/CH_4 fractionator increases the tolerance versus CO_2 precipitation by a factor of about 100. Depending on the actual conditions in a real project, CO_2 removal upstream of the N_2 rejection system can by either completely avoided or at least limited to a part stream of the feed stock. Significant commercial advantages can be expected, as the additional investment cost for the partitioned column is small in comparison to an amine system.

The novel process is especially suitable for on-spec sales gas recovery in a gas loop, where nitrogen is used for enhanced oil recovery.

Biographies

After graduation at the Technical University Munich as PhD in Applied Physics, Dr Heinz Bauer joined Linde Engineering 30 years ago. He worked in different positions in process design as well as sales and marketing. His focus has been on all kind of cryogenic gas processing including cracked gases, synthesis gas, refinery off-gases and of course natural gas. Presently he is responsible as Vice President for technology of LNG & gas processing plants.

After her research visit at the University of California, Berkeley, Claudia Gollwitzer extended her stay in the US to collaborate with the process and plant engineering team of Linde Process Plant Inc, Tulsa, OK. As a graduate of the Technical University of Munich, she writes her diploma thesis at Linde Engineering in the field of LNG and gas processing. In her thesis she studies the NRU-Process concerning CO_2-tolerance by means of a powerful in house optimization tool.

After graduation at the Technical University Berlin in Environmental Process Engineering, Martin Gwinner joined Linde Engineering 20 years ago. He worked as a process engineer for various plant types including flue gas purification, sulphur recovery and natural gas processing. In the last 15 years he focused on all kind of cryogenic natural gas treatment processes, especially LNG production and Nitrogen Removal.

References

[1] PERP Program – Gas Processing and NGL Extraction: Gas Conditioning, New Report Alert, Nexant Chem Systems, May 2006

[2] Interstate Natural Gas Quality, Specifications & Interchangeability, Center for Energy Economics, December 2004

[3] http://www.mtrinc.com/nitrogen_removal.html

[4] http://www.aet.com/nreject.htm

[5] http://www.offshore-technology.com/projects/natuna/

[6] http://www.basf.com/group/corporate/en/innovations/research-verbund/facts-and-figures/milestones/index

[7] http://www.montz.de/sites/products/dividing.fr.html

MERCURY REMOVAL FROM GASEOUS AND LIQUID HYDROCARBONS

Neil Eckersley, UOP LLC, Des Plaines, IL, USA

A Honeywell Company

Introduction

Mercury is a naturally occurring element found in small but measurable concentrations in many oil and gas fields throughout the world. Mercury is most frequently detected in its elemental form and tends to be prevalent in natural gas processing and LNG (Liquefied Natural Gas) facilities. Due to advances in detection systems, mercury can now be accurately measured down to single digit nanogram levels in the case of gases and to ppbwt levels in the case of liquid hydrocarbons. The removal and capture of mercury is important for a number of reasons:

- Process plants with brazed aluminium heat exchangers are susceptible to corrosive attack by mercury
- Product streams eg. Condensates, LPG, NGL and naphtha in the case of refineries, are less valuable to producers when "distressed" by mercury
- Many refinery and petrochemical catalysts are poisoned by mercury
- Mercury is toxic and should be removed on health and safety grounds.

In supplying purification solutions to the oil & gas processing industry, UOP has developed a number of approaches for removing mercury using regenerative HgSIV products and non-regenerative metal oxide and metal sulphide technology in both gas and liquid phase streams. This paper focuses on the non-regenerative applications. In recent years, mercury levels have increased from typical highs of 30 or 40 μg/NM^3, to levels exceeding 1,000 μg/NM^3 in the Pacific Rim area. As gas processing plants become more flexible, it becomes possible to process more gas of questionable quality. Mercury-containing gas, natural gas liquids and condensates derived from gas fields of variable quality, require treatment to a rigorous standard in order to protect gas plant assets, pipelines and complex processing equipment.

Gas processors require safe, reliable and effective mercury removal process solutions in order to achieve requisite sales gas quality specifications. In the last decade this has lead to more plant operators relying on different mercury removal unit (MRU) technologies to meet their various contractual obligations. This paper will describe several commercial case histories and discuss the plant specific drivers that lead to the adoption of fixed bed, non-regenerable MRU technology.

The accurate measurement of various mercury species in process streams is key to determining the selection of the most appropriate purification technology. A number of analytical techniques are available to measure mercury down to ppb levels and it is important for process plant operators to consider how mercury is measured in order to remove mercury to ultra low levels. The latest proven analytical methods for the measurement are reviewed and discussed.

Mercury analysis and typical global levels

UOP's Technical Service teams have performed a number of mercury surveys over the years on process streams in the gas and liquid phase, on a diverse range of product streams in Asia Pacific, Europe, North and South America, The Middle East and Africa. Table 1 details mercury levels that UOP analysts have reported globally in the gas phase.

Table 1. Reported elemental mercury levels detected in natural gas

Region	Mercury concentration / µg/NM^3
North Africa	1-100
North America	1-20
South America	1-105
South East Asia	10-2,000
Middle East	1-10
Europe	1-50

Gas phase surveys have been carried out using two different analysis systems:

The Tekran Model 2600 utilising cold vapour atomic fluorescence spectroscopy and the Lumex RA 915+ utilising atomic absorption spectroscopy with Zeeman correction. The Tekran system is typically used to measure mercury in gases and the Lumex system has been used to measure mercury in both gaseous and liquid streams. Accurately measuring mercury in the field presents a complex challenge and requires a significant degree of analytical understanding; including not only sample analysis but also sample collection. Metallic sample collection containers should be avoided when considering liquid samples since mercury will readily plate out on metal surfaces. Sample line conditioning is crucial, since mobile mercury will tend to accumulate on inner wall pipe-work. The presence of different mercury species and other chemical interference can also impact the measurements taken.

Mercury removal using non-regenerable adsorbent technology - metals or activated carbon?

In addition to forming amalgams with several metals, mercury is very mobile and will adsorb onto pipeline surfaces and other equipment commonly found in gas processing plants. Mercury can then desorb back into gas streams, passing through contaminated pipelines and prolonged periods of time can elapse between the installation of an upstream mercury removal unit (MRU) and the complete purging of a pipeline.[1]

MRU products have long been used to take advantage of the affinity that mercury has with certain surfaces. In recent years this has meant that more advanced metal based adsorbents have been developed to remove mercury from a variety of hydrocarbon process streams. These modern metallic based products have tended to replace the older sulphur impregnated activated carbon technologies that have traditionally been specified into Liquified Natural Gas (LNG) plants. Metal based MRU products tend to benefit from a larger operational envelope when compared to carbons which are only effective in treating dry gas, meaning carbon is only optimally effective in processing gas downstream of any dehydration section. The extensive micro-porous nature of activated carbon means that capillary condensation is a problem when operating at or close to the dew point of a gas. [2] Sulphur promoted carbon is also prone to sulphur dissolution when exposed to "wet" gas streams. This leaching of sulphur off the carbon leads to sulphur slip from the MRU, potentially damaging downstream equipment and reducing useful mercury capacity. At the same time, more gas processors are requesting that mercury removal takes place either at the well head or close to the front of the gas plant, where the gas can be "wet". **The thinking here is to prevent the migration of mercury to various locations within a gas plant and avoid any subsequent partitioning into processed natural gas and condensate streams[3].**

The handling of mercury-contaminated pipe-work, classified as hazardous waste, needs to be considered in the context of existing occupational health guidelines. Although mercury has a high boiling point it also has a relatively high vapour pressure. This combined with its inherent toxicity means that extreme caution is required when handling mercury contaminated pipe-work. The American Conference of Governmental Industrial Hygienists (ACGIH) has assigned mercury vapour a threshold limit value (TLV) of 0.025 mg/m^3 of air as a time weighted average (TWA) for a normal 8-hour workday and a 40-hour workweek.[4] This TLV has itself halved since 1991, when the previous assigned level was 0.05 mg/m^3.

When "up-front" mercury removal is required on raw gas, the challenge becomes one of how to treat natural gas that is not necessarily "bone-dry". Apart from the basic requirement to ensure that adequate mercury removal can be achieved on gases at or close to their dew points, from the point of capital expenditure, it is important to ensure that MRU reactor volumes are minimised where possible. This is particularly important when positioning an MRU off-shore where spatial constraints are critical. The contact times required by sulphided carbon products often lead to a larger than practical MRU footprint. The drive to reduce capital budgets and avoid large volumes of spent material has lead to gas processors examining technologies other than sulphided carbon.

The flexibility of metal based MRU technology

UOP has developed a range of non-regenerable adsorbents to improve on existing MRU technologies, to be used on gaseous and liquid streams commonly found in and around natural gas processing facilities. Gas streams containing thousands of micrograms of mercury per normal cubic metre (NM^3) and liquid streams containing thousands of parts per billion by weight (ppbwt) of mercury can be successfully treated using UOP 'GB' adsorbents. Instead of older technology carbon, transition metal oxides and sulphides are utilised. The active component of the GB adsorbent products is a metallic sulphide and the GB products are supplied either in their oxide form and are sulphided in-situ by simple contact with the process gas or liquid to be treated, or are supplied pre-sulphided. UOP GB products have the ability to remove mercury from sulphur-containing or sulphur-free gaseous and liquid hydrocarbons, in a variety of positions within any gas plant. The products have been developed to be compatible with modern mercury recovery routes. Mercury is removed and recovered from the discharged adsorbents via vacuum distillation and sold into specialist applications for re-use. The remaining active metal is sent for recovery via a smelting process and then sold back onto the open market. This process ensures that the MRU product is handled in an environmentally friendly way.[5] In gas phase service, the adsorbents have been developed to treat both wet and dry gases, without the same wet gas limitations of other non-regenerable products. A commonly suggested flow-scheme location for a UOP GB product is shown in **Figure 1**. The UOP "GB" product range is currently successfully treating liquid hydrocarbon streams in addition to gaseous fluids, in many gas plants and 3 case studies will be examined in this paper. Sulphur is anchored to the metallic substrate, preventing subsequent dissolution and slippage onto downstream equipment.

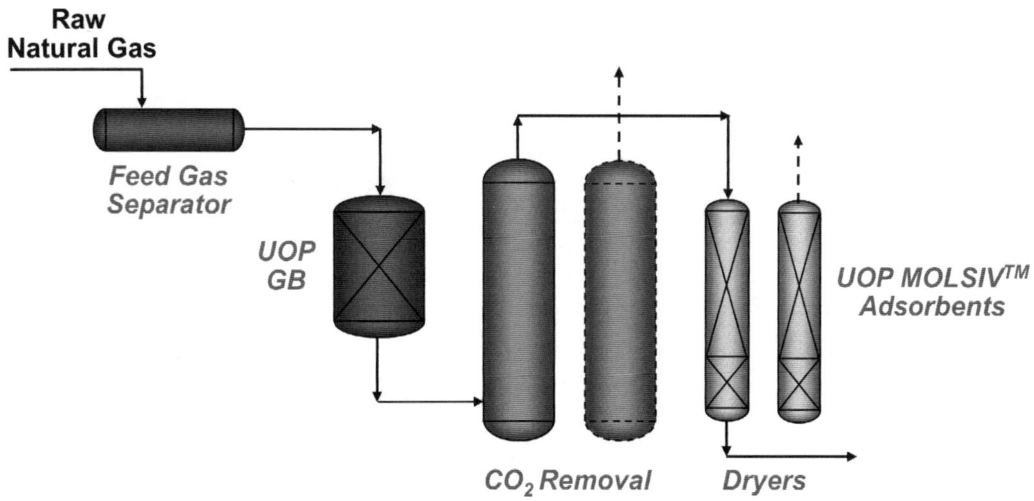

Figure 1

Mercury in gas processing plant liquids

UOP has conducted several studies to better understand the distribution of mercury within gas processing facilities. Mercury does not only enter gas plants with raw natural gas but can also come in through (for example) off-shore condensate streams. Since condensates can represent valuable refinery and petrochemical plant feed-stocks, careful consideration needs to be applied in purifying them to meet downstream process plant equipment and catalyst specifications. **Case study 2** below, discusses how a condensate stream with a high attendant level of mercury (2000 ppbwt) has been successfully treated with UOP technology.

Apart from entering gas processing facilities via condensate streams, mercury can also be transferred from in-coming natural gas to subsequently produced ethane, propane, butane and LPG as liquids are separated from gas and isolated within the gas plant to provide sales streams to downstream domestic and process plant consumers. Petrochemical customers taking gas plant liquids need to be aware that mercury specifications in traded feed-stocks are becoming tighter and financial penalties are becoming more common and punitive as awareness grows in terms of the need to better control the quality of process unit hydrocarbons. **Case study 3** below examines the fate of gaseous ethane prior to being sold by the gas processor to a customer operating an ethylene cracker. The ethane mercury specification as it passes to the cracker is <0.01 µg/NM3 and the influent mercury entering the ethylene plant C2 feed MRU is 10 µg/NM3. Specification is met using a UOP GB product.

Of particular concern to the customers who buy gas processing plant-derived liquids, is the effect of mercury on downstream process catalysts. Precious metal catalysts operated at relatively low process temperatures (<200 °C) are susceptible to mercury poisoning, resulting in deactivation and premature change-out. Even if mercury is driven off these catalysts via heat treatment, sintering of the sensitive catalyst can result with a permanent loss in activity compromising the process economics of the unit in question.

Clearly there are several ways to remove mercury from natural gas processing plants. Capital expenditure limitations, equipment protection requirements, environmental factors and the needs of downstream customers all need to be considered. The following case studies illustrate some of the factors involved in choosing which mercury removal system is most applicable for a given process plant.

Mercury removal case studies

Case 1: PTT Thailand, UOP GB-562

The PTT Gas Separation Plant No. 5 **(GSP-5)** is located in Map Ta Phut, Rayong, Thailand was commissioned and started up in 2004. The on-shore facility processes raw gas via a pipeline from off-shore gas fields in the Gulf of Thailand. The raw gas entering GSP-5 is conditioned to remove CO_2, H_2O and Hg and the total gas flow treated is 530 MMSCFD. The MRU is designed to protect an NGL recovery plant, incorporating a cryogenic unit that includes a brazed aluminium heat exchanger. The initial charge of mercury removal adsorbent used from plant start-up was sulphur impregnated activated carbon. The gas plant MRU configuration consists of two parallel reactors each designed to process 265 MMSCFD of gas. The MRU is located upstream of the amine plant and the dryers and is positioned to treat raw gas as it enters the facility. Historically the gas entering the MRU has contained some liquid hydrocarbon with a TEG carry-over component. The initial charge of activated carbon experienced a premature mercury breakthrough after 2 years in service. Table 2 provides a summary of the MRU process conditions.

Table 2. PTT GSP-5 MRU process conditions

Gas treated	Natural Gas
Gas flow-rate	265 MMSCFD
Operating pressure	48 Kg/cm^2
Operating temperature	18 °C
Hg influent range	50-200 $\mu g/NM^3$
Hg effluent specification	<0.01 $\mu g/NM^3$

Figure 2

After the activated carbon was discharged from the reactors, the MRU was re-filled with UOP GB-562 absorbent. Figure 2 shows a photograph of the GB-562 parallel flow reactors at GSP-5.

For 3 years since the GB-562 was commissioned in March 2008, the plant has continuously recorded effluent mercury levels below the required specification and maintained the start of run pressure drop which was a priority for the customer. This successful performance has been achieved despite the GB-562 treating a liquid entrained, water saturated natural gas.

Case 2: PTT Thailand, UOP GB-346S

The PTT Gas Separation Plant No. 4 **(GSP-4)** is located in Khanom, Surathani, in the south of Thailand, 700 kms from Bangkok. The facility was commissioned and started up in 1997. GSP-4 receives gas via a 24" pipeline from an off-shore platform operated by PTT EP. PTT installed a dew point control unit upstream of the existing gas plant, chilling the feed gas down to below 16°C in order to knock-out liquid condensate. This condensate contains 2000 ppbwt mercury and approximately 200 ppmwt water. UOP GB-346S has been successfully used in an MRU upstream of a dehydration unit and cryogenic unit, since 2009.

The gas plant MRU configuration on the condensate stream consists of one single up-flow reactor designed to process 6,962 kg/hour of liquid. The MRU is located upstream of a UOP dehydration unit utilizing molecular sieve technology. Table 3 provides a summary of the MRU process conditions.

Table 3. PTT GSP-4 MRU process conditions

Liquid treated	Natural Gas Condensate
Liquid flow-rate	6,962 kg/hour
Operating pressure	31 bar G
Operating temperature	12 °C
Hg influent	2000 ppbwt
Hg effluent specification	<1 ppbwt

Figure 3

Figure 4 shows a photograph of the GB-346S single flow reactor at the GSP-4 facility. To ensure on-spec condensate and complete removal of mercury, from start-up, GB-346S is supplied in it's pre-sulphided form and as such reduces effluent Hg levels to <1 ppbw.

The design life of the GB-346S is 4 years and since coming on-stream, the plant has continuously recorded mercury levels below the required specification for PTT's condensate customers. The project payback period to install GB-346S was just 6 months or 13% of the initial charge design lifetime and PTT is now able to supply liquid to more customers.

Case 3: PTT Chemical Thailand, UOP GB-564

The PTT Chemical **I-1 Olefins** plant is located in Map Ta Phut, Amphoe Mueang, Rayong, Thailand. Commissioned and started up in 1989, the facility processes Natural Gas and Ethane from several on-shore PTT Gas plants and provides ethane as a feedstock to the PTT Chemical ethylene cracker. The ethane must first pass through an MRU and then through an Acid Gas Removal Unit (AGRU) to reduce CO_2 levels. The ethylene cracker has a mercury specification on the ethane feed, of <0.01 µg/NM^3 to protect cryogenic equipment. The AGRU has an H_2S specification to avoid high pressure drop and foaming issues with it's mono ethanol amine (MEA) liquid. In 2008, PTT Chemical experienced problems with H_2S slipping to the AGRU, causing solid particles to form, leading to high pressure drop and foaming. The existing sulphur impregnated activated carbon in the MRU could not adequately reduce H_2S levels to the 0.1ppmv level required by the AGRU. UOP was asked to provide a solution that would remove H_2S and Hg, to meet both the AGRU H_2S specification and the ethylene cracker Hg specification.

In January 2009 UOP provided PTT Chemical with GB-564 to replace the activated carbon and in doing so converted what was an MRU into an MRU and SRU (Sulphur removal unit). Table 4 details process conditions for the GB-564 MRU/SRU.

Table 4. PTT Chemical MRU/SRU process conditions

Gas treated	Ethane
Gas flow-rate	73,000 kg/hour
Operating temperature	35-40 °C
Operating pressure	13 Kg/cm^2
Hg Influent	10 µg/NM^3
Hg Effluent spec	<0.01 µg/NM^3
H_2S Influent	2-5 PPMV
H_2S Effluent spec	<0.1 PPMV

Figure 4

After the activated carbon was discharged from the reactor, the MRU was re-filled with UOP GB-564 adsorbent. Figure 4 shows a photograph of the GB-564 reactor at PTT Chemical.

On installing the GB-564, PTT Chemical was able to operate the MRU as an MRU/SRU and co-remove H2S as well as Hg. The GB-564 was designed to operate without any operator input, to meet downstream H_2S specifications for a 12 month period until the upstream source gas plants were able to optimize their own AGRU system. In addition the GB-564 was designed to remove mercury to the desired outlet specification for a period exceeding 10 years.

Discussion

If mercury is allowed to reside in a gas processing plant un-checked, it can cause severe damage to process equipment and potentially compromise the health and safety of plant operators. UOP has developed a technology to remove mercury from both gas and liquid phase streams at various locations within a gas plant. GB products afford maximum mercury removal, complete plant asset protection and the cleanest possible environment for gas processing operational staff. This latest technology of non-regenerative products can offer long lasting adsorbent life cycles, resulting in an overall improvement in plant process economics, minimizing the $ spend per kg of Hg removed. It is important that MRU products can simultaneously remove both hydrogen sulphide and mercury and they can operate as efficiently in the liquid phase as the gas phase, to increase the value of liquid products that are sold on to downstream chemical and petrochemical customers.

Acknowledgements

The author would like to acknowledge the kind cooperation of PTT Public Company Limited, in the writing of this paper and the patience of Fiona, Katie and George.

References

[1] AIChE Paper Jointly authored by UOP and Equistar, Mercury removal from cracked gas a liquid streams, 2004 Ethylene Producers Conference
[2] Biscan, D.A., Gebhard, R.S., Matviya, T.M., Impact of process conditions on mercury removal from natural gas using activated carbon, presented at the 8th International Conference On Liquefied Natural Gas, 1986
[3] B. Edmonds, R.A.S Moorwood and R.Szcepanski., Mercury Partioning In Natural Gases and Condensates, presented at the GPA European Chapter Meeting London, March 1996
[4] 2009 TLVs and BEIs - ACGIH
[5] Private correspondence with Begemann Milieutechniek B.V. Mercury waste reclaimers

(22a) Fundamentals of Alarm Management

Peter G. Herena, Kenexis Consulting Corporation, Elgin, IL

When the information era began thirty years ago, the focus for the refinery control specialist was placing as much data as possible into the hands of engineers and operators. The philosophy remained prevalent though the volume of transferred data leaped by orders of magnitude every decade. Eventually refinery operators began to realize that the sheer volume of data was, in part, causing the problem; and thus the field of Alarm Management was born.

But how does one go through the process of identifying the most critical alarms, and once that is known, what are the methods used to get the most important information in the hands of the correct person? In this presentation the fundamentals of Alarm Management will be discussed, including assessing hazards, identifying data transmission bottlenecks and understanding how our brains are wired to recognize and respond to information. Specific examples will be cited and current good practices will be described.

See more of this Session: Process Control Optimization and Tutorial
See more of this Group/Topical: Topical 7: 14th Topical on Refinery Processing

(22b) Who's Controlling Your Industrial Control System? and What Makes You So Sure? An Information Security Primer for Systems Professionals

Todd Haverkos, Iconium Security, Elgin, IL

In the wake of Stuxnet – a game changing piece of malware specifically targeting air gapped industrial control systems infrastructure – electronic security has rocketed up the worry list for control systems professionals. Todd will discuss the tools in the current arsenal of modern attackers, how they evade detection, and the ecosystem surrounding the underground economy that fuels for-profit malicious software and hackers as well as discussing specifics of Stuxnet that reverse engineers are discovering as they work to decompile and decrypt the intent behind the highly advanced class of threat that Stuxnet represents. Todd will discuss what specifically a control systems professional should be thinking about with respect to information security in their day-to-day work, and how to keep systems from becoming the next victim (or at least detect whether a system is a current victim).

See more of this Session: Process Control Optimization and Tutorial
See more of this Group/Topical: Topical 7: 14th Topical on Refinery Processing

(22c) A Look Into Advanced Controls for Automotive Engines

Aaron Matthews, University of Wisconsin-Madison, Chicago, IL

This presentation features an in-depth look at engine controls for gasoline, diesel, and hybrid drivetrains. The most important control tasks are then identified and their implications on engine performance are analyzed. Examples are given showing driver input and controller output sensitive to environmental conditions. There are also examples of advanced hardware configurations with a brief list of pros/cons for each. The most important control tasks are then identified and their implications on engine performance are analyzed. The presentation also discusses emissions testing procedures for transient and stationary engines.

See more of this Session: Process Control Optimization and Tutorial
See more of this Group/Topical: Topical 7: 14th Topical on Refinery Processing

Importance of Data Reconciliation on Improving Performances of Crude Refinery Preheat Trains

Edward M. Ishiyama[1,2], Simon J. Pugh[1], D. Ian Wilson[2], William R. Paterson[2] and Graham T. Polley[3], (1) Process Engineering Technology, IHS-ESDU, London, United Kingdom, (2) Chemical Engineering and Biotechnology, University of Cambridge, Cambridge, United Kingdom, (3) Chemical Engineering, Universidad de Guanajuato, Guanajuato, Mexico

Abstract

Performance of a preheat train is evaluated through plant monitoring data (temperature and flow measurements); here the reliability of data reconciliation plays a significant role. Most preheat trains do not monitor all stream temperatures (or flow rates) and frequently guess the missing data through 'short-cut' methods. Wrong choice of 'short-cut' methods such as the use of linear interpolation is shown to result in misleading conclusions.

This paper introduces a systematic data reconciliation approach through three steps. 1) Generation of missing data through a modified preheat train simulator based on *Energy Fuels*, 2009, *23*(3), pp 1323–1337). 2) Filtering unreliable data through a 'trusted' heat balance. 3) Grouping heat exchanger monitoring data into different time sections to identify trends under different operating periods.

The superiority of the new data reconciliation methodology is illustrated through a series of case studies. The reconciled data are then utilized to make key operational decisions to improve preheat train performance.

1. Introduction

Preheat trains are networks of heat exchangers, usually complex, with flow splitters and bypass streams, which transfer heat between the distillation tower products and its crude oil feed. These units are prone to fouling. Fouling is the formation of unwanted deposits on the heat transfer surfaces with time. The fouling behaviour is dynamic and is a combination of several factors (*i.e.* a combination of particulate, chemical reaction, crystallization and corrosion fouling in the preheat train). The complex nature of the fouling behaviour was illustrated by Ishiyama *et al.* (2010), where refinery fouling analysis and laboratory experimental reports were summarized (reproduced as Figure 1). Figure 1 is a plot between actual plant heat exchanger fouling rates against wall shear stress. Note that all exchangers in this plot are shell-and-tube units with crude on the tube-side. The solid line is the correlation proposed by Joshi *et al.* (2009) based on plant fouling data collected by Shell Global Solutions. This correlations show that the average rate of fouling decreased with increasing tube-side wall shear and the influence of temperature was not significant to the observed fouling rates. On the same plot are hollow symbols (two data sets) and filled symbols (5 data sets), representing fouling rate observed in pilot plant experiment and 5 other refineries. Each data point present exchangers operated at different temperatures (detailed in Ishiyama *et al.*, 2010), and would raise the concern that the use of shear stress alone is insufficient in evaluating fouling behaviours in certain exchangers in various refineries.

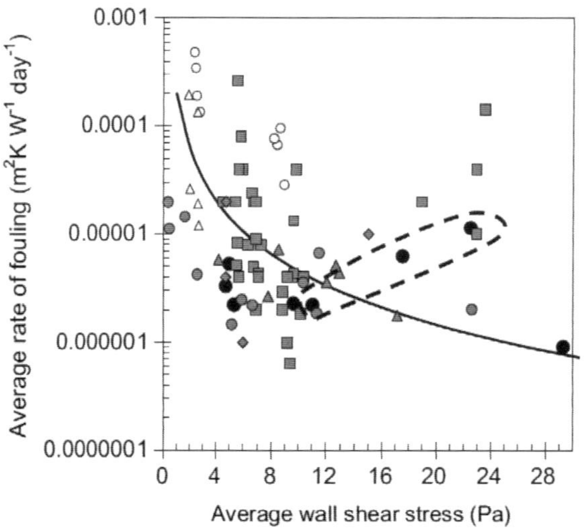

Figure 1: Plot of average fouling rate against estimated wall shear stress. Solid locus: correlation reported by Joshi *et al.* (2009). Solid symbols are the fouling rates obtained from 5 refineries; open triangle and open square are pilot plant fouling data obtained from Wood River tube flow and Exxon tube flow, respectively. (both pilot plant data sets reported by Knudsen *et al.*, 1999). (Reproduced from Ishiyama *et al.*, 2010).

Evaluations of fouling rates are based on the temperature and flow measurements available in the plant, and the accuracy and the reliability of these plants monitoring data reflects the accuracy of the calculated fouling rates. Most of the time, not all stream temperatures are measured and/or the measured stream temperatures are erroneous. In this paper we denote data reconciliation as fitting plant monitoring data to specific models in order to get data that follow mass and energy conservation laws.

The fouling resistance of an exchanger, R_f, is given by,

$$R_f = 1/U - 1/U_{cl} \qquad (1)$$

Here U is the overall heat transfer coefficient and U_{cl} is the overall heat transfer coefficient at the clean state, respectively. Both U and U_{cl} are based on the tube external surface area. Note that U_{cl} is calculated for each operating condition, not the design (clean) value based on the design operating condition.

Exchanger heat duty, Q, is evaluated by

$$Q = UA_o F \,\text{LMTD} \quad (2a)$$

$$= m_c C_{pc} (T_{c,out} - T_{c,in}) \quad (2b)$$

Here, Q, is the heat duty, A_o is the heat transfer area (based on the tube external surface), F is the temperature correction factor and LMTD is the logarithmic mean temperature difference. m is the mass flow rate, C_p is the specific heat capacity and $(T_{c,out} - T_{c,in})$ is the temperature driving force. Subscript 'c' denote crude stream.

With missing temperature or flow measurements it is impossible to evaluate the value of U from equation (1, 2a,b) alone. The main aim of this document is to demonstrate the methodology practiced in the data reconciliation and network simulation software tool (*smart*PM^{TM}) which is being developed at IHS ESDU, UK. The method is based on guessing the fouling resistances of each exchanger; the guessed fouling resistances are used in the network simulator to obtain the reliable plant measurements. The initial guess for the fouling resistance is very important to obtain an accurate data reconciliation; this is based on obtaining reliable temperature guesses for missing/unreliable data sets.

2. Missing temperatures and/or flow measurements – assuming available measurements are reliable

When the plant consists of missing measurements, the missing values are generated through a constructed simulator, which could either be used for a single heat exchanger or a heat exchanger network of any structure. The steps are divided into four parts.

1. Initial mass balance is performed to generate any missing stream mass flow rates.
2. Initial guess for the missing temperature is made.
3. Then these values are used to calculate the fouling resistance guesses.
4. These fouling resistance values are then used as initial values and the temperatures of the streams generated through the simulation. The steps are summarized in Figure 2.

The initial temperature guess is done through linear interpolation of the known temperatures associated with the exchanger group.

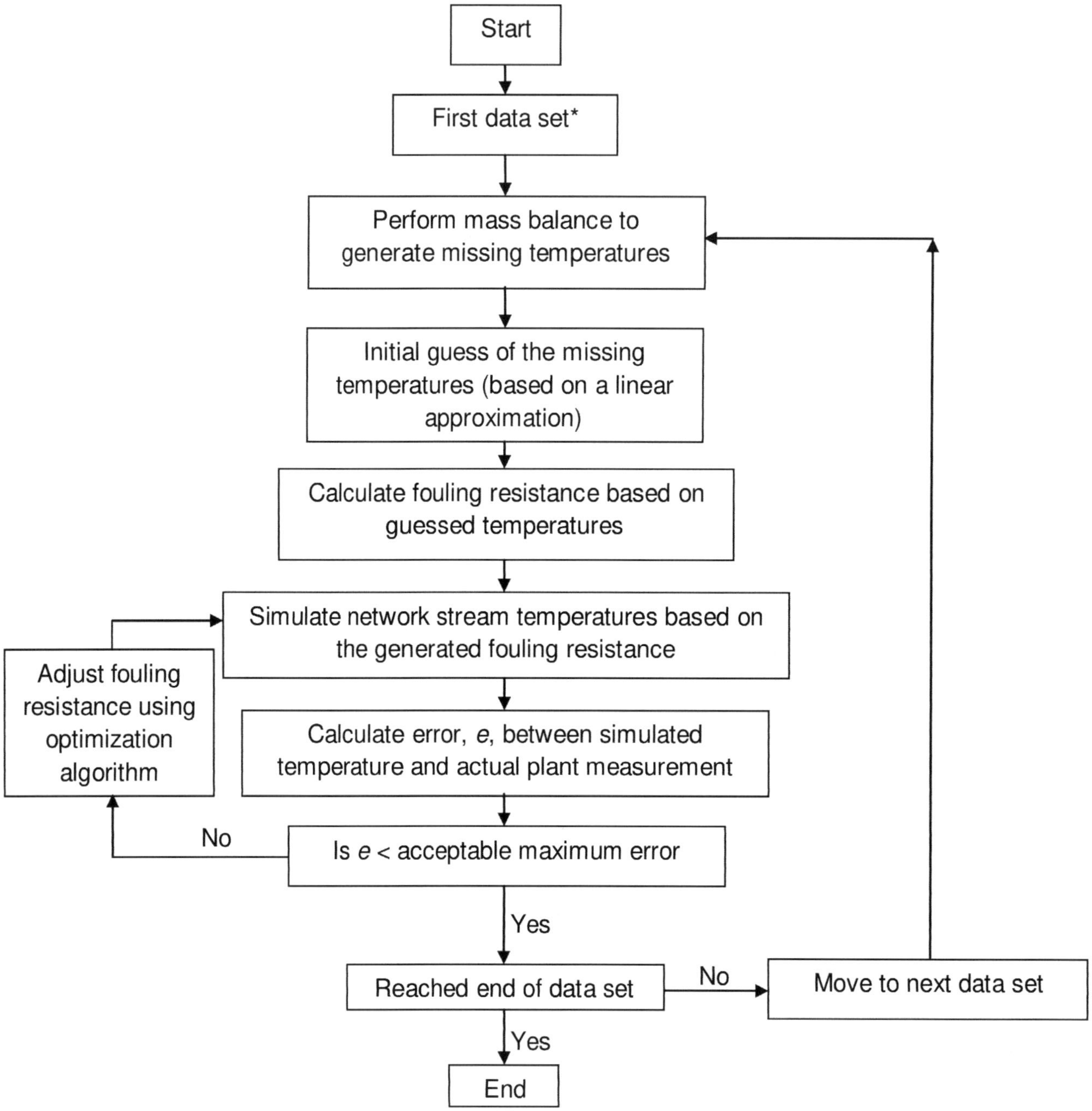

Figure 2: Data reconciliation steps with missing intermediate flow/temperature measurements. *'Data set' refers to plant monitoring data

3. Missing temperatures and/or flow measurements – some available measurements are unreliable

In cases where the available measurements could be unreliable (this is likely to be the case in most refineries), the data reconciliation process needs further analysis. It is necessary to have information on what measurements could be trusted. For instances, measurements at the coil inlet, desalter inlet/outlet, storage tank and product/pump-around streams coming from the distillation column are considered reliable.

The reliable data are then used to provide the basis for mass and energy balances. One approach is to assume that there is linear variation between temperatures of the reliable streams and get an approximate initial guess. If there is not more than one reliable temperature on a stream then the temperature is inferred comparing the clean operating condition (based on the simulated value) to the current operating condition. For example, assume a preheat train consisting of reliable measurements at the preheat train inlet and the coil inlet. The clean operating condition could be obtained through performing a network simulation taking all exchangers are clean.

Unmeasured temperatures are guessed based on an idealized location of the stream between the crude storage tank and the furnace inlet (both temperatures are always measured). Assume that the temperature increment linearly with distance from the tank to the furnace, represented by the linear form, $y = m x + c$, where x gives the location and y the temperature at location 'x'. The value of x is not important in guessing the stream temperature at the fouled state as this could be eliminated as

$$y_2 = \left(\frac{m_2}{m_1}\right) y_1 + \left(c_2 - \frac{m_2}{m_1} c_1\right)$$

Here subscripts 1 and 2 represent the clean and the fouled states, respectively.

By using the initial temperature guess and obtaining the initial fouling resistance, the results are then fine tuned to minimized measured and simulated temperature drop differences across heat exchangers using *smart*PM simulation. The approximate search method assumes the temperature profile to change linearly as it moves along the length of the preheat train. The rigorous search then uses an optimization algorithm to find the missing/unreliable data based on reliable measurements.

Case study

Figure 4 is a section of the preheat train of a well instrumented refinery processing heavy crude oil. All temperatures are measured. In the first case study the temperatures labelled T_1 to T_8 in the streams are assumed to be missing and in the second case study the temperatures labelled T_1 to T_{10} in the streams are assumed to be missing. Data reconciliation process carried out in two ways. 1) data reconciliation is based on linear approximation of temperature and local heat balances. 2) data reconciliation is carried out using the methodology described under sections 2 and 3.

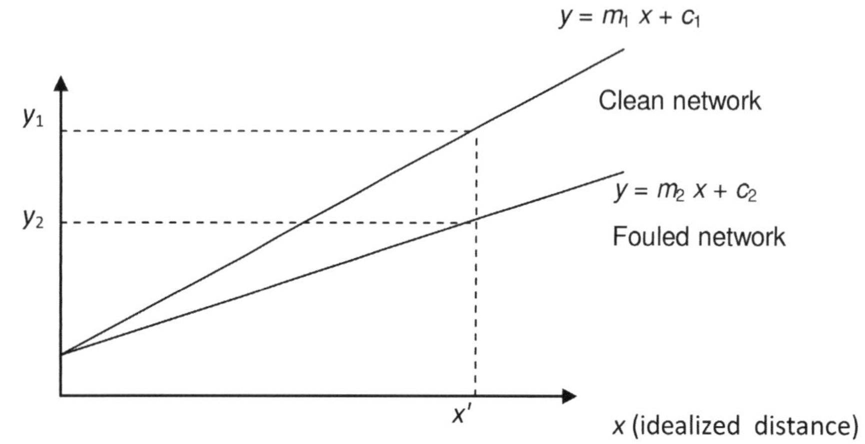

Figure 3: Illustration of evaluating the initial guess for the unmeasured stream temperature

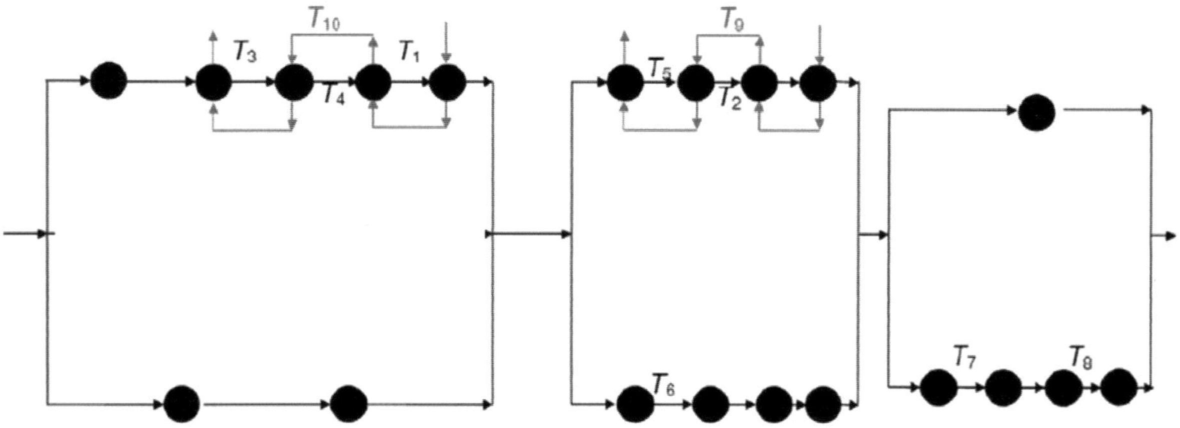

Figure 4: Case study preheat train. The refinery is well instrumented. The labels T_1 to T_{10} are the stream temperature measurements that are the subject of this study.

Discussion

The fouling resistance profile at a given time instance is compared in Figure 5 with the actual plant fouling resistance. The actual plant fouling resistance could be obtained as all plant measurements are available. The assumption of having temperatures T_1 to T_{10} missing is for illustration purpose only.

In Figures 5 (a,b), the data set appearing as 'I' shows an over-prediction of the fouling resistance. Thus telling the user to clean an exchanger earlier than it should have. For the data set denoted as 'II' fouling is under predicted, hence masking the actual effect of fouling in the unit. This would provide misleading information not to clean a unit, when it should have been cleaned.

(a)

(b)

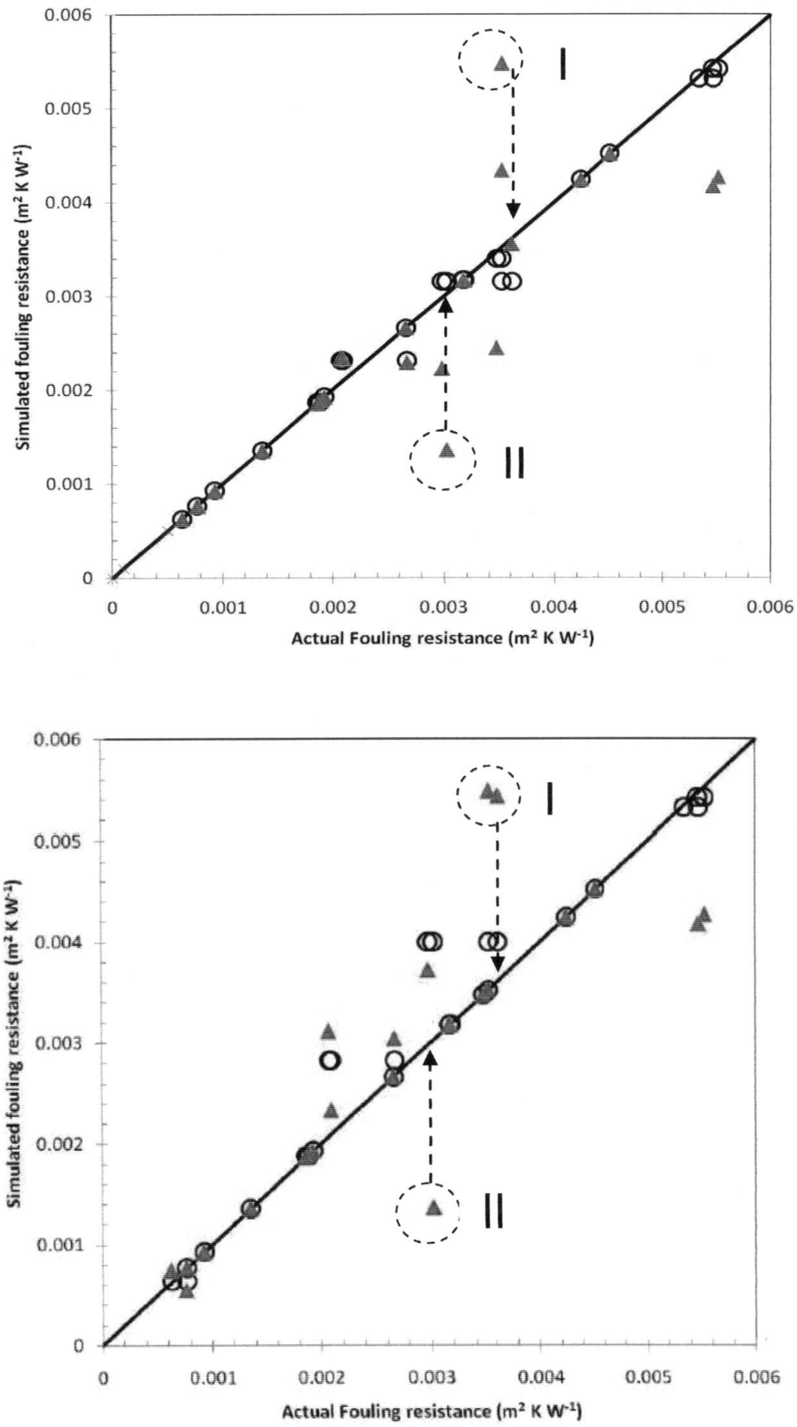

Figure 5: Fouling resistance at a given time instance for the case study exchangers. Filled triangles present results based on simple linear temperature approximation alone and the hollow circles present result obtained through this study. (a) assumes temperatures T_1 to T_8 are missing and (b) assumes temperatures T_1 to T_{10} are missing.

Conclusions

Data reconciliation is a vital process before analysing plant monitoring data. A method of reconciling data with missing/unreliable stream temperatures and flow rates is illustrated through a newly developed software tool (*smart*PM^{TM}) at IHS based on the work done at Cambridge University. The importance of data reconciliation is demonstrated through a case study network where incorrect data reconciliation is illustrated to result in wrong decisions on the degree of fouling of certain units.

References

Ishiyama E.M., Paterson W.R. and Wilson D.I. (2009). Platform for techno-economic analysis of foulng mitigation options in refinery preheat trains. *Energy and Fuels*, 23(3) 1323 – 1337.

Ishiyama E.M., Heins A.V., Paterson, W.R., Spinelli L. and Wilson D.I. (2010). Scheduling cleaning in a crude oil preheat train subject to fouling: Incorporating desalter control. *Applied Thermal Engineering*, 30 1852-1862.

Joshi H.M., Shilpi N.B. and Agarwal A. (2009). Relate crude oil fouling research to field fouling observations, in: OCnference Proceedings of the 8th International conference of Heat Exchanger Fouling and Cleaning, Schladming, Austria.

Knudsen J.G., Lin D. and Ebert W.A. (1999). The determination of the threshold fouling curve for a crude oil, in : Bott, T.R. (editor), Understanding Heat Exchanger Fouling and Its Mitigation. Begell House Inc., Castelveccio Pascoli, Italy, pp 265-272.

(22e) Tutorial - Measuring Hydrocarbon Liquid Levels In Highly Dynamic Petroleum Processes

Steve Newton, Emerson, Sandy, UT and **Sarah Parker**, Emerson, Chanhassen, MN

Measuring Hydrocarbon Liquid Levels in Highly Dynamic Petroleum Processes

Measuring liquid levels in highly dynamic processes can be very challenging. Rapid level changes, changing liquid densities, boiling liquids, and interface fluids can combine to make it difficult to get an accurate and/or stable measurement. This presentation will discuss several of these difficult applications and the solutions found using the various measurement abilities of the guided wave radar. We will outline best practices for the mechanical installation and the level instrument and explore how to best set up the measurement in order to improve loop tuning.

In some challenging applications, initial liquid levels were changing so rapidly that the level transmitters could not track the changes in real time. Also, level measurements were not believable since liquid densities were also changing. The loops could not be properly tuned because of the volatility of the level measurements, which in turn caused liquids to boil and levels to be unstable. The units were frequently tripping due to high or low alarms, and the plant could only be operated in manual.

Using the full capabilities of the guided wave radar allowed accurate stable level measurements in a variety of highly dynamic applications. Using probe-end-projection, damping rates, and using other advanced gage settings we were able to stabilize boiling liquid level measurements, which allowed proper loop tuning to stabilize the process. The plant now operates without upsets or trips. The end result was stable process control with very little process variation.

See more of this Session: Process Control Optimization and Tutorial
See more of this Group/Topical: Topical 7: 14th Topical on Refinery Processing

(22f) Optimization of A Complex System of Crude Oil Distillation Units Using NLP and Metamodels

Diana C. Lopez[1], Luis J. Hoyos[1], Sara M. Chaparro[2] and Jes˙ s A. Castro[1], (1)Instituto Colombiano del Petróleo ICP, ECOPETROL S.A., Piedecuesta, Colombia, (2)Ambiocoop Ltda., Piedecuesta, Colombia

The objective of this paper is to show the optimisation of a complex system of Crude Oil Distillation Units (CDU) belonging to a Colombian refinery of ECOPETROL S.A. This system is composed by crude oil blending scheme and all distillation units installed in the refinery. Distillation units are made up by 5 atmospheric towers, 5 atmospheric furnaces and 4 vacuum towers; crude oil blending is integrated for 14 Colombian crude oils, which were, at first, mixed in 4 mixtures (pre-mixtures) which, at second, were used for preparing the crude oil diets.

Optimisation results were optimum feedstock composition and operating conditions for the CDU system and was carried out through Non Linear Programming (NLP) Model developed by the research institute belonging to ECOPETROL S.A. NLP model optimized the system economic profit and uses Metamodels approach to represent Atmospheric Distillation Towers (ADTs) and Assays cutting with Operating Modes to model the Vacuum Distillation Towers (VDTs). This Model has been used for short, middle and long term planning.

In NLP model, crude oil blending restrictions were mass balances, mixing rules, property boundaries like acidity, sulfur content, specific gravity. Atmospheric distillation process was modeled through Metamodels calculating volumetric flows, properties and temperatures for each ADT product and pumparound; product properties were API gravity, sulfur, neutralization number, conradson carbon, nickel, vanadium, total nitrogen. Metamodels were based on rigorous simulations of distillation units in PRO/II.

Optimization model was formulated in GAMS 23.3, was solved with CONOPT and had a user interface developed in Microsoft Office Excel 2007. NLP model found new crude oil diets and process variable points for each CDU with better economic results than those obtained with the feed compositions of normal short term planning and operation in the refinery. NLP model was able to evaluate new crude oil blending schemes such as segregation of crude oil streams in the current pre-mixtures and inclusion of new pre-mixtures for the refinery; for these new cases, the model calculated the best performance of distillation system.

See more of this Session: Process Control Optimization and Tutorial
See more of this Group/Topical: Topical 7: 14th Topical on Refinery Processing

(28a) PX Recovery by Crystallization

Weihua Jin, GTC Technology, Houston, TX

Paraxylene (PX) is one of the fastest growing major petrochemical products in the world. PX is used as a feedstock for terephthalic acid (PTA), which in turn is used to make polyester film, fiber, and chips. Demand for paraxylene and PTA is growing rapidly, especially in the developing regions of the world. This brings attention to new manufacturing methods, which can produce paraxylene at lower cost and which can be brought on stream more quickly to match the market cycles in demand. While historical crystallization technologies have been denigrated as obsolete and not competitive, recent technology improvements have proven the notion that crystallization is not competitive untrue.

Today's paraxylene crystallization technologies include numerous innovations that improve the operation of crystallization processes and reduce the production cost. This paper will present details of the outlining the benefits of crystallization technologies for paraxylene production.

Fundamentals and recent innovations in crystallization will be highlighted. Special insight will be given from the operator's perspective, and the strong relevance and application possibilities of crystallization in modern aromatics complexes will be shown.

See more of this Session: Advances In Aromatics and Petrochemicals Processing I
See more of this Group/Topical: Fuels and Petrochemicals Division - See Also Topicals 4, 6, and 7

(28b) Advances In Cumene/Phenol Technology - Improving Phenol Plant Profitability

Robert J. Schmidt, R&D Aromatics & Derivatives Development Center, Sr. Development Associate, UOP LLC - A Honeywell Company, Des Plaines, IL

Cumene/phenol producers are facing several challenges today. Feedstock and energy costs are driving up production costs and putting a squeeze on operating margins. End users are demanding higher quality products while newer plants coming on-stream have better cost positions due to favorable raw material supply or larger scale of economies.

Recent advances in cumene/ phenol technologies provides an example of what producers might do to lower their operating costs and deliver the highest quality of products available to the marketplace. This presentation will focus on recent improvements in cumene and phenol technology and how they impact plant profitability.

- *Overview/update of global cumene/phenol market*
- *Advances in cumene process technology for cumene production*
- *Advances in phenol process technology*
- *Potential future developments*

See more of this Session: Advances In Aromatics and Petrochemicals Processing I
See more of this Group/Topical: Fuels and Petrochemicals Division - See Also Topicals 4, 6, and 7

Process Development of Preparing Dinonylphenol-Substituted Methane by Acid Resin-Packed Bed Reactor

Zongli Zhu, Zhiqiang He, Xiaoxiao Huang, Yun Fang*

School of Chemical and Material Engineering, Jiangnan University, Wuxi, Jiamgsu 214122, China

Abstract The process to prepare skeleton of Gemini surfactants is so complex and expensive that the production and application of Gemini surfactants is hindered although they have attracted widespread attention for their unique properties. In this paper an acid resin-packed bed reactor was used to valuate the acid resin catalyzed condensation of nonylphenol and formaldehyde to synthesize the skeleton of Gemini surfactants, dinonylphenol-substituted methane such as bis-(2-hydroxy-5-nonylphenyl)-methane. And the process optimizing and period evaluation of the resin was carried out. The experimental results show that the conversion of nonylphenol based on formaldehyde could reach above 94.3% and the selectivity of dinonylphenol- substituted methane is 92.1% when the reaction in the molar ratio of nonylphenol to formaldehyde 4:1 at 60 °C for 6 h.

Keywords dinonylphenol-substituted methane; acid resin; nonylphenol; formaldehyde;

1 Introduction

Recent years have seen significant advances in the field of Gemini surfactants. These surfactants [1-2] are now the subject of increasing interest in academic and industrial research laboratories, because they appear to be superior in many applied fields to the contrastive conventional surfactants that were made up of one hydrophilic head and one hydrophobic tail. Gemini surfactants have many unique properties such as low critical micellar concentrations (cmc), high viscoelasticities, and enhanced abilities for lowering the oil-water interfacial tension. In addition, they have better wetting and lime-soap dispersing properties [3-4]. But the process to synthesize skeleton of Gemini surfactants is so complex and expensive that less of commercial Gemini surfactants could be found in market. It is reported that surfactants based on dual coupling of nonylphenol are useful as emulsifying and dispersing agents. Grossmann et al. [5] have reported in 1975 to synthesize sulfuric acid semi-esters of oxalkylated polynuclear alkyphenols but the process is so complex and concentrated hydrochloric acid as catalyst is corrosive. Baumann et al. [6] synthesized bis-(4-nonyl-2-phenol)-methane in 1995 used oxalic acid dehydrate as catalyst, but the reaction temperature is high and the catalyst can not be reused. Xu et al. [7] synthesized bis-(2-hydroxy-5-nonylphenyl)-methane in 2008, reported the product yield was 60%. So it is meaningful to improve the reaction conversion rate and decrease the cost.

2 Experimental

2.1 Materials

Nonylphenol supported by Sasol Chemical Company (Nanjing, China); formaldehyde and sodium dodecylbenzenesulfonate purchased from Sinopharm Chemical Reagent Company (Shanghai, China); acid resin made by self; Deionized water obtained from a Millipore Milli-Q water purification system (Millipore, USA).

2.2 Synthesis of dual coupling of nonylphenol

22 g (0.1 mol) nonylphenol, 0.75 g (0.025 mol) formaldehyde, 0.082 g sodium dodecylbenzenesulfonate and a certain amount of petroleun ether are put into a flask and heated with stirring and introduction of a nitrogen stream to 60 °C. Then 6 g acid resin is added into the flask. The mixture is stirred at this temperature for 6 h. The resulting mixture is nature cooled to room temperature. Then the acid resin is removed by the suction filter, and the product is washed with distilled water to remove the residual formaldehyde and sodium dodecylbenzenesulfonate. After that, removing residual water and solution by vacuum drying. The product is purified by column chromatography. A final product was characterized by FT-IR and ESI-MS.

3. Results and discussion

3.1 Structure identification of dual coupling of nonylphenol

The FT-IR spectrum of the condensation product are shown in Figure 1. The absorption peaks of trisubstituted benzene =C-H out-of-plane bending vibration absorption peaks at υ = 826.9 cm^{-1} is feature spectra, indicating that condensation has occurred.

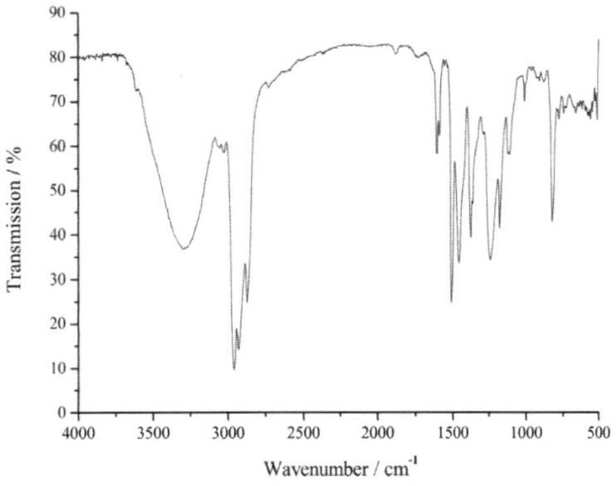

Fig. 1 The FTIR spectrum of dinonylphenol-substituted methane

Typical ESI-MS spectra of the products are given in Figure 2. The ESI–MS technique was employed to measure the molecular weight of the objective product in this work. Under the negative ion mode, only the anionic $[C_{31}O_2H_{47}]^-$ should be detected (Fig. 2). The mass-to-charge ratio (m/z) of product is 451.3, which matches well with the theoretical calculation result (451) of the molecule ion weight for $[C_{31}O_2H_{47}]^-$.

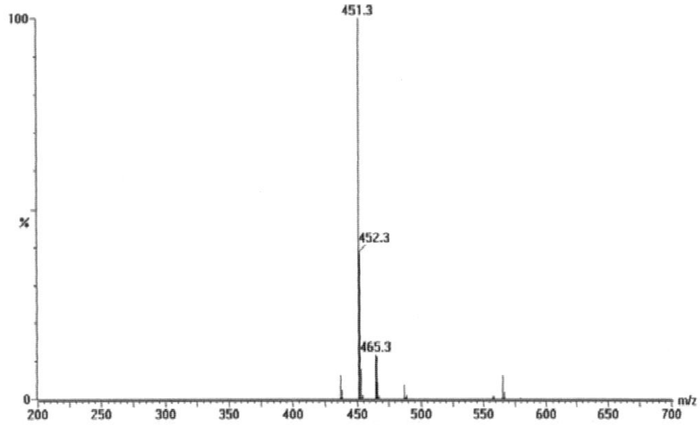

Fig. 2 The ESI–MS spectrum of $[C_{31}O_2H_{47}]^-$ at negative ion mode

3.2 Effect of reaction parameters on the condensation

In batch model, reaction time influenced the reaction conversation was examined from 2 h to 8 h, and the highest conversation was obtained at 6 h then declined. The effect of mole ratio of nonylphenol and formaldehyde on conversation was also investigated. When the mole ratio was above 4 the conversion curve turned flat, thus the suitable mole ratio was chosen as 4:1. Reaction temperature influenced the selectivity of target product was tested at 50, 60 and 70 °C, the highest reaction selectivity was observed at 60 °C.

3.3 Continuous evaluation of the resin

In packed bed model, when the reactor diameter is 21 mm and the resin filling height is 210 mm, the reaction parameters kept as the molar ratio of nonylphenol to formaldehyde 4:1 at 60 °C; the conversion and selectivity data at different residence time were showed in figure 3 and figure 4.

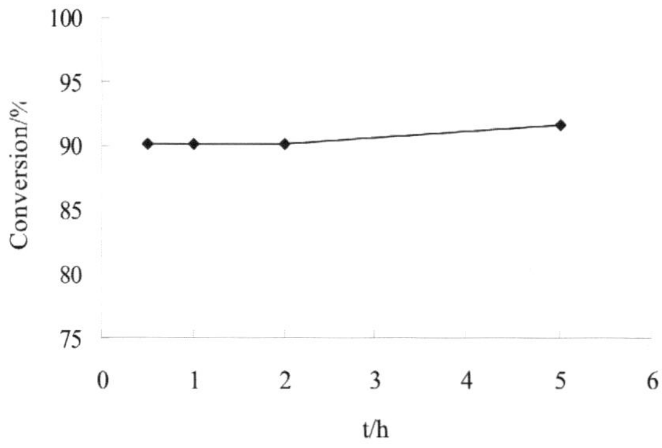

Fig.3 Effect of residence time on conversion.

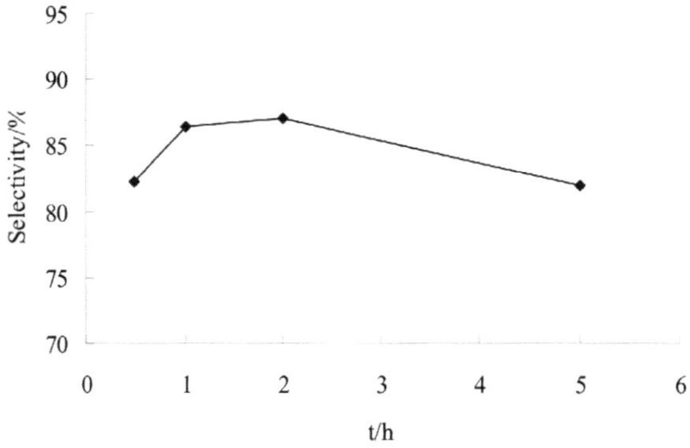

Fig.4 Effect of residence time on selectivity.

The effect of residence time on conversion is shown in Figure 3. When the reaction time is 0.5 h, the conversion is above 90% and with the time extended the conversion remained almost unchanged. The corresponding selectivity is shown in figure 4. In 2 h, selectivity is increased and then declined. Therefore, the optimum residence time is 2 h.

4 Conclusions

An acid resin-packed bed reactor was used to valuate the acid resin catalyzed condensation of nonylphenol and formaldehyde to synthesize the skeleton of Gemini surfactants dinonylphenol-substituted methane. The experimental results of batch model show that the conversion of nonylphenol based on formaldehyde could reach above 94.3% and the selectivity of dinonylphenol-substituted methane is 92.1% when the reaction in the molar ratio of nonylphenol to formaldehyde 4:1 at 60 °C for 6 h. And continuous evaluation of the resin was studied. The experiment results of acid resin-packed bed show that at 2 h the conversion of formaldehyde could reach 90.1% and the selectivity of dual coupling of nonylphenol is 87.1%.

Reference

[1] Menger F M, Littau C A: Gemini surfactants: synthesis and properties. J Am Chem Soc 1993, 113:1451-1452.

[2] Zana R, Benrraou M, Rueff R: Alkanediyl-CX,-w-bis(dimethylammonium bromide) surfactants. 1. Effect of the spacer chain length on the critical micelle concentration and micelle ionization degree. Langmuir 1991, 7:1072-1075.

[3] Zana, R.; Talmon, Y: Nature 1993, 362, 228.

[4] Zana R: Critical micellization concentration of surfactants in aqueous solution and free energy of micellization. Langmuir 1996,12:1208-1211.

[5] Max Grossmann,Frankfurt(Main), Ruth Faber, Burgkirchen, Heinz Uhirig, Steinbach, Taunus: Dispering and wetting agents for use in organic biocides. US: 3919429, 1975-12-11.

[6] Hans-Peter Baumann, Ettingen. Organic compounds. US: 5470355, 1995-12-28.

[7] Xu Nian, Li Gang: Synthesis of Bis(2-hydroxy-5-alkylphenyl)methane. Chinese Journal of Organic Chemistry 2008, 28(10):1780-1784.

(28d) Studies of Structural and Electronic Properties of Polyethyelene Glycol-Polystyrene-Palladium Complex

Dodi Heryadi, Supercomputing Laboratory, King Abdullah University of Science and Technology, Thuwal, Saudi Arabia

This study presents an investigation into orientation of polyethyelene glycol, polystyrene, and palladium inside water. Molecular mechanics, semi-empirical and ab initio studies are utilized to further understand the structural and electronic properties of this complex. The solvent model employed in this work is the polarizable continuum model (PCM). The results show the potential in designing palladium catalysts with enhanced reactivity.

See more of this Session: Advances In Aromatics and Petrochemicals Processing I
See more of this Group/Topical: Fuels and Petrochemicals Division - See Also Topicals 4, 6, and 7

Modeling Catalyst Deactivation in Dehydrogenation of Ethylbenzene to Styrene

Jalal Babaei, Morteza Baghalha*
*Department of Chemical and Petroleum Engineering, Sharif University of Technology, Tehran 11365-9465, Iran
E-mail address: baghalha@sharif.edu

Abstract

Long-term deactivation of a commercial K-promoted Fe_2O_3 catalyst used in the dehydrogenation of ethylbenzene to styrene monomer was studied using 30 months of operating data of three industrial SM radial fixed bed adiabatic reactors. The Langmuir-Hinshelwood, Hougen-Watson model (LHHW) was used for reaction kinetics modeling. The kinetics of coking from ethylbenzene and styrene and the short-term deactivation function due to the quick coke formation were also considered in the model. The kinetics of long-term deactivation of the catalyst for ethylbenzene dehydrogenation to styrene between 550 and 630 °C were determined by allowing the effects of temperature and time on the catalyst activity. Using the plant data, 5 unknown parameters of the long-term activity function, related to the slow loss of potassium from the catalyst, were optimized. The optimized value of the order of the deactivation law was determined to be 3. The effect of internal mass-transport limitation was found to be minimal.

Keywords: Dehydrogenation, Ethylebenzene, Styrene monomer, Deactivation, K-Fe_2O_3 catalyst

1. Introduction

Styrene is the simplest and by far the most important member of a series of aromatic monomers. Also known commercially as styrene monomer (SM), it is a versatile monomer extensively used for the manufacture of plastics, including crystalline polystyrene, rubber – modified impact polystyrene, expandable polystyrene, acrylonitrile – butadiene – styrene copolymer (ABS), styrene – acrylonitrile resins (SAN), styrene – butadiene latex, styrene – butadiene rubber (SBR) and unsaturated polyester resins [1]. The styrene process was developed in the 1930s by independently and simultaneously BASF in Germany and by Dow Chemical in the USA. An important fraction of the world styrene monomer production is obtained by catalytic dehydrogenation of ethylbenzene (EB), potassium promoted Fe-catalyst is typically used [2], in the presence of steam. Steam is used to supply heat for this endothermic reaction, to inhibit coke formation, and to reduce the partial pressure of the products and thus increase the equilibrium conversion [3].

An adiabatic fixed bed reactor with either radial or axial flow pattern is used in industrial plants. Axial flow adiabatic reactors have been analyzed by Sheel and Crowe [4], Clough and Ramirez [5] and Sheppard et al. [3]. These authors carried out studies of simulation and optimization of existing ethylbenzene dehydrogenation reactors.

On the other hand, the radial flow reactor allows lower pressure drops and good flow distribution. This type of reactor is especially suitable for processes requiring low pressures to increase the equilibrium conversion. Radial flow reactor also exhibits an additional advantage over the axial flow reactor where a smaller catalyst pellets could easily be used in the bed [6]. For all these reasons, radial reactors are being preferred to perform the dehydrogenation of ethylbenzene.

Sheel and Crowe [4] considered the activity of the catalyst constant, although actually varying with both time and bed depth, because insufficient data were available over a long period of time.

Lim et al. [7] calculated a deactivation factor of commercial catalysts in adiabatic radial flow beds (two reactors in series) using neural network modeling. Actual plant data were used in their study to test their model.

Shibata and Kiyoura [8] mentioned the effects of adding potassium as a structural promoter on the catalyst properties; including, decreasing the catalyst apparent bulk density and avoiding shrinkage of the catalyst tablet upon calcinations during the catalyst preparation procedure. As a selective promoter, potassium also reduces cracking reactions to benzene and carbon that occur mostly in n-type oxides with acidic sites. By adding potassium, p-type oxide with negative surface potential is produced that retards the formation of benzene from side reactions [9].

Hirano [10] was the first to claim that the active sites in this catalyst are $KFeO_2$. He also mentioned that in situ catalyst deactivation with CO_2 lead to deactivation of these sites. However, he found no relationship between the amount of coke deposited and the deactivation rate, leading the author to the conclusion that coke was not a major cause for the catalyst deactivation in his experiments. This may be due to the fact that coke deposition is a short term deactivation mechanism for this type of catalyst. Devoldere and Froment [11] studied coke formation and gasification in the catalytic dehydrogenation of ethylbenzene.

Lee [9] and Herzog and Rase [12] reported that a shift in the potassium promoter toward the centre of the catalyst pellets occurs. There is also a potassium migration toward the lower temperature region in the reactor exit; since dehydrogenation is an endothermic reaction. They suggested that the catalyst deactivation mostly arises from migration and loss of potassium with process time. Fundamental properties of $K\text{-}Fe_2O_3$ catalysts were studied by Muhler et al. [13]. These authors found that the catalytic activity was represented by the surface Fe^{3+} and K^+ ions which are incompletely coordinated. $KFeO_2$ phase was proposed to be the active site. This phase would be stable at the reaction temperature, but is not detectable at room temperature. $K_2Fe_{22}O_{34}$ phase is postulated to represent the storage phase for potassium.

In spite of the industrial importance, kinetics of the deactivation of the potassium-promoted iron oxide catalyst has not been investigated in detail. The purpose of this study is to examine changes in the catalyst activity with process time (activity decay law) over a period of about 30 month by using industrial plant data and to investigate the validity of mechanism and kinetics of dehydrogenation reactions reported in literature adopted in the current study.

2. Experimental

Commercial catalyst of this system is the potassium-promoted iron oxide. In typical industrial operation, the activity of this catalyst decreases gradually with process time and the catalyst must be replaced with new material after about 2 to 3 years. To compensate for the catalyst deactivation during the catalyst life cycle, the reactor inlet temperature is gradually increased in order to maintain a rather constant conversion of EB.

The industrial data for the adiabatic radial-flow reactor for the styrene production was obtained from a SM unit. This plant has three ethylbenzene dehydrogenation reactors in series, as shown in Figure 1. First reactor has only a dehydrogenation bed. In the older technology, the outlet flow of the first reactor would heat up in a heat exchanger before entering the 2nd reactor. However, in the newer SMART technology that has been used in this plant, an oxidation bed is used at the beginning of the 2nd and 3rd reactors to heat the partially-converted EB stream and to eliminate part of hydrogen in this stream which shifts the equilibrium to produce more styrene. Through this technique, EB conversions up to 80% would be achieved. The geometric data of reactors are reported in Table 1.

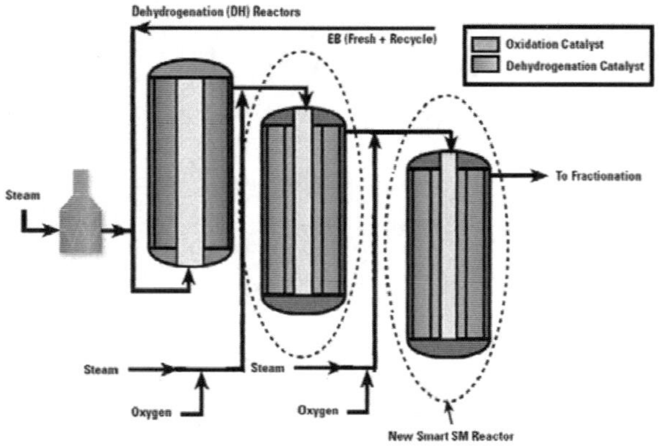

Figure 1. Configuration of the adiabatic radial flow reactor system

Table 1. Geometrical Parameters of Reactors

First Reactor	
Inner diameter of dehydrogenation bed (m)	0.99
Outer diameter of dehydrogenation bed (m)	2.108
Average height of the bed (m)	6.08
Second Reactor	
Inner diameter of oxidation bed (m)	0.91
Inner diameter of dehydrogenation bed (m)	1.350
Outer diameter of dehydrogenation bed (m)	2.470
Average height of the bed (m)	6.353
Third Reactor	
Inner diameter of oxidation bed (m)	1.150
Inner diameter of dehydrogenation bed (m)	1.520
Outer diameter of dehydrogenation bed (m)	2.570
Average height of the bed (m)	6.393

The obtained plant data consist of the reactors input-output components flow rates and temperatures on the 5th day of each month during a period of 30-months plant operation. During this period, a fresh commercial K-Fe_2O_3 catalyst was used in the dehydrogenation reactors at the start of operation. The catalyst had a cylindrical shape with 3 mm in diameter and 3.5 mm in length. The physical properties of the dehydrogenation (as well as the oxidation) bed is reported in Table 2. At the end of this period (month 30), the catalysts were highly deactivated. As a result, the operation was ceased at the time and the catalysts were replaced. The feed to the 1st reactor was a mixture of fresh and recycled stream of ethylbenzene plus steam which is illustrated in Figure 2, as molar flow rates vs. the operation time. In the EB mixed stream to the 1st reactor, there exist some impurities, as reported in Figure 3.

Table 2. Physical Properties of Catalysts

Dehydrogenation bed catalysts	
Catalyst bulk density (ρ_b)	1325 kg/m3
Catalyst diameter	3.5 mm
Catalyst bed porosity	0.35
Oxidation bed catalysts	
Catalyst bulk density (ρ_b)	1250 kg/m3
Catalyst diameter	3 mm
Catalyst bed porosity	0.4

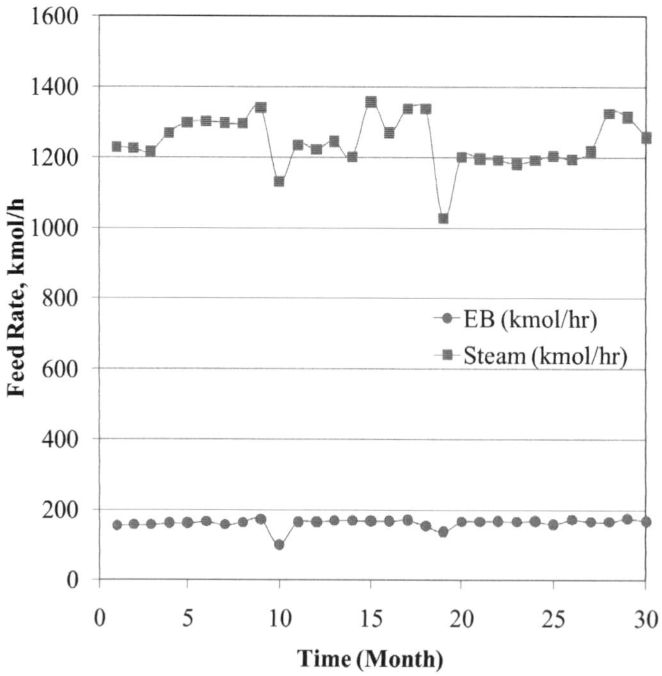

Figure 2. Inlet ethylbenzene and steam to the first reactor

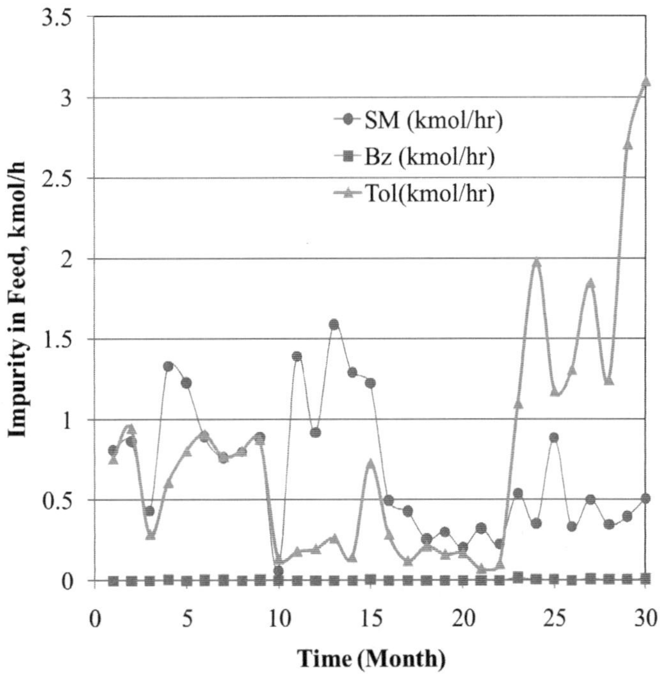

Figure 3. Impurities in the inlet stream to the first reactor

For the 2nd and 3rd reactors, the molar flow rates of the diluting steam (DS), the oxygen-rich gas (ORG), and the stream temperatures of the mixture are reported in Figures 4, 5 and 6, respectively. Steam as a diluting agent reduces the temperature of the exit stream from the first reactor, so it would prevent a possible explosion due to the rapid oxygen – hydrogen reaction in the catalytic oxidation bed of the 2nd (and also the 3rd) reactor.

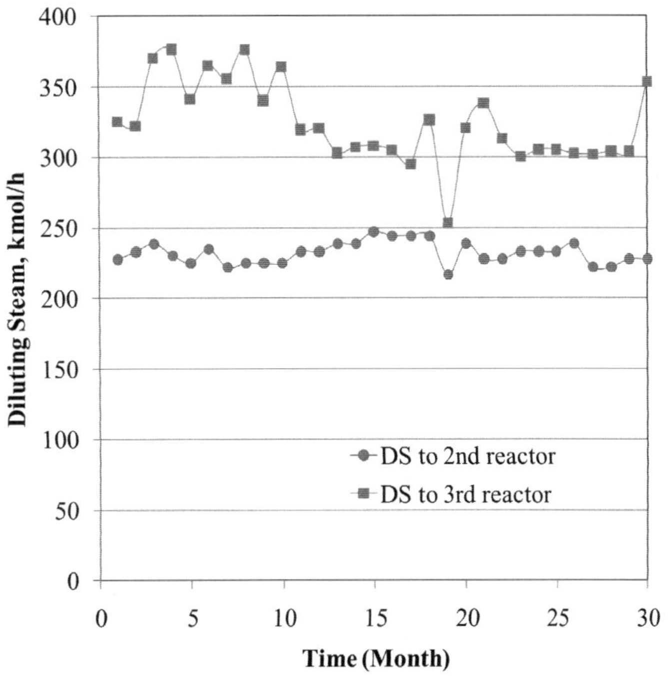

Figure 4. Comparison of diluting steam (DS), input to the second and the third reactors

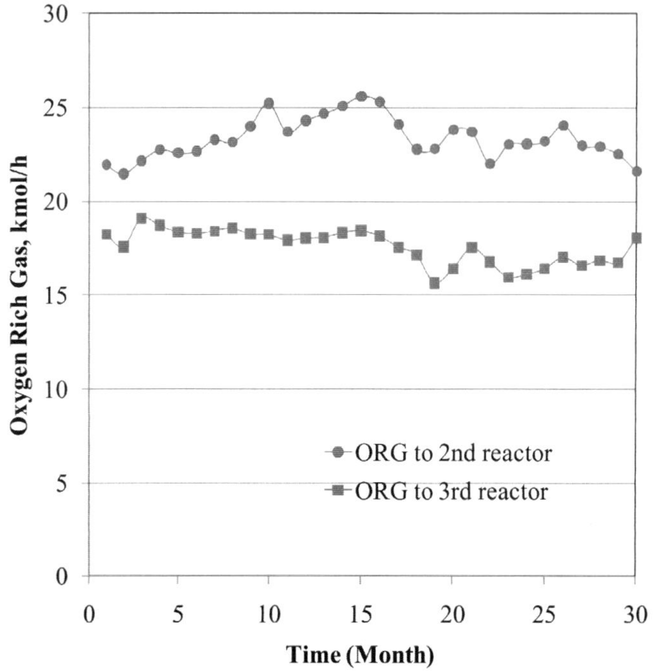

Figure 5. Comparison of oxygen rich gas (ORG) flow input to the second and the third reactors

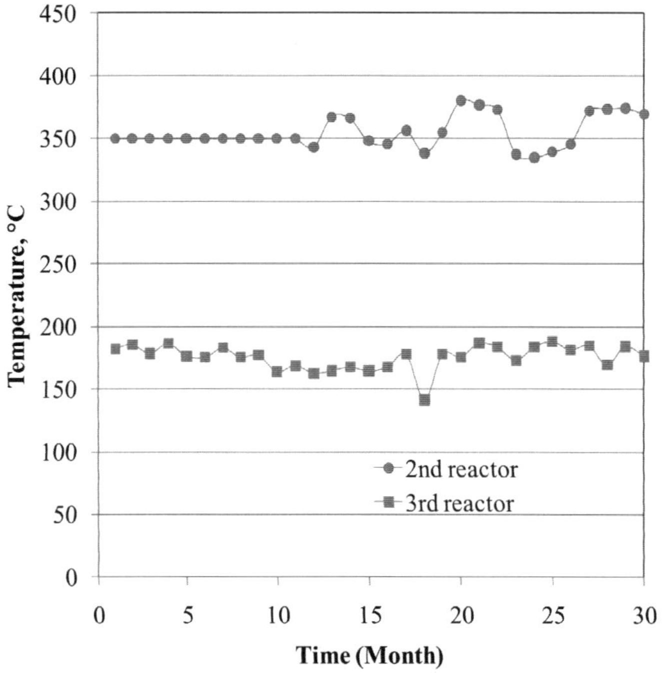

Figure 6. Comparison of temperature of DS (Diluting Steam) and ORG (Oxygen Rich Gas) mixed flow to the second and the third reactors

The amount of steam used in the SM unit is an important economic parameter which affects the total energy cost of the plant. The steam-to-oil (SOR) ratio should be kept around 2 for the proper and optimal operation of the SM unit [5]. As shown in Figure 7, the actual month to month SOR data of our industrial reactors (considering all the 3 reactors combined) are also close to 2. Figure 7 shows that SOR values decreased as the operation had continued. This observation was also reported by Lim et al. [7]. However, they argued that SOR must be increased vs. the operation time in order to increase the profit function.

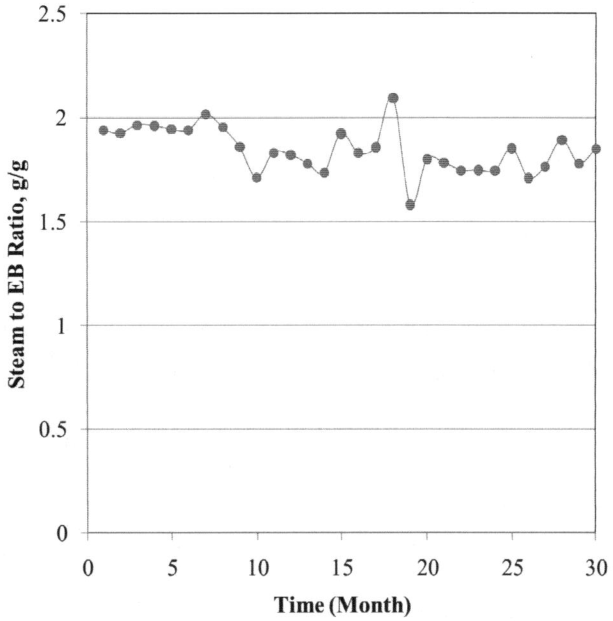

Figure 7. The total Steam (sum of all the 3 reactors) to EB Ratio (SER, in g/g) implemented in the SM plant

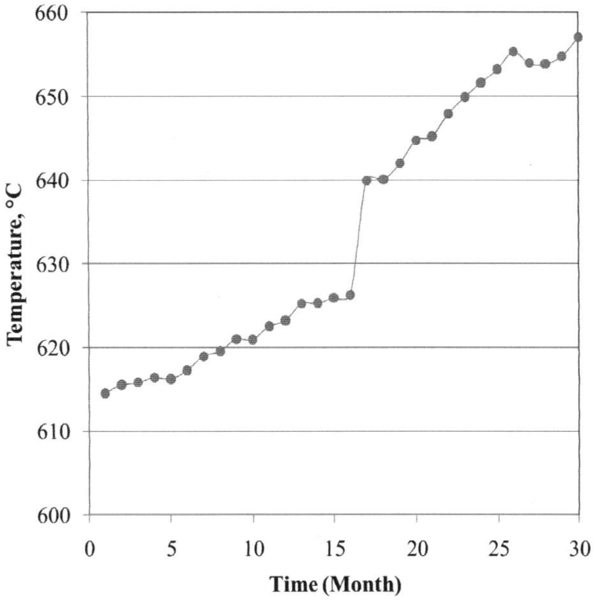

Figure 8. Operational input temperature data of feed stream to the first reactor

As mentioned previously, the inlet temperature to the first reactor must be raised with the operation time in order to compensate for the catalyst deactivation. The actual plant data for the inlet temperature to the 1st reactor, as shown in Figure 8, very well depicts this requirement.

3. Reaction sets

3.1. Catalytic reactions

The generally accepted catalytic EB dehydrogenation reactions [14] are as follows:

$$EB \underset{k_1^{-1}}{\overset{k_1}{\rightleftarrows}} SM + H_2 \tag{R1}$$

$$EB \xrightarrow{k_2} Bz + C_2H_4 \tag{R2}$$

$$EB + H_2 \xrightarrow{k_3} Tol + CH_4 \tag{R3}$$

$$SM + 2H_2 \xrightarrow{k_4} Tol + CH_4 \tag{R4}$$

Many researchers [4, 5, 15, 16, and 17] have implemented power law kinetics to represent the apparent rate equations for the catalytic dehydrogenation of EB to ST. In such models, the internal diffusion effects of the catalyst, as well as the catalyst size and type on rate of reactions are considered in the rate coefficients. The first study on the intrinsic rate equations for the catalytic dehydrogenation of EB was proposed by Carra and Forni [18]. They assumed only one type of active sites which would generate all products in parallel reactions; whereas Hirano [10] postulated a variety of specialized reaction sites for the styrene, benzene and toluene reactions. In the present study, we implemented the heterogeneous approach and hence used the LHHW kinetics model to drive the reaction kinetics. The corresponding rate equations for reactions R1 to R4 are [14]:

$$r_1 = \frac{k_1 K_{EB}(P_{EB} - \frac{P_{SM}P_{H_2}}{K_{eq}})}{(1 + K_{EB}P_{EB} + K_{H_2}P_{H_2} + K_{SM}P_{SM})^2} \tag{1}$$

$$r_2 = \frac{k_2 K_{EB} P_{EB}}{(1 + K_{EB}P_{EB} + K_{H_2}P_{H_2} + K_{SM}P_{SM})^2} \tag{2}$$

$$r_3 = \frac{k_3 K_{EB} P_{EB} K_{H_2} P_{H_2}}{(1 + K_{EB}P_{EB} + K_{H_2}P_{H_2} + K_{SM}P_{SM})^2} \tag{3}$$

$$r_4 = \frac{k_4 K_{SM} P_{SM} K_{H_2} P_{H_2}}{(1 + K_{EB}P_{EB} + K_{H_2}P_{H_2} + K_{SM}P_{SM})^2} \tag{4}$$

These rate equations correspond to the rate-determining steps on dual sites and involve an adsorbed hydrocarbon and molecularly adsorbed hydrogen. The kinetic parameters of these reactions are reported in Table 3.

Table 3. Values of the LHHW Kinetic Parameters for the reaction numbers 1 to 4 (Lee and Froment [14])

Pre-exponential Factor of Catalytic Reaction i [kmol/(kg-cat h)]	
A1	$4.594*10^{9}$
A2	$1.060*10^{15}$
A3	$1.246*10^{26}$
A4	$8.024*10^{10}$
Pre-exponential Factor of Adsorption Species j [bar^{-1}]	
A_{EB}	$1.014*10^{-5}$
A_{SM}	$2.678*10^{-5}$
A_{H2}	$4.519*10^{-7}$
Activation Energy [kJ/mol]	
E_1	175.38
E_2	296.29
E_3	474.76
E_4	213.78
Adsorption Enthalpy [kJ/mol]	
$\Delta H_{a,EB}$	-102.22
$\Delta H_{a,SM}$	-104.56
$\Delta H_{a,H2}$	-117.95

3.2. Water reforming and water-gas shift reactions

Two water reforming and one water-gas shift reactions from pseudo homogenous model were implemented.

$$CH_4 + H_2O \xrightarrow{k_5} CO + 3H_2 \qquad (R5)$$

$$C_2H_4 + 2H_2O \xrightarrow{k_6} 2CO + 4H_2 \qquad (R6)$$

$$CO + H_2O \xrightarrow{k_7} CO_2 + H_2 \qquad (R7)$$

The kinetic parameters of these reactions were adopted from sadeghzadeh et al. [16], as reported in Table 4.

Table 4. Kinetic Parameters for Water – Gas Reactions (Sadeghzadeh et al. [16])

Reaction Number	A	E (kJ/kmol)
5	0.12	103996
6	-3.21	65723
7	21.24	76382

3.3. Thermal EB dehydrogenation reactions

Thermal radical-type reactions cannot be ignored in the derivation of the kinetic model for EB dehydrogenation. They occur in the zones without catalyst or in the voids of zones that contain only inert solids or in the void fraction of the catalyst bed itself. The three main thermal EB dehydrogenation reactions and their corresponding rate equations are as follows.

$$EB \underset{k_{t,1}^{-1}}{\overset{k_{t,1}}{\rightleftarrows}} SM + H_2 \quad \text{(R8)}$$

$$EB \xrightarrow{k_{t,2}} Bz + C_2H_4 \quad \text{(R9)}$$

$$EB + H_2 \xrightarrow{k_{t,3}} Tol + CH_4 \quad \text{(R10)}$$

$$r_{t,1} = k_{t,1}(P_{EB} - \frac{P_{SM}P_{H_2}}{K_{eq}}) * \frac{\varepsilon_b}{\rho_b} \quad (5)$$

$$r_{t,2} = k_{t2}P_{EB} * \frac{\varepsilon_b}{\rho_b} \quad (6)$$

$$r_{t,3} = k_{t,3}P_{EB} * \frac{\varepsilon_b}{\rho_b} \quad (7)$$

The assumed [14] kinetic parameters for these reactions are given in Table 5.

Table 5. Pre-exponential Factors and Activation Energies for the Thermal Reactions

Reaction Number	$A_{t,i}$ [kmol/(m^3 . hr. bar)]	$E_{t,i}$ [kJ/kmol]
8	$2.2215*10^{16}$	272.23
9	$2.4217*10^{20}$	352.79
10	$3.8224*10^{17}$	313.06

3.4. Oxidation bed reaction

For the catalytic oxidation reaction of hydrogen (R11) in the oxidation bed of reactors 2 and 3, the rate equation of Hoiberg [19] was implemented in the present study.

$$H_2 + 1/2O_2 \xrightarrow{k_{11}} H_2O \quad \text{(R11)}$$

$$r_{11} = k_{11}P_{O_2}^{0.8}\exp(\frac{-E_{11}}{RT}) \text{ kmol/m}^3\text{h} \quad (8)$$

In Eq. (8), K_{11} = 5310000 kmol/m^3.h.bar$^{0.8}$ and E_{11} = 50210.37 kJ/kmol [19].

4. Reactor Model

4.1 The basic equations

Based on Kareeri et al. [25], the reactor shown in Figure 1 could be classified as CF-π radial flow fixed bed reactor. In the CF-flow configuration, the gas is fed to the center pipe and travels in radial direction outward from the center of the pipe to the annular channel. If the axial flow directions in the annular channel and in the center pipe are opposite, it is classified as the π-flow type. Kareeri et al. [25] concluded that when the ratio of the center pipe cross-sectional area to the annular channel cross sectional area is less than unity, the CF-π configuration has the lowest flow deviation from a plug flow reactor. This ratio is less than 1 in all the three industrial reactors that were considered in the present work. Therefore, plug flow in the radial direction is a proper assumption for the reactor model.

Applying the steady state mass and heat balance to a differential control volume of the reactor [16] leads to the following set of differential equations for the dehydrogenation bed (Eqs. 9 and 10) and the oxidation bed (Eqs. 11 and 12).

$$\frac{dF_i}{dr} = \sum_{i,j=1}^{10} \nu_{ij}\eta_j r_j.\rho_b.A_r \quad (9)$$

$$\frac{dT}{dr} = \frac{\sum_{j=1}^{10}(-\Delta H_j).\eta_j r_j.\rho_b.A_r}{\sum_{i=1}^{11} F_i C_{Pi}} \quad (10)$$

$$\frac{dF_i}{dr} = \sum_{i=5,8,11} \upsilon_{i.oxi} r_{oxi}.A_r \quad (11)$$

$$\frac{dT}{dr} = \frac{-\Delta H_{oxiation}.r_{oxidation}.A_r}{\sum_{i=1}^{11} F_i C_{Pi}} \quad (12)$$

Where $F_j(R_0) = F_{j0}$ and $A_r = 2\pi r Z$. The enthalpies of reaction and the heat capacities are calculated from smith [26], based on the ideal gas law. The pressure drop in both the dehydrogenation and the oxidation beds were estimated by the Ergun equation [17].

$$\frac{dP}{dr} = \frac{-10^{-5} f}{d_p \rho_g}\left(\frac{m}{2\pi r Z}\right)^2 \quad (13)$$

Where, ρ_g is the density of gas mixture (kg/m^3), m is the total mass flow (kg/h), d_p is diameter of spherical catalyst, r is the length of radial bed, and Z is the height of reactor. f is the friction factor that can be calculated from Eq. (14).

$$f = \frac{1-\varepsilon}{\varepsilon^3}\left(1.75 + 150\frac{1-\varepsilon}{\mathrm{Re}}\right) \quad (14)$$

Where, ε is the bed porosity and $\mathrm{Re} = \frac{d_p m}{\mu_g (2\pi r Z)}$ is the particles Reynolds number. μ_g is the viscosity of the gas mixture (kg/m.s). ρ_g, the density of gas mixture can be calculated from Eqs 15 and 16.

$$\frac{1}{\rho_g} = \sum_i^{Components} \frac{w_i}{\rho_i} \quad (15)$$

$$\rho_i = \frac{PM_w}{RT} \quad (16)$$

η_j in Eqs. 9 and 10 is the effectiveness factor for reaction Rj in the dehydrogenation bed. In the following subsection, the method of calculation of the effectiveness factor is discussed.

4.2 The effectiveness factor

For solid catalytic reactions, when the reaction rate is much faster than the diffusion rate of the reactants into the solid structure (inlet of bed), it is needed to introduce an effectiveness factor, η, into the reaction rate equation. The effectiveness factor was calculated only for the first equilibrium reaction R1. For the other reactions, it was verified that the values of the corresponding effectiveness factors were close to one. To obtain the effectiveness factor for an equilibrium first order reaction kinetics, the so-called Thiele Modulus can be defined [20] according to,

$$M_T = L\sqrt{\frac{K'''\rho_c}{D_e X_{eq}}} = L\sqrt{\frac{K'''\rho_c}{D_e(\frac{K_{eq}}{1+K_{eq}})}} \tag{17}$$

Where $\rho_c = \frac{\rho_b}{1-\varepsilon_b}$ and L is R/2 for a cylindrical catalyst. Then the η versus M_T curves for all forms of rate equations closely follow the curve for the first-order reaction, as in Eq. (18).

$$\eta_{FirstOrder} = \frac{\tanh(M_T)}{M_T} \tag{18}$$

The diffusion coefficients were calculated from the Fuller et al. formula [21].

$$D_{AB} = \frac{0.00143T^{1.75}}{PM_{AB}^{\frac{1}{2}}[(\sum_v)_A^{\frac{1}{3}} + (\sum_v)_B^{\frac{1}{3}}]^2} \tag{19}$$

Where D_{AB} is the binary diffusion coefficient, cm^2/s; T is the temperature in K; M_A, M_B are the molecular weights of A and B in g/gmol; and M_{AB} is $2[(1/M_A) + (1/M_B)]^{-1}$; P is the pressure in bar. $\sum_v$ is calculated for each component by summing atomic diffusion volumes. Knudsen diffusivity in gases in a straight cylindrical pore can be calculated from the kinetic theory.

$$D_{KA} = \frac{4}{3}r\sqrt{\frac{2RT}{\pi M_A}} = 9700r\sqrt{\frac{T}{M_A}}\ (\mathrm{cm^2/s}) \tag{20}$$

Where $r = \frac{2V_g}{S_g} = \frac{2\varepsilon_s}{S_g\rho_c}$, S_g is the total surface area in cm^2/g, ρ_c is the pellet density in g/cm^3, and ε_s is the catalyst internal void fraction. Our previous work [22] on the characterization of the styrene catalyst has revealed that the value of S_g for the fresh dehydrogenation catalyst was 1.72 m^2/g.

In contrast to a homogeneous medium, a porous catalyst contains nonuniform pore structures which intersect with others to form a network where the fluid may follow the tortuous path. To take into account the texture properties of the porous catalyst, the effective diffusivity D_{eA} for component A diffusing through a porous catalyst can be evaluated through Eqs. (21) and (22).

$$\frac{1}{D_A(r)} = \frac{1}{D_{AB}} + \frac{1}{D_{KA(r)}} \tag{21}$$

$$D_{eA} = \frac{\varepsilon_s D_A(r)}{\tau} \tag{22}$$

τ is the tortuosity factor that was estimated based on the Bruggemann correlation [23], as presented in Eq. (23). The void fraction of the dehydrogenation catalyst is 0.35 [24].

$$\tau = \varepsilon_s^{-0.5} \tag{23}$$

5. Deactivation

The catalyst deactivation causes the catalytic reactions R1 to R4 proceeds slower. Hence, the corresponding rate equations, Eqs. (1) to (4) could be prorated by multiplying with an activity parameter, "a", where "a" is less than 1. Deactivation of the styrene catalysts is a complex process. Several deactivating mechanisms are interrelated and take place concurrently. Probably a good distinction can be made on the basis of reversibility of deactivation [2]. In industrial practice one discriminates between short-term reversible deactivation and long-term irreversible deactivation. Site blocking by carbon deposition or CO_2 poisoning (not formation of K_2CO_3 which would be stable under reaction conditions) may cause the short-term deactivation. Regeneration with steam results in the potassium- catalyzed gasification of carbon deposits and the structural regeneration of the $KFeO_2$ phase [13]. It is assumed that the short-term catalyst deactivation is mostly dominated by the coke formation (with a_c represents the catalyst activity as it is affected by the coke formation) and the long-term catalyst deactivation is governed by the potassium loss (with a_k represents the catalyst activity as it is affected by the potassium loss). Hence, the catalyst activity, "a", can be represented by Eq. (24).

$$a = a_C . a_K \qquad (24)$$

5.1 Short-term deactivation

The amount of coke deposited on a typical iron oxide catalyst depends on several factors. Besides the specifications of the catalyst and the associated promoters, important operating restrictions are steam to EB ratio (also called SOR or steam to oil ratio), the pressure and the temperature. Although a higher temperature could be achieved with addition of more steam in favor of equilibrium conversion to styrene, side reactions may also become significant under such conditions [27]. The empirical Voorhies [28] correlation for coking in the catalytic cracking of gas oil has been widely accepted and generalized beyond the scope of the original contribution. Yet, such an equation completely ignores the origin of the coke. The rate of coking must depend on the composition of the reaction mixture. Froment and Bischoff [29] were the first to relate these factors quantitatively to the rate of coking and to draw the conclusions from it as far as kinetics and reactor behavior are concerned. Coke can be formed either by a reaction path parallel to the main reaction resulting EB, or by a reaction path consecutive to the main reaction resulting SM. The deactivation functions for coking and for the main reaction are considered the same, according to a study on the dehydrogenation of 1-butene into butadiene system [30] with an observation that the presence of hydrogen acts as an inhibitor of coke formation.

The kinetic modeling of the intrinsic rate of coke formation is accomplished along the approach of Froment [29].

$$r_c(P_j,T,C_c) = \frac{dC_c}{dt} = r_c^o(P_j,T)\Phi_c(C_c) = \frac{k_{EB}K_{EB}P_{EB} + k_{SM}K_{SM}P_{SM}}{(1+K_{EB}P_{EB} + K_{SM}P_{SM})^{n_c}}(1-\alpha C_c)^{n_c} \qquad (25)$$

α=32.12 is the deactivation parameter, kg of catalyst/kg of coke [30]. Because coverage of the active sites is involved, this deactivation function can only be of the linear type [11]. n_c in Eq. (25) represents the number of active sites deactivated by the formation of the coke precursor; hence $n_c = 1$ is assumed here [11].

Devoldere and Froment [11] first developed a detailed kinetic model for the coke formation and gasification during ethylbenzene dehydrogenation. The model for the coke formation is based on a two step mechanism: coke precursor formation (monolayer or chemical adsorption) and coke growth (multi layer or physical adsorption). Due to lack of experimental data on the catalyst coke content, a simple one step coke formation model was assumed in the present work. The catalyst coke content is affected by the operating conditions during coke formation and gasification and how the dynamic equilibrium coke content varies with the feed conditions along the industrial 3-bed adiabatic reactor. The kinetics of coke gasification was assumed as follows [11, 31].

$$H_2O + C_f \underset{k_{-c1}}{\overset{k_{c1}}{\rightleftarrows}} C_f(O) + H_2 \tag{Rc1}$$

$$C_f(O) \xrightarrow{k_{c2}} CO + C_f \tag{Rc2}$$

Where $K_{H_2O} = \dfrac{k_{c1}}{k_{-c1}}$, $K_{H_2O} = 2.25*10^6 \exp(\dfrac{-42600}{RT})$, $k_2 = 4.85*10^6 \exp(\dfrac{-48200}{RT})$. Hence, the rate of gasification is expressed as in Eq. (26).

$$r_G(p_j,T,C_C) = \frac{k_2}{1+K_{H_2O}\dfrac{P_{H_2O}}{P_{H_2}}} C_{tG} = \frac{k_2}{1+K_{H_2O}\dfrac{P_{H_2O}}{P_{H_2}}} C_c^{n_G} \tag{26}$$

C_{tG} is the total sites for gasification. The value for the exponent n_G, expressing the relationship between the total number of active sites for gasification and the catalyst coke content, was found to be close to 1/3 [11]. The gasification reaction occurs at the edges of the carbon, which is oxidized by water. The net rate of coke formation can be calculated as,

$$\frac{dC_C}{dt} = r_C^o(p_j,T)\Phi_c(C_C) - r_G(p_j,T,C_C) \tag{27}$$

The "dynamic equilibrium coke content" is attained when the net rates of coke precursor formation and coke growth are zero.

$$\frac{dC_C}{dt} = \frac{k_{EB}K_{EB}P_{EB} + k_{SM}K_{SM}P_{SM}}{(1+K_{EB}P_{EB}+K_{SM}P_{SM})}(1-\alpha C_C) - \frac{k_2}{1+K_{H_2O}\dfrac{P_{H_2O}}{P_{H_2}}} C_c^{\frac{1}{3}} = 0 \tag{28}$$

After calculation of the dynamic equilibrium coke content (Cc), as suggested for Dumez and Froment [30] for the dehydrogenation of 1-butene into butadiene, an exponential deactivation function was implemented to calculate the effect of coking on the catalyst activity.

$$a_C = \exp(-\alpha_C C_C) \tag{29}$$

The short-term catalyst deactivation due to coke formation quickly reaches a rather constant level in the first few days of the plant operation. After that, a slow long-term deactivation process due to the potassium loss from the catalyst matrix slowly but continuously reduces the catalyst activity to the end of the catalyst life.

5.2 Long-term deactivation

To model the slow long-term catalyst deactivation due to the potassium loss, a_K was assumed to change as a function of time and temperature, according to [24]:

$$-\frac{da_K}{dt} = k_d \exp[-\frac{E_d}{R}(\frac{1}{T}-\frac{1}{T_R})]a^d \tag{30}$$

$$a_K = \{k_d(d-1)\exp[-\frac{E_d}{R}(\frac{1}{T}-\frac{1}{T_R})]t+1\}^{\frac{1}{1-d}} \tag{31}$$

6. Results and Discussion

All the kinetic parameters for the reactions were adopted from the relevant literature, as listed in Tables 3 to 5. The only unknown parameters in our kinetics modeling were related to the long-term activity function in Eq. (31). Assuming different catalytic sites for styrene, benzene and toluene formation, three deactivation rate constants for use in Eqs. 30 and 31, including k_d (SM), k_d (Bz) and k_d (Tol), were assumed as three adjustable parameters. The activation energies for the deactivation reactions in relation to the three reactions R1 to R3 were assumed to be equal in value, E_d, which was considered as the 4th adjustable parameter. The order of decay law, d in Eq. (30), was given constant values of one, two, and three. For each value of d, the 4 adjustable parameters were optimized to minimize the objective function (OF), which is the RMS error between the plant data and the predictions for the 1st 20 months operation period, according to Eq. (32).

$$\mathrm{OF} = \sqrt{\frac{1}{180} * \sum_{i=1}^{20} \left(\frac{X_i - X_{\exp}}{X_{\exp}}\right)^2 + \sum_{i=1}^{20} \left(\frac{y_i - y_{\exp}}{y_{\exp}}\right)^2 + \sum_{i=1}^{20} \left(\frac{T_i - T_{\exp}}{T_{\exp}}\right)^2} \quad (32)$$

The values of the optimized parameters, as well as the OF values are reported in Table 6. For each value of d, the RMS errors for the prediction of the last 10 months of plant operation are also reported in Table 6. As reported in this table, the value of d=3 produced the best results in terms of minimizing the OF and the RMS error for the prediction of the last plant operation.

Table 6. Optimized Parameters for Various Order of Decay Law

d	E_d (kJ/mol)	k_d (SM)	k_d (Bz)	k_d (Tol)	OF (fit) for months 1 to 20	Error (Prediction) for months 21 to 30
3	401	$8.8*10^{-8}$	$8.5*10^{-8}$	$4.5*10^{-7}$	0.147	0.105
2	260	$7.78*10^{-7}$	$6.32*10^{-7}$	$5.14*10^{-6}$	0.157	0.193
1	131	$5.63*10^{-6}$	$4.35*10^{-6}$	$3.90*10^{-5}$	0.178	0.453

Using the values of the optimized parameters, the ethylbenzene conversions at the exit of each reactor vs. the months of operation were calculated, as depicted in Figure 9. As seen in this figure, there is a good agreement between these predictions and the plant data. In Figure 10, the effectiveness factors for reaction R1 vs. the length of all the three reactors are shown. The values close to 1 indicates that the ethylbenzene mass transport in the large macropores of the catalyst is not the rate-limiting process.

The predicted temperature and pressure profiles along the length of all the three reactors for the 1st month are shown in Figure 11 and Figure 12, respectively. The plant data at the reactors outlet are also shown in Figures for comparison. As seen, there is a good agreement between the experimental and the predicted values.

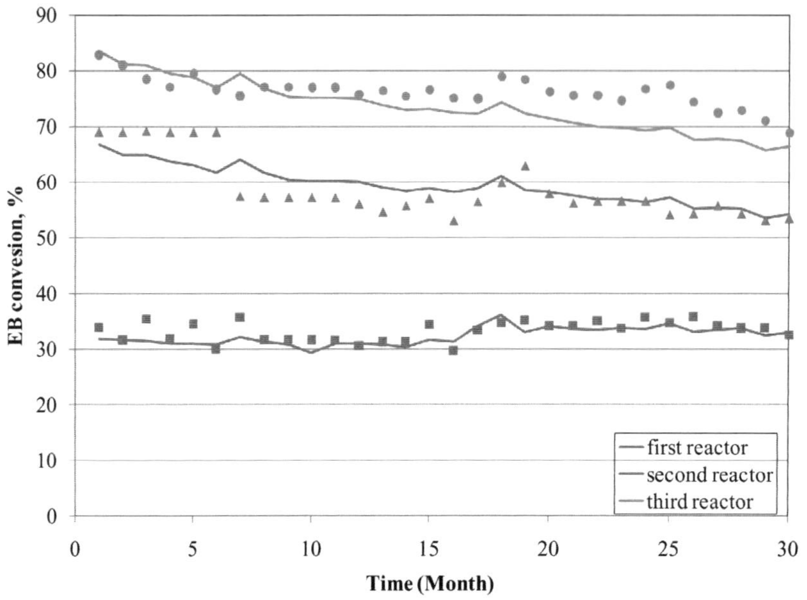

Figure 9. Predicted EB conversion according to the third order decay law (lines), as compared to the plant data for the first (□), the second (Δ) and the third reactors (○)

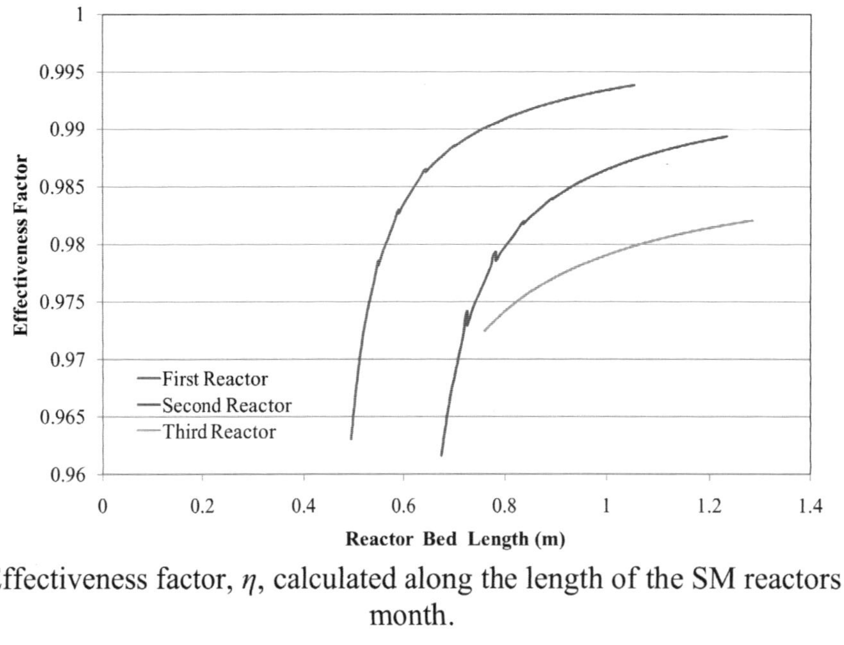

Figure 10. Effectiveness factor, η, calculated along the length of the SM reactors for the first month.

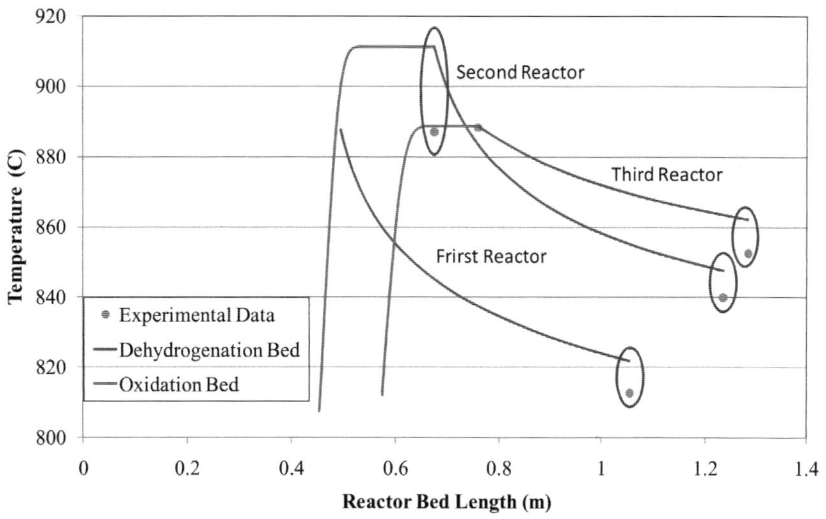

Figure 11. Temperature profiles of three reactors along the dehydrogenation and oxidation bed for the first month.

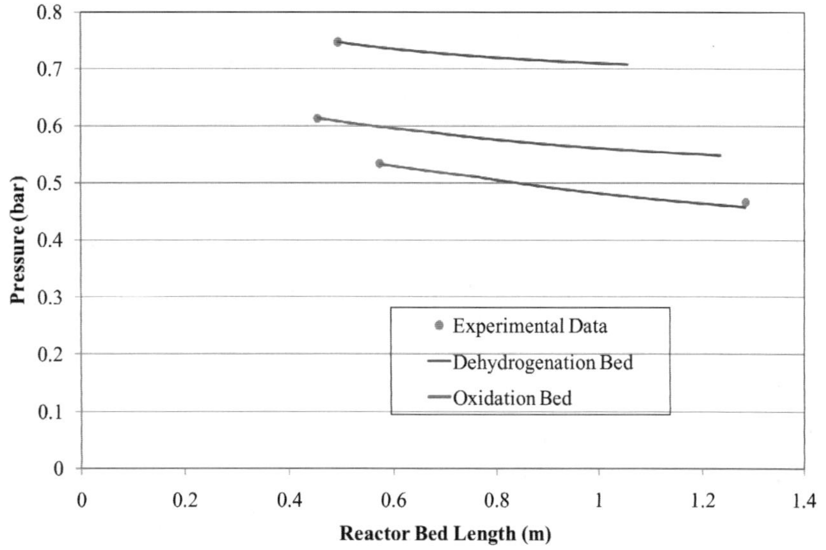

Figure 12. Pressure profiles of three reactors along the dehydrogenation and oxidation bed for the first month.

7. Conclusions

The Langmuir-Hinshelwood, Hougen-Watson model (LHHW) was used for reaction kinetics modeling of the catalytic dehydrogenation of ethylbenzene to styrene monomer. The kinetics of coking from ethylbenzene and styrene and the short-term deactivation function due to the quick coke formation were considered in the model. The kinetics of long-term deactivation of the catalyst for ethylbenzene dehydrogenation to styrene between 550 and 630 °C were determined by allowing the effects of temperature and time on the catalyst activity. Using 30 months of operating data of three industrial SM radial fixed bed adiabatic reactors, 5 unknown parameters of the long-term activity function, related to the slow loss of potassium from the catalyst, was optimized. The optimized value of the order of the deactivation law was determined to be 3.

References:

1. Kirk-Othmer, *Encyclopedia of Chemical Technology*, (4th ed.), Vol 22, Page 476
2. Meima, G.R.; Menon, P.G. "Catalyst Deactivation Phenomena in Styrene Production", *Appl. Catal., A: General* 212 (2001) 239–245.
3. Sheppard, C. M.; Maier, E. E. "Ethylbenzene Dehydrogenation Reactor Model", *Ind. Eng. Chem. Proc. Des. Dev.* 25 (1986) 207-210.
4. Sheel, J. G. P.; Crowe, C. M. "Simulation and Optimization of an Existing Ethylbenzene Dehydrogenation Reactor", *Can. J. Chem. Eng.* 47 (1969) 183-187.
5. Clough, D. E.; Ramirez, W. F. "Mathematical-Modeling and Optimization of Dehydrogenation of Ethylbenzene to Form Styrene", *AIChE J.* 22 (1976) 1097-1105.
6. Savoretti, A. A.; Borio, D. O.; Bucala, V.; Porras, J. A. "Non-adiabatic Radial flow Reactor for Styrene Production", *Chem. Eng. Sci.* 54 (1999) 205-213.
7. Lim, H.; Kang, M.; Park, S.; Lee, J. "Modeling and Optimization of a Styrene Monomer Reactor System Using a Hybrid Neural Network Model", *Ind. Eng. Chem. Res.* 43 (2004) 6441-6445.
8. Shibata, K.; Kiyoura, T. "Effect of Potassium Promoter on Iron Oxide Catalysts for Dehydrogenation of Ethylbenzene to Styrene", *Bull. Chem. Soc. Jpn.* 42 (1969) 871-874.
9. Lee, E. H. "Iron-Oxide Catalysts for Dehydrogenation of Ethylbenzene in Presence of Steam", *Catal. Rev. Sci. Eng.* 8 (1973) 285-305.
10. Hirano, T. "Roles of Potassium in Potassium-Promoted Iron-Oxide Catalyst for Dehydrogenation of Ethylbenzene", *Appl. Catal.* 26 (1986) 65-79.
11. Devoldere, K. R.; Froment, G. F. "Coke Formation and Gasification in the Catalytic Dehydrogenation of Ethylbenzene", *Ind. Eng. Chem. Res.* 38 (1999) 2626-2633.
12. B.D. Herzog; H.F. Rase, "In Situ Catalyst Reactivation: Used Ethylbenzene Catalyst with Agglomerated Potassium Promoter Dehydrogenation", *Ind. Eng. Chem. Prod. Res. Dev.* 23 (1984) 187-196.
13. Muhler, M.; Schutze, J.; Wesemann, M.; Rayment, T.; Dent, A.; Schlogl, R.; Ertl, G. "The Nature of the Iron Oxide-Based Catalyst for Dehydrogenation of Ethylbenzene to Styrene.1. Solid-State Chemistry and Bulk Characterization", *J. Catal.* 126 (1990) 339.
14. Lee, W.J.; Froment G. F. "Ethylbenzene Dehydrogenation into Styrene: Kinetic Modeling and Reactor Simulation", *Ind. Eng. Chem. Res.* 47 (2008) 9183–9194.
15. Abdalla, B. K.; Elnashaie, S. S. E. H.; Alkhowaiter, S.; Elshishini, S. S. "Intrinsic Kinetics and Industrial Reactors Modeling for the Dehydrogenation of Ethylbenzene to Styrene on Promoted Iron-Oxide Catalysts", *Appl. Catal. A-Gen.* 113 (1994) 89-102.
16. Sadeghzadeh, J.; Kakavand, M.; Farshi, A.; Abedi, M.H.; "Modeling of Radial Flow Reactors of Oxidative Reheat Process for Production of Styrene Monomer", *Chem. Eng. Technol.* 27 (2004) 139-145.
17. Abdollahi, F.; Mostoufi, N.; Sotudeh-Gharebagh R. "Optimization of Radial Flow Reactors of Styrene Production", *International Journal of Chemical Reactor Engineering,* 5 (2007) Article A75.
18. Carra, S.; Forni, L. "Kinetics of Catalytic Dehydrogenation of Ethylbenzene to Styrene", *Ind. Eng. Chem. Proc. Des. Dev.* 4 (1965) 281-285.

19. Hoiberg, J.A.; Lyche, B.C.; Foss, A.S., "Experimental Evaluation of Dynamic Models for a Fixed-Bed Catalytic Reactor", *AIChE J.* 17 (1971) 1434-1447.
20. Levenspiel, O. "*Chemical Reaction Engineering*", 3rd ed.; Wiley: New York, 1990.
21. Reid, R. C.; Prausnitz, B. E.; Poling, T. K. "*The Properties of Gases and Liquids",* 4th ed.; McGraw-Hill: New York, 1987.
22. Baghalha, M.; Ebrahimpour, O. "Structural Changes and Surface Activities of Ethylbenzene Dehydrogenation Catalysts During Deactivation", *Applied Catalysis A:* General. 326 (2007) 143–151.
23. Vafai, K. Handbook of Porous Media, 2nd ed.; CRC Press: New York, 2005.
24. Fogler, H. S. "*Elements of Chemical Reaction Engineering"*, 4th ed.; Prentice Hall: Englewood Cliffs, NJ, 2006.
25. Kareeri, A.A., Zughbi, H.D., Al-Ali, H.H., "Simulation of Flow Distribution in Radial Flow Reactors", *Ind. Eng. Chem. Res.* 45 (2006) 2862-2874.
26. Smith, J. M.; Van Ness, H. C. "*Introduction to Chemical Engineering Thermodynamics"*, 6th ed.; McGraw-Hill: New York, 2001.
27. Matsui, J.; Sodesawa, T.; Nozaki, F. "Activity Decay of Potassium-Promoted Iron Oxide Catalyst for Dehydrogenation of Ethylbenzene", *Appl. Catal.* 51 (1989) 203–211.
28. Voorhies, A. "Carbon Formation in Catalytic Cracking", *Ind. Eng. Chem.* 37 (1945) 318.
29. Froment, G. F.; Bischoff, K. B. "*Chemical Reactor Analysis and Design"*, 2nd ed.; John Wiley & Sons: New York, 1990.
30. Dumez, F. J.; Froment, G. F. "Dehydrogenation of 1-Butene into Butadiene Kinetics, Catalyst Coking, and Reactor Design", *Ind. Eng. Chem. Proc. Des. Dev.* 15 (1976) 291.
31. Katta, S.; Kealrns, D. L. "Study of Kinetics of Carbon Gasification Reactions", *Ind. Eng. Chem. Fundam.* 20(1981) 6-13.

Alternative Fuels Research Laboratory

Authors: Angela D. Surgenor, NASA Glenn Research Center, Cleveland, OH
Jennifer L .Klettlinger, NASA Glenn Research Center, Cleveland, OH
Leah M. Nakley, Sierra Lobo, Cleveland, OH
Chia H. Yen, University of Toledo, Cleveland, OH

Abstract

NASA Glenn has invested over $1.5 million in engineering, and infrastructure upgrades to renovate an existing test facility at the NASA Glenn Research Center (GRC), which is now being used as an Alternative Fuels Laboratory. Facility systems have demonstrated reliability and consistency for continuous and safe operations in Fischer-Tropsch synthesis and thermal stability testing. This effort is supported by the NASA Fundamental Aeronautics Subsonic Fixed Wing project.

The purpose of this test facility is to conduct bench scale Fischer-Tropsch (F-T) catalyst screening experiments. These experiments require the use of a synthesis gas feedstock, which will enable the investigation of F-T reaction kinetics, product yields and hydrocarbon distributions. Currently the facility has the capability of performing three simultaneous reactor screening tests, along with a fourth fixed-bed reactor for catalyst activation studies.

Product gas composition and performance data can be continuously obtained with an automated gas sampling system, which directly connects the reactors to a micro-gas chromatograph (micro GC). Liquid and molten product samples are collected intermittently and are analyzed by injecting as a diluted sample into designated gas chromatograph units.

The test facility also has the capability of performing thermal stability experiments of alternative aviation fuels with the use of a Hot Liquid Process Simulator (HLPS) in accordance to ASTM D 3241 "Thermal Oxidation Stability of Aviation Fuels (JFTOT method). An Ellipsometer will be used to study fuel fouling thicknesses on heated tubes from the HLPS experiments.

A detailed overview of the test facility systems and capabilities are described in this paper.

Introduction

Rising crude oil prices, energy independence, fluctuating global climate changes, and improving combustion performance while meeting advanced emission requirements are all major concerns in the area of commercial aviation. Reducing capital costs associated with synthetic jet-fuel production from non-conventional sources is also of concern. NASA Glenn's Alternative Fuel research effort focuses primarily on improving combustion performance and reducing emissions in advanced jet engine designs that would use alternative fuels including but not limited to F-T fuels. In the 1920's the F-T fuel process was first developed by Franz Fisher and Hans Tropsch at the Kaiser Wilhelm Institute for Coal Research in Germany prior to World War II. This process was initially developed to help Germany's need for fuel while in the middle of a petroleum shortage. The initial process for producing F-T fuels was focused on converting coal into synthetic fuel, since Germany had an abundant coal supply.

Fischer - Tropsch jet fuel is produced from reacting synthesis gas (carbon monoxide and hydrogen) over the surface of a catalyst. F-T jet fuel is comparable to Jet A-1 fuel and has been shown to have superior properties in engine testing. Research has proven significantly reduced particulate emissions without effecting engine performance, excellent thermal stability at elevated temperatures, and superior low temperature properties. Since F-T fuels do not contain aromatics or sulfur contaminants, they are typically considered higher quality, and have shown to burn cleaner than petroleum-based fuels. This in turn results in lower emission concentrations of CO, unburned hydrocarbons, and harmful particulates being released into the atmosphere.

Although there are many positive aspects of F-T fuels, there are also technical, economic, and strategic challenges that come along with this type of fuel. Challenges that still need to be overcome include the interchangeability of F-T fuels with crude-oil derived kerosene-type fuels; sealing compatibility of fuel systems already conditioned in crude-oil derived kerosene-type fuels with succeeding exposure to F-T fuels containing no aromatics; demand and supply of F-T fuels at prices comparable to crude-oil derived kerosene-type fuels; and the modification of existing fuel specifications to allow for general approval of F-T kerosene-type fuels (ref 1). The standards set by Sasol in South Africa for supplying alternative fuel to commercial aircraft is a major factor driving today's technology and investments in the direction of producing synthetic F-T fuels.

Research hardware at the NASA Glenn Research Center consists of three -1 liter, high pressure, Fischer -Tropsch continuous stirred tank reactors (CSTR). Inside of these reactors, gaseous hydrogen and carbon monoxide are mixed in the presence of a catalyst and heated under pressure to cause the F-T synthesis reaction. The F-T reaction products are collected as molten wax, liquid oil, water, and light gas. Catalysts used in the F-T process are either iron or cobalt based materials with promoters and supports such as silica and alumina. The gas, liquid, and molten wax samples taken from the F-T reactors are thoroughly analyzed by Gas Chromatographs (GCs) to investigate F-T reaction kinetics and product distributions.

A Hot Liquid Process Simulator (HLPS) enables thermal stability experimental studies to be conducted with conventional and alternative aviation fuels, as well as blended fuel derivatives. This system provides a rapid thermal stability screening test for candidate fuels before conducting expensive experiments in combustion test facilities.

Facility Overview

The Alternative Fuels Laboratory located at the NASA Glenn Research Center in Cleveland, Ohio is the only test facility at NASA Glenn solely used for Alternative Fuels Research. Figure 1 shows an external view of the Alternative Fuels Laboratory.

Figure 1: External view of Alternative Fuels Laboratory

The Alternative Fuels Laboratory consists of a control room, power supply room, and a high bay test cell. The control room contains a portion of facility control systems, an operator's control station, and the primary data acquisition system. A separate room within the control room contains a Hot Liquid Process Simulator (HLPS), an Ellipsometer and associated data acquisition hardware for Fuel Stability Testing. The power supply room contains various mechanical systems, ventilation controls and facility power panels. The 840 ft^2 test cell contains the Fischer-Tropsch Reactor hardware, system components, an activation reactor, a mini-lab housing three GCs, and ventilation system. The test cell also includes a mezzanine area in which individual reactor gas control panels and a catalytic combustor are installed. The test cell also contains individual reactor control systems which are operated by Programmable Logic Controllers (PLCs). Figure 2 shows all three CSTRs at floor level.

Figure 2: Three CSTR reactors units at floor level, control panels and combustor on mezzanine level.

A small, partially enclosed miniature laboratory (mini-lab) used for sample analysis is also located directly inside of the test cell. The mini-lab consists of a sink, cabinetry, countertops, three gas chromatographs, a computer and printer. Figure 3 shows the mini-lab enclosure.

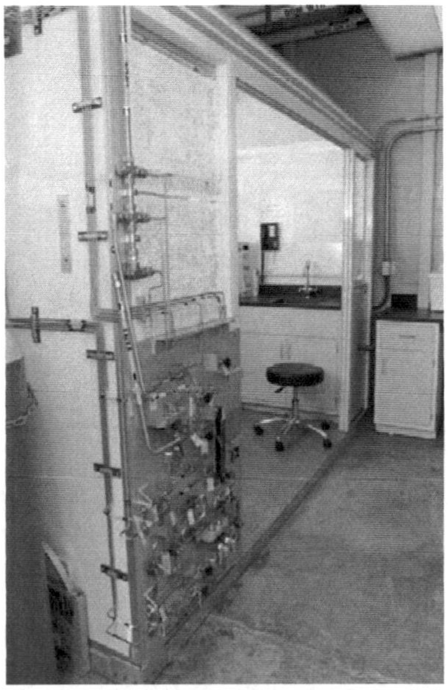

Figure 3: Mini-Lab Enclosure

Building Utilities

Gaseous nitrogen, compressed shop air, gaseous argon, and domestic water are supplied directly into the building from surrounding test facilities. The 2400 psig gaseous nitrogen supply is used to pressurize and purge the F-T reactor systems and auxiliary equipment with flow capabilities from 0-120 SLPH for each reactor. Compressed shop air is also available at 125 psig in multiple locations throughout the laboratory. Ultra high purity hydrogen, ultra high purity carbon monoxide, and nitrogen/argon gas feed systems are also supplied to the test facility with a flow range from 0-120 SLPH for each reactor.

Facility Research Systems

The alternative fuels research hardware consists of three CSTRs, a fixed-bed reactor and a catalytic combustor system. All three CSTR reactors have identical specifications. The basic reactor units were procured from Pressure Products Industries (PPI), model FC-1 series, 1-liter reactors. The CSTR reactor vessels are rated for up to 6,000 psig (408 atm) at 343°C (650°F) (ref. 2). Current operating pressures for the CSTR reactors are limited to 450 psig at 280°C, based on the system designs employed by Glenn Research Center and by the operational requirements for running F-T synthesis studies. Figure 4 shows a CSTR unit. (ref.2)

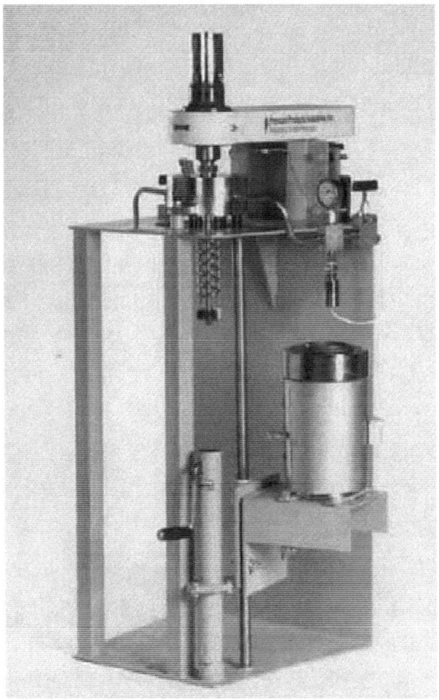

Figure 4: One (1) liter capacity CSTR (ref. 2)

A block flow diagram in figure 5 depicts the F-T rig process configuration. Each integrated reactor system includes individually controlled gas feed systems, a continuous stirred reactor unit, and individual sampling ports for removing refinery gas, liquid, and molten wax samples. Each CSTR is equipped with a dedicated programmable logic controller (PLC) for control and operation monitoring.

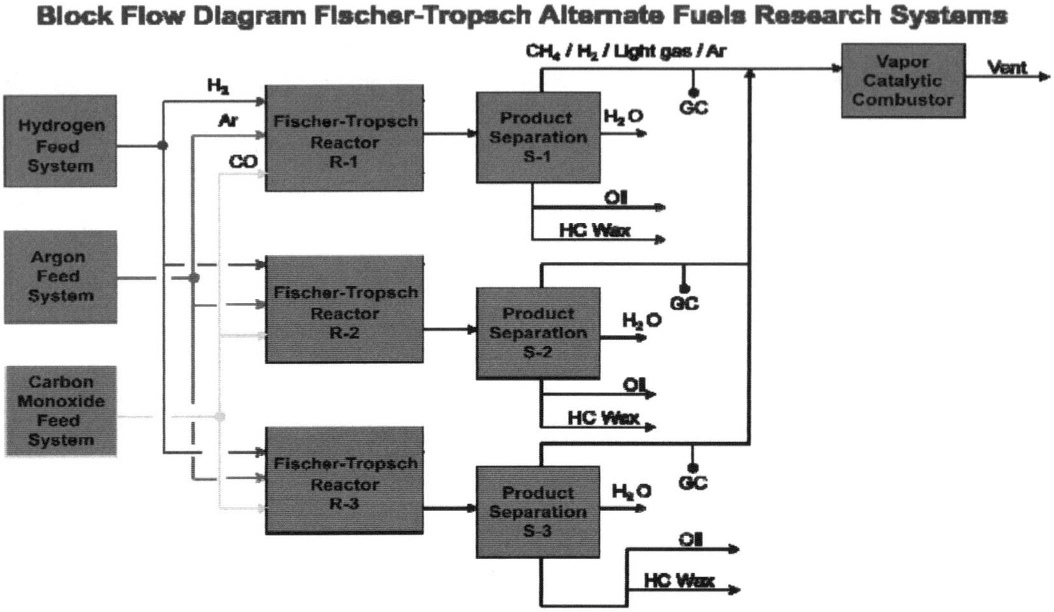

Figure 5: Block Flow diagram of F-T process configuration

The internal components of each F-T reactor consist of an agitator, a gas feed sparger and sample filter. The F-T reactor itself is constructed of 316 SS material and has internal body dimensions of 3"dia x 10.25" long and has a capacity of one liter. The reactor is custom fitted with an internal agitator driven by a ½ HP AC inverter duty variable speed explosion proof motor, typically controlled to run at 500 – 1250 rpm. The reactor agitator can be operated as low as 100 rpm for catalyst activation purposes.

The agitator shaft is hollow in order for the vapor phase product to be re-circulated from top to bottom in the reactor. There are two sets of agitator blades on the shaft. The first set of pitched impellers is equally level with the gas/liquid product interface to force downward gas action and to reduce foaming. The second set of impellers consists of six vertical turbine blades located at the bottom of the shaft. A gas feed mixture consisting of CO, H_2, and Ar is injected into the reactor below the turbine blades to maximize contact and to break up gas feeds into small diameter bubbles. An internal baffle is installed in the reactor to improve contact efficiency and to avoid swirl.

Liquid product draw-off is taken from the top of the reactor liquid volume across a sintered metal filter. The filter selection is based on the catalyst to be tested. Internal reactor temperatures can be controlled through reactor wall heaters by a dedicated PLC for each reactor. Reactor pressure is individually controlled by a manual back-pressure regulator on each reactor. Figure 6 shows a cut-away view of the internal reactor components. (ref. 2)

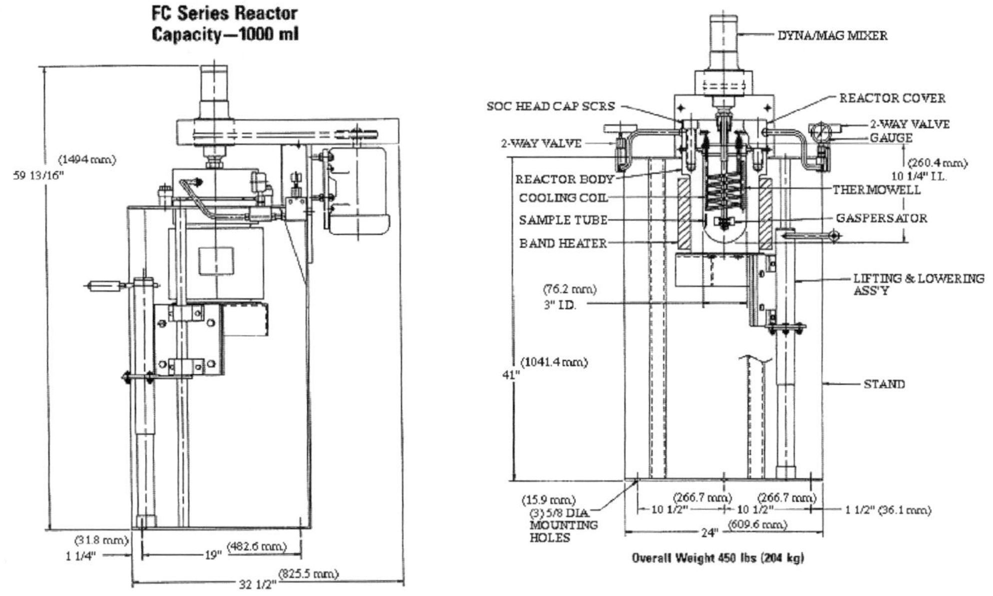

Figure 6: Cut-away view of internal components of CSTR from Pressure Products Industries (PPI) – (ref. 2)

Each F-T reactor assembly has three stainless steel cylindrical traps installed downstream of the reactors, which are used to collect F-T liquid products as three different fractions including molten waxes, light oil and water condensate. Each trap is 500 cc in capacity. Two traps are heated with electric heat tracers which are controlled by the reactor PLCs at the following conditions: hot trap at 200 °C (wax); warm trap at 100 °C (light oil). Water by-product is collected in the cold trap which is wrapped with a copper cooling coil where chilled ethylene glycol is circulated to maintain an internal temperature of 0 °C. Liquid sample level is maintained inside the reactor by manually drawing off liquid samples daily. To prevent wax from plugging the sample lines, all liquid lines are heated with heat tracing tape to approximately 175 °C.

Fixed-Bed Reactor R-4

The R-4 reactor is a fixed-bed reactor designed by NASA GRC based on requirements of a similar unit used by the Center for Applied Energy Research at the University of Kentucky. This separate reactor is used for the activation of cobalt catalysts. The reactor is designed to process catalyst materials up to 225 psig and 430°C with controlled flows of gaseous hydrogen, carbon monoxide, and either argon or nitrogen as an inert gas. Independent gas flow rates to the reactor can range from 0-120 SPLH. The average amount of catalyst loaded into the reactor is 10-20 grams, although the amount varies according to the test parameters. The reactor is capable of holding approximately 50 grams of catalyst, depending on catalyst density.

The reactor is fabricated from a 1" OD, seamless 316 Stainless Steel tube with a wall thickness of 0.083" and is 12" long. A catalyst bed is supported by packed glass wool that is inserted inside of the reactor and is replaced for every test. The reactor is heated by an external electric mighty-band coil heater rated up to 940W.

The reactor instrumentation monitors reactor internal and wall temperatures along with pressure. Two type K thermocouples are inserted internally from the bottom of the reactor and are at fixed lengths. One thermocouple is near the bottom of the catalyst bed (~ ¼" above the glass wool), while the other is near the center of catalyst bed. Multiple thermocouple locations verify that the catalyst bed is uniformly heated to enable full catalyst activation. Figure 7 shows a diagram of the fixed-bed reactor with glass wool and catalyst loading.

Figure 7: Fixed-Bed Reactor R-4: Glass wool and catalyst loading

The fixed-bed reactor is controlled by both a PLC and Labview software, which regulates gas flows, temperatures, and automatic alarms and shutdowns for high temperatures and pressures. The reactor temperature control includes a ramp & soak PID controller and an optional manual control with the use of the Labview control software. Figure 8 shows pictures of the fixed-bed reactor (R-4) with and without an insulation jacket, which is typically used during operation.

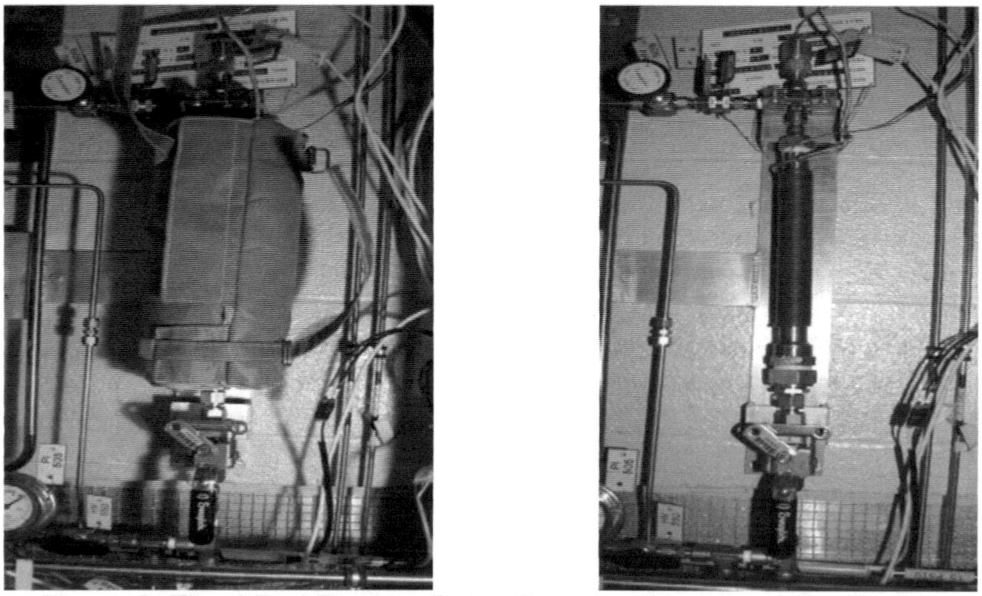

Figure 8: Fixed-Bed Reactor R-4 with and without insulation jacket

Catalytic Combustor

Catalytic Combustor

A catalytic combustor system is installed on the exit side of the reactors to oxidize refinery gases generated during the Fischer-Tropsch reaction prior to venting into the atmosphere. The combustor system is an in-house GRC design which has been successfully utilized in other facilities at NASA Glenn. This combustor takes electrically preheated air from the 125 psig facility shop air system and mixes it with the refinery gases prior to the introduction into the catalytic combustor. The combustor operating temperature is controlled nominally between 750 – 850 °C by varying the air feed flow that is sent to the combustor. The percent excess air used in the oxidation process ranges from 480 – 540% (above stoichiometric). The combustor exhaust gases are air quenched down to 150-260°C prior to release into the atmosphere through the facility vent stack. The combustor operates below 15 psig and is less than 6" in diameter. This gas clean-up system effectively eliminates CO, CH_4, and residual light hydrocarbon emissions for a more environmentally friendly discharge into the atmosphere. Figure 9 shows the catalytic combustor.

Figure 9: Catalytic Combustor

Hot Liquid Process Simulator (HLPS)

The Alternative Fuels Laboratory also houses a Hot Liquid Process Simulator (HLPS), model HLPS-400 manufactured by Alcor. The HLPS analyzer is used to determine fuel breakpoint according to ASTM D3241 specification test called JFTOT (Jet Fuel Thermal Oxidation Test). Breakpoint can be defined as the highest temperature at which a fuel can pass the JFTOT.

The HLPS uses the idea of running fluid through a resistance heated tube-in-shell heat exchanger while monitoring flow, temperature, and pressure. An air saturated fuel sample is loaded into a high-pressure reservoir. The fuel reservoir is pressurized to ~500 psig with gaseous nitrogen and is set to a constant flow rate. At the outlet of the reservoir the fuel flows through a pre-filter and then over the heated tube. The heated tube is controlled by a ramp and soak controller which increases tube temperature to the specified set-point. At the outlet of the heated tube, a small filter is in place to capture any particulates in the heated fuel. The HLPS analyzer measures differential pressure (DP) over this filter. At the end of 150 minutes test, the tube is assessed by tube discoloration in a visual tube rater which is an internally lit black box consisting of a standard ASTM color chart. The tube is optically compared to the color chart and is assigned a color number ranging from 1 to 4 (1 is metallic silver, 2 is slightly tan, 3-4 are shades of brown). A color rating of three or above or a DP greater than 25 mmHg constitutes a failed test. This test is repeated in 5°C increments until a failed test is observed. The HLPS analyzer is equipped with a visual tube rater to rank tube discoloration; a differential pressure transducer to quantify heat induced particulate formation; and a heater system to handle a wide range of fluids including highly viscous materials. Figure 10 shows a diagram of the Hot Liquid Process Simulator unit. Figure 11 shows the test tube apparatus cutout views.

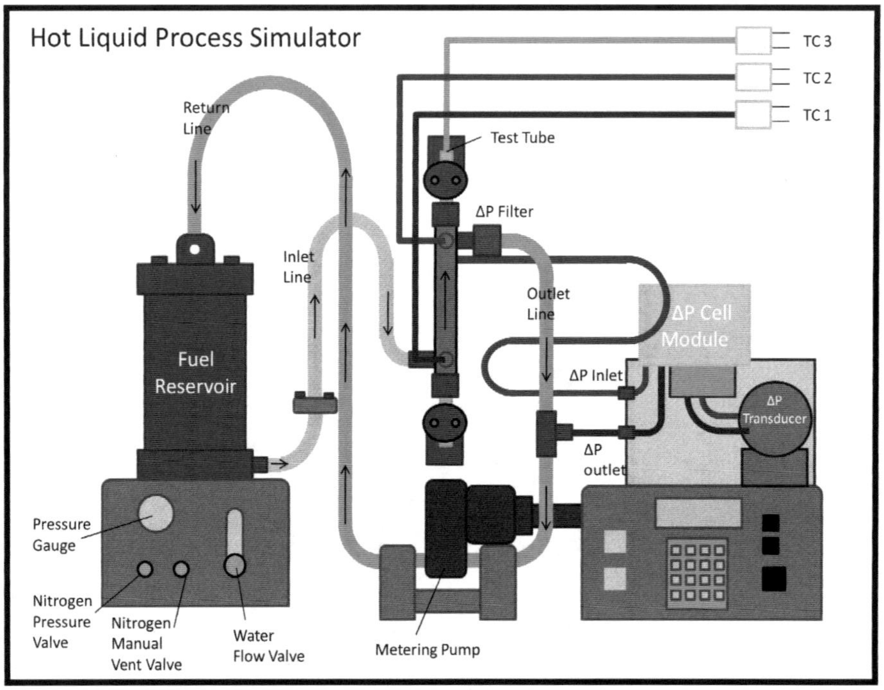

Figure 10: Hot Liquid Process Simulator

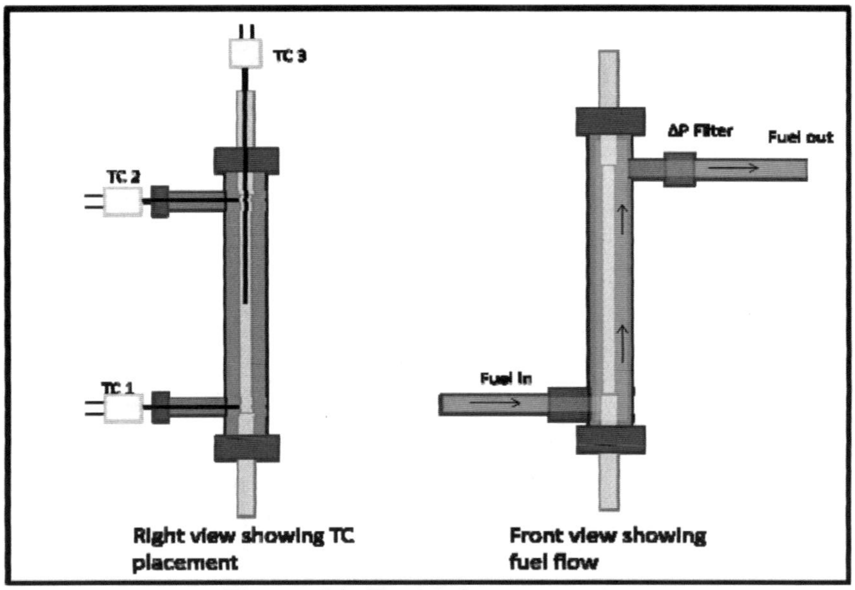

Figure 11: Test tube apparatus

The HLPS components are mainly constructed of stainless steel. The HLPS reservoir can hold up to 1L of fuel and is capable of testing fuel fouling tendencies at conditions up to 650°C. Various types of heater tubes can be used in order to reach temperature requirements for each individual test. JFTOT tests require the use of an Aluminum tube, which has an upper limit of 380°C. Steel or stainless steel tubes are also available for tests requiring higher temperatures of up to 650°C. The flow rate can be varied from 0.25 to 5mL/min for each test, although JFTOT requires a set flow rate of 3mL/min. Typical tests use approximately 600 ml of fuel per test, and take approximately four hours to assess one temperature point. The number of tests needed per fuel sample depends on what fuel is tested, and where the actual breakpoint temperature is found.

Gas Analysis Systems

The Gas Analysis equipment is located in the mini-lab enclosure located inside of the test cell. Gas Analysis equipment consists of an Agilent 3000A Micro Gas Chromatograph (Micro GC) and two Agilent 6890 Gas Chromatographs (GCs). The Micro GC is used to analyze refinery gas samples taken directly from the F-T reactors on a continuous basis. One of the Agilent 6890 GCs is designated for oil analysis, and another is designated for wax analysis.

All three gas chromatographs separate and identify compounds in the given sample. Agilent's EZChrom software is used to control the GCs and generate a detailed report for all samples sent to or injected into the gas chromatographs. The chemical compositions and quantities are used to study the overall mass balance, product distribution, and reaction kinetics. The gas chromatographs are capable of measuring compositions of light refinery gas mixtures (H_2, CO, CH_4, $C_1 - C_4$, CO_2, and N_2), light oils $C_4 - C_{44}$, and heavy hydrocarbons $C_{11} - C_{80}$ (alkanes and alkenes).

Refinery gas samples are delivered to the Agilent 3000A Micro GC through stainless steel tubing, which directly connects the product gas line to the inlet of the GC. A gas liquid separator is attached to the inlet of the Micro GC. This serves as a flow controller and filter for samples before they enter four internal capillary columns and pass through a Thermal Conductivity Detector (TCD) for component identification. The Agilent 3000A Micro GC contains four internal columns which all have unique separation capabilities and therefore require specific carrier gases to move the samples through the columns. Columns currently being used consist of a "MS 5A" column which uses Argon as a carrier gas, and a "Plot U", "Alumina Plot", and "OV-1" columns which helium as a carrier gas. Calibration is performed daily using one of two catalyst defined calibration gas mixtures. Figure 12 shows the Agilent 3000A Micro GC.

The Agilent 6890 GC's are used to analyze liquid hydrocarbon products that are manually collected from the reactors. Light oil samples and wax samples are analyzed by two separate GC's each equipped with an Automatic Liquid Sampler for direct injection into a column. Once a sample is injected into the GC, helium carrier gas moves the sample through either an Agilent "DB-5 "or "HP5" column before it is sent to a hydrogen Flame Ionization Detector (FID) to be identified . Calibration is performed periodically using applicable chemical standards. Figure 13 shows the Agilent 6890 GC.

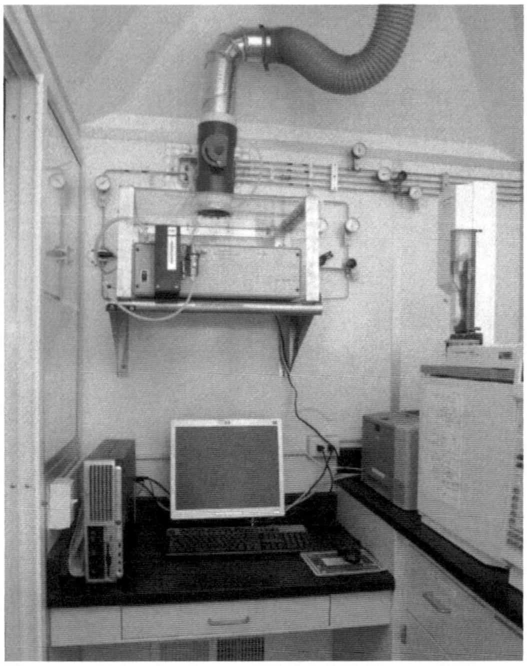
Figure 12: Agilent 3000A Micro GC

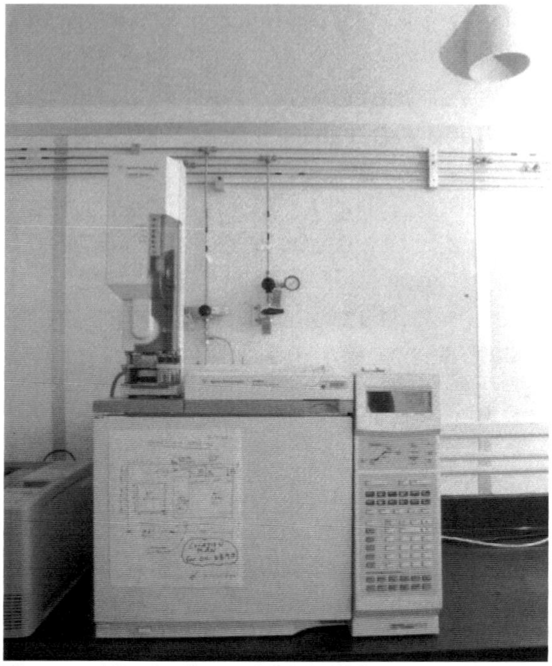
Figure 13: Agilent 6890 GC

Facility Control Systems

Facility process systems are operated by two main control systems: Programmable Logic Controllers (PLCs) and Labview software which utilizes National Instruments hardware. Each F-T reactor is independently controlled and monitored by a designated PLC. The fixed-bed reactor, catalytic combustor and all other components are controlled by the Labview software. PLCs are primarily used to control the three main reactor units and do not save data.

EZChrom software is used to control Gas Chromatograph operations, generate sample data reports, and provide automated sampling and analysis sequences. The EZChrom control system operates all gas chromatographs according to user defined parameters.

The HLPS can be operated either by using a push button control panel located directly on the HLPS unit or by the Labview program installed on the HLPS computer in the laboratory. The software for the HLPS operation was designed by Alcor-PAC and can only be modified by Alcor.

Instrumentation/Emergency Shutdowns

The F-T reactor systems include numerous types of instrumentation used to monitor and control facility and system operations. Temperatures, pressures, and flows are reported by the PLCs and Labview and are stored as data files. Both the PLCs and Labview contain integrated alarm features to ensure orderly and safe shutdowns. Facility and/or reactor shutdown sequences are instantaneously initiated when any programmed alarm or shutdown point is reached. The current Labview configuration for instrumentation consists of 278 Labview channels which are available for data acquisition. The expansion of additional modules and channels is available.

Temperature measurements are obtained with type-K thermocouples, while pressure measurements are taken with high accuracy diaphragm-type pressure transducers. Gas supply flow rates are controlled by mass flow controllers for each individual gas supply system. Refinery gas flow rates are measured by the use of wet test meters. All components are calibrated on a regular basis to ensure accuracy.

Data Acquisition

One computer has been designated as the primary data acquisition system and is directly connected to the National Instruments hardware and internal facility network. This computer executes the Labview software which collects data from the facility instrumentation. Data is transferred from the PLCs to the Labview computer for data logging purposes. The Labview software does not have control authority over the reactor PLCs other than to initiate emergency shutdowns. The second computer in the control room is used for monitoring the gas chromatographs processes, data retrieval, and monitoring individual reactor operations.

The EZChrom software is solely used for GC data acquisition. All liquid and gas sample results are stored in the GC computer.

The thermal stability data obtained from the Hot Liquid Process Simulator and is saved directly on the HLPS computer through the Labview program.

Test Duration

Fischer-Tropsch research consists of long duration reactor tests which run continuously 24 hours a day, 7 days a week. Once F-T testing is initiated, the facility runs in an automated, unattended mode with the exception of daily sample collection. Approximately two months of the year is used for facility maintenance, upgrades, and instrument calibration.

Thermal Stability tests conducted with the HLPS unit take approximately 4 hours per sample. Multiple sample tests are needed to determine the actual thermal stability of the fuel being analyzed. A maximum of two tests can be performed per day, and it typically takes about 1-3 days of testing to determine the actual "breakpoint" of a fuel. Thermal Stability tests can be performed at any time and do not have a designated maintenance shutdown period.

Future Installation

A steam generator system which can be used to vaporize de-ionized water for research purposes has already been designed for future installation. The design has incorporated a Chromolox vertical steam boiler, model CHPES which is ASME Section I "S' stamped pressure vessel. The system will be able to generate 60 pounds per hour of steam at 235 psig. It is a commercial off the shelf process system that consists of a condensate flash vessel, piping, valving, pump, electric vaporizing heater, associated instrumentation and controls with factory preset shutdowns for safe operation.

A de-ionized water (DI) system which will provide conditioned water to the steam generator system and the mini-lab has also been planned and designed for future installation. The de-ionized system is an electrically powered, commercial off the shelf unit that will be provided by Siemens Water Technologies. This system will provide between 0.15 - 3 gpm of de-ionized water flow. It will provide De-ionized water at 50 – 200 KΩ-cm to the steam generator and de-ionized Industrial-Grade water at > 1 MΩ-cm to the mini-lab.

Facility Capability Summary

F-T reactors	
Reactor capacity	1 Liter
Total # of reactors	3
Materials of Construction	316 SS
Operating pressure	0-450 psig
Operating temperature	0- 280°C
Gas feeds	UHP H2, UHP CO, N2 and Argon
Gas feed flow rates	0-120 SLPH per feed
Shop Air	125 psig
Reactor motor speed	100-1250 rpm
Catalyst type	Cobalt or Iron
Fixed Bed Reactor	
Material of Construction	316 SS
Operating pressure	0-225 psig
Operating temperature	0- 430°C
Gas feeds	UHP H2, UHP CO, N2 and Argon
Gas feed flow rates	0-120 SLPH per feed
Hot Liquid Process Simulator	
Materials of Construction	316 SS
Reservoir capacity	1 Liter
Amount of fuel used per test	600 ml
Pressure	500 psig
Maximum temperature	650 °C
Flow rate range	0.25 to 5 ml/min
Single test duration	4 hours
Gas Chromatography	
Agilent 3000A Micro GC (gas)	TCD, C1 to C4 compounds
Agilent 6890 GCs (liquid)	FID, C6 to C80 compounds

Concluding Remarks

The F-T reactors are capable of determining the conversion of syngas into synthetic crude from non-conventional resources. By using the F-T reactors to develop novel catalysts coupled with the development of the innovative F-T process, new pathways can be identified to reduce future fuel production costs. F-T process promotes fuel production from non-petroleum sources such as coal, natural gas, and renewable biomass. Alternative jet fuels thermal stability can be assessed using HLPS in this facility.

Future research plans include the testing of novel catalysts with unique types of supports and promoters. Catalyst developments will be performed at NASA Glenn along with Universities and external businesses specializing in catalyst development.

References

1. http://aiche.confex.com/aiche/s07/preliminaryprogram/abstract_80933.htm
2. http://www.pressureproductsindustries.com/reactors/FCSeries.html

(32b) Study of Fischer-Tropsch Synthesis (FTS) In a Batch Reactor with TiO_2 Supported Co Catalyst

Xiaojun Lu[1], Diane Hildebrandt[1] and David Glasser[2], (1)Center of Material and Process Synthesis, University of the Witwatersrand, Johannesburg, South Africa, Johannesburg, South Africa, (2)Centre of Material and Process Synthesis, University of the Witwatersrand, Johannesburg, South Africa, Johannesburg, South Africa

INTRODUCTION: Laboratory scale research for low temperature FTS is normally carried out in a fixed bed or a slurry bed reactor. A batch reactor in the gas-solid reaction regime is seldom used by researchers in either evaluating a catalyst or studying the mechanism of the FTS It is however of value in investigating the FTS as it creates an even reactant distribution environment for all the catalyst pellets in the reactor, and eliminates the effect of the solvent, which is used in a slurry bed. The continuous mode reactors (PFR and CSTR) are operated at steady state in most cases, while the batch reactor is operated at unsteady state and the pressure of the reactor and partial pressure of the reactants and products change with the extent of the reaction. This may offer extra information for us to understand the behavior of FT reaction.

METHODS: The Fischer-Tropsch reaction was conducted in a batch reactor on a TiO_2 supported cobalt catalyst (10%Co/90%TiO_2, BET area 28.6 m^2/g, average pore diameter 35.8 nm) in a gas-solid regime to present another way of looking at the Fischer-Tropsch Synthesis. The batch operation was initiated after the reactor was first operated at steady state in CSTR mode. The reaction conditions applied were that of typical low temperature FTS for cobalt catalyst with a reaction temperature 210 ^oC, starting pressure 20 bar(g), $H_2/CO = 2$ in the feed. The reaction duration was varied from 20 minutes to 22.5 hours

RESULTS: The conversion at various reaction durations was tracked and the reaction rate was compared to that in the CSTR mode. The concentrations of the reactants were correlated to reaction time and the reaction rate could be expressed as first order with H_2 concentration. CH_4 selectivity was investigated and the olefin to paraffin ratios for the light hydrocarbons was compared at different reaction durations and to that for the steady state in the CSTR mode. The product distribution for C_1-C_9 is given and the results were obscure as an ascending trend was observed with the increase of carbon number. The pressure of the reactor was monitored.

THE GRAPHICS (Figure 1 and Figure 2) ON THIS PAGE WERE NOT AVAILABLE ON THE E-MEDIA VERSION OF THIS DOCUMENT

Fig. 1: The pressure in the reactor at different reaction durations with the corresponding CO conversions (starting at 20bar)

Fig. 2: *The reaction rates in the CSTR and batch operation modes*

The reactor system pressure at different reaction durations in the batch reactor are given in Figure 1. The corresponding CO conversions are given as well. The reaction rates for the CSTR and batch operational modes are presented in Figure 2.

DISCUSSION & CONCLUSIONS: For the results presented in Figure 1, if one assumes that all the hydrocarbons condense, the calculated pressure in the batch would be significantly higher than the measured. Thus one can only presume that at least a fraction of the water condenses. We have attempted to simulate this behaviour. For the reaction rates in CSTR mode and Batch mode presented in Figure 2, an apparent increase is observed when the reactor was switched from CSTR to batch mode. As there would no expectation of a change of the catalyst properties, this is believed to be caused by the accumulation of the products in the reactor.

See more of this Session: Production of Fuels and Chemicals From Fischer-Tropsch Hydrocarbons I
See more of this Group/Topical: Topical 6: 11th Topical Conference on Gas Utilization

(32c) A Comprehensive Heat Transfer Model for the Fischer-Tropsch Tubular Fixed Bed Reactor

Xiaowei Zhu[1], Xiaojun Lu[2], Diane Hildebrandt[2] and David Glasser[3], (1)Center of Material and Process Synthesis, University of the Witwatersrand, Johannesburg, South Africa, (2)Center of Material and Process Synthesis, University of the Witwatersrand, Johannesburg, South Africa, Johannesburg, South Africa, (3)Centre of Material and Process Synthesis, University of the Witwatersrand, Johannesburg, South Africa, Johannesburg, South Africa

INTRODUCTION: The chemistry of the Fischer-Tropsch synthesis can be described as the polymerization of single carbon units on a catalytic site. The FTS reaction is highly exothermic and has an adiabatic temperature rise of up to 1750K. Efficient heat removal is of crucial importance because too high temperatures may lead to a loss in selectivity, possible thermal runaways and rapid deactivation of the catalyst. The investigation of the heat transfer and accurate heat transfer model are crucial when a tubular fixed bed reactor needs to be designed for this synthesis. The current heat transfer models are for classical packed bed reactors. When a fixed bed reactor is used for FTS, there may be liquid involved. A study from a colleague (Xiaojun Lu et al 2010) showed that water might be both in liquid and vapor phases under FT reaction conditions. The interaction between reaction, heat transfer, VLE, condensation, and boiling is very complex and is not covered in the current models. Thus, it would not be appropriate to apply these models in this situation. What we are attempting to do in this paper is to set up a heat transfer model, considering the heat transfer by the conduction mechanism and the condensation and evaporation of the material, to offer a better simulation for FT reactors.

METHODS: In this study, a tubular reactor with an inner diameter of 23 mm is used to investigate the radial heat transfer in a fixed bed FTS reactor with TiO_2 supported cobalt catalyst (10% Co/90% TiO_2, BET area 28.6 m^2/g, average pore diameter 35.8 nm). Thermo-wells were built into the reactor at different radial positions as indicated in Figure 1. Thermocouples could be moved up and down to measure the temperature at different axial positions. Multiple heating jackets offered a flat axial temperature profile (<0.5C) under no flow conditions. FTS reaction was performed at typical low temperature FTS conditions. Different control temperatures and flow rates were chosen

The model being developed currently regards the heat transferred in the bed as a combination of the conduction mechanism and the heat flux due to the radial flow of the materials in the reactor as a consequence of condensation and evaporation that is governed by vapour liquid equilibrium.

THE GRAPHICS (Figure 1 and Figure 2) ON THIS PAGE WERE NOT AVAILABLE ON THE E-MEDIA VERSION OF THIS DOCUMENT

Fig.1: A temperature profile of the catalytic bed when the FT reaction took place (Twall=200oC, P=20bar, SV=2.25 NL/(h-gcat))

Fig. 2 The k_{eff} of the catalyst bed at different FT reaction rates

RESULTS: A two dimensional temperature profile of the catalyst bed is presented in Figure 1. When wall control temperature was at 200°C, the maximum temperature rise in the bed was as high as 27°C with a SV of 2.25 NL/(h-gcat). In Figure 2, calculated effective thermal conductivity coefficients (k_{eff}) of the catalyst bed are plotted in respect of the reaction rates that were derived at different reaction conditions. A higher reaction rate results in a lower k_{eff}.

DISCUSSION & CONCLUSIONS: The measured temperature profiles of the bed show that the temperature rise in the catalyst bed is very sensitive to the wall control temperature, and a small increment (from 200°C to 205°C) of T_{wall} could make the maximum temperature rise increase from 27°C to 45°C. An obvious conclusion from the k_{eff} values derived for the different temperatures suggests that an assumption of constant heat transfer characteristics for the whole bed is not sufficient and even incorrect The complexity of the vapour liquid equilibrium and the condensation and evaporation of the products in different parts of the bed limits the application of the classic heat transfer model from the literature. The preliminary simulation shows promising results when using the model taking both the conduction and the heat flux carried by the side flow of the stream with condensation and evaporation into consideration.

References:

Xiaojun Lu, Diane Hildebrandt, and David Glasser

Study of Fischer-Tropsch Synthesis (FTS) in a Batch Reactor with TiO2 supported Co catalyst Abstract AIChE Spring Meeting 2010

See more of this Session: Production of Fuels and Chemicals From Fischer-Tropsch Hydrocarbons I

See more of this Group/Topical: Topical 6: 11th Topical Conference on Gas Utilization

Fischer-Tropsch synthesis: Effect of start-up solvents in a slurry reactor

Venkat Ramana Rao Pendyala, Gary Jacobs, and Burtron H. Davis*

Center for Applied Energy Research, University of Kentucky, 2540 Research Park Dr. Lexington, KY-40511, USA
davis@caer.uky.edu

Introduction:

Since many decades the Fischer–Tropsch synthesis (FTS) is an industrially established process for the production of mainly linear hydrocarbons with broad chain length distributions from synthesis gas or syngas [1]. Considering the continuously increasing crude oil price in recent years and severe environmental regulations concerning the sulphur content, it represents an alternative route for the production of petrochemical substitutes as well as high-quality fuels [2]. This interest is mainly driven by the increasingly stringent environmental legislation on transportation fuels and the possibility to monetize abundant and remote natural gas reserves as well as renewable biomass sources with independence from petroleum supply. FT derived (or synthetic) diesel fuel, which displays substantially high cetane number (typically above 70) than that obtained in conventional refineries from crude oil while being virtually free of environmentally harmful sulfur and have very low aromatic contents. With successful application of FTS technologies at large scale by Sasol and Shell, energy industries are considering using this technology as an alternative route to compensate the depleting resources of crude oil [3]. Great efforts have been made to develop efficient catalysts and to explore new technology in order to meet the demands of the economical feasibility of FTS process for industrial application.

Depending on the H_2/CO ratio of the feed, FTS is normally carried out with either cobalt- or iron based catalysts. Iron catalysts are often selected for converting coal-derived, CO-rich syngas, due to the fact that the high water gas shift (WGS) activity adjusts the H_2/CO ratio upward and low cost, flexible operation conditions as well as reasonable products distribution [4]. At present, most of slurry phase FTS use the precipitated iron catalyst promoted with potassium and copper [5]. On the other hand, cobalt catalysts are typically used in gas-to-liquids (GTL) technology, as they have low intrinsic WGS activity, suitable for converting H_2-rich, natural gas-derived syngas.

Fixed and fluid beds as well as slurry bubble columns are reactor configurations employed for commercial FTS. The main advantages of the slurry reaction over other technologies are that (a) heat removal is easier, and consequently, control of the overall exothermic reaction is better; (b) it is cheaper and simpler to build, operate, and maintain; and (c) specific volume is smaller (for similar production) [6]. On the other hand, this type of reactor has low one-pass conversion; the catalyst erodes with time and ages faster [6]. Yet the performances of these types of reactor may be different, due to the fluid dynamic and dimensional characteristics of two systems. In fact it is common practice to initially load a slurry reactor with a start-up solvent, to suspend the fresh catalyst, before starting the reactor.

Start-up solvent with different molecular weight may have different mass and heat transport effects on FT catalysts. A light solvent may have better mass transport than a heavy solvent, but it may easily escape from the slurry reactor. When a high space velocity and a relatively high temperature were used in in-situ reduction and FT reaction, the slurry level in a continuously stirred tank reactor (CSTR) can rapidly decrease if light solvent is used and this results in a rapid decrease in the conversion. The effect of initial wax medium on a low alpha iron FTS catalyst was studied by Gormley, et al. [7]. They found that the initial wax medium had little effect on the initial activity, but a major effect on rate of deactivation. Satterfield and stinger [8] determined the effect of liquid composition on the slurry FT synthesis of a fused magnetite catalyst. They observed that phenanthrene, n-octacosane, triphenylmethane performed well and Fomblin YR deactivated rapidly. The objective of this present study was to determine the effect of start-up solvent on the rate of the reaction and selectivity of the FT synthesis over the cobalt based catalysts as well as potassium promoted precipitated iron catalysts by using a slurry reactor.

Experimental:

Precipitated iron catalysts were prepared using a ferric nitrate solution obtained by dissolving iron (III) nitrate nonahydrate (1.17M) in deionized water, and then tetraethylorthosilicate (TEOS) was added to provide the desired Fe:Si ratio. A flow of the TEOS and iron nitrate mixture was added to a CSTR precipitation vessel together with a stream of ammonium hydroxide (15.6M) that was added at a rate to maintain a pH of 9.0. The final filter cake was dried for 24 h in an oven at 110 °C with flowing air. For this study, the Fe:Si catalyst base powder was then impregnated with the proper amount of aqueous KNO_3 solution to produce the desired composition of Fe:Si:K = 100:4.6:1.25 (atomic ratio). The catalyst was dried at 110 °C overnight followed by impregnation and calcination at 350 °C in an air flow for 4 h.

Condea Vista Catalox (high purity γ-alumina, 150 m^2/g) was used as the support. The catalyst was prepared by a slurry impregnation method. In this method, which follows a Sasol patent [9], the ratio of the volume of solution used to the weight of alumina was 1:1, such that approximately 2.5 times the pore volume of solution was used to prepare the loading solution. Two impregnation steps were used, each to load 12.5% of Co by weight. Between each step the catalyst was dried under vacuum in a rotary evaporator at 333K and the temperature was slowly increased to 373 K. After the second impregnation/drying step, the catalyst was calcined under air flow at 673 K. The promoter was added by the incipient wetness impregnation technique and the precursor utilized for noble metal addition was tetraammineplatinum (II) nitrate. After Pt addition, the sample was dried and calcined again, with the same conditions as before.

The FTS experiments were conducted in a 1 L CSTR equipped with a magnetically driven stirrer with turbine impeller, a gas-inlet line, and a vapor outlet line with a stainless steel fritted filter placed external to the reactor. A tube fitted with a SS fritted filter extending below the liquid level of the reactor was used for withdrawing reactor wax (i.e., rewax), thereby maintaining a relatively constant liquid level in the reactor. Separate mass flow controllers were used to control the flow rates of hydrogen and carbon monoxide. Carbon monoxide was passed through a vessel containing lead oxide on alumina to remove any traces of iron carbonyl. The gases were premixed in an equalization vessel prior to entering the reactor. The reactor temperature was maintained constant (± 1 °C) by a temperature controller. After the iron catalyst was activated, syngas was introduced at a rate of 6 NL/h g-Fe. Reaction conditions were 270 °C, 1.3MPa, $H_2/CO = 0.7$, and a stirrer speed of 750 rpm.

Results and discussion:

Polywaxes (polyethylene) are fully saturated homopolymers of ethylene that exhibit a high degree of linearity and crystallinity. The physical properties of various start-up solvents were shown in Table-1. These have narrow molecular weight distributions with a typical polydispersity (Mw/Mn) of 1.08. Product densities at 25 °C range from a low of 0.80 g/cc for C-30 oil to a high of 0.98 g/cc for polywax 3000. With increasing the molecular weight of start-up solvent, increases the average carbon number, viscosity, and melting point. In application, these exhibits sharp melt point, fast recrystallization, low melt viscosity, excellent heat stability and resistance to chemical attack. All these polywaxes are solids at room temperature; however C-30 is in the form of liquid.

The effect of start-up solvents during FT synthesis over the cobalt based catalysts as well as potassium promoted precipitated Fe catalysts was studied. To study the effect of start-up solvent, similar activation and reaction conditions were maintained in all the cases. The effect of start-up solvents on syngas conversion for potassium promoted iron catalyst were shown in Figure 1. At similar conditions (Temperature, Pressure and GHSV) by varying the start-up solvent, all the cases showed a similar induction period in which the catalyst effected increasing conversion from a low initial point to a stable conversion. After reaching the steady-state conversion level all these start-up solvents were exhibited essentially the same conversion. These results are clearly indicated that there is no significant effect on activity by using different startup solvents in a slurry reactor for iron catalysts. Gormley et al. [7] observed on a low-alpha iron catalyst that, increasing the molecular weight of initial wax deactivates more rapidly than lighter wax. However, for the cobalt catalysts the syngas conversion was found to be varying with solvent media, which might be due to the particle size of the iron catalysts being very small relative to the cobalt alumina particles used. An unsupported iron catalyst is easily broken up into fine particles in a CSTR. Supported cobalt catalysts have greater particle sizes and better attrition property than the iron catalysts; thus it may have different transport properties than the iron catalyst.

Figure 2 shows the syngas conversion of the cobalt based catalysts for different start-up solvents. All the solvents have shown a similar initial conversion, but by using the C-30 oil the syngas conversion was further increased in next two days. Polywax-3000 solvent showed an initial high activity, but it decreased in 100 hrs and then it reached the steady state level. Polywax-2000 showed a little high activity than polywax-3000. Polywax-500 and C-30 oil solvents have shown a high activity than other solvents and the deactivation rate is less comparatively high molecular weight solvents. The activity of these solvents increases the following trend: Polywax-3000 < Polywax-2000 < Polywax-500 < C-30 oil. Here, it followed a trend with decreasing the molecular weight of the solvent syngas conversion was found to increase. Chao and Lin [10] have reported that differences in solubilities, mass transfer coefficients and diffusion coefficients in waxes of varying molecular weight. They concluded that solubility of the gases decreases as the wax becomes heavier. If the solubility is the reason for this, it should be happen for both the cases iron as well as cobalt catalysts. In this study, for the iron catalysts there is no effect of start-up solvent, whereas for the cobalt catalysts activity was found to be increase with decreasing the molecular weight of the stat-up solvent. An explanation based on solubility differences is ruled out.

Table-2 shows the effect of start-up solvent on selectivity for an iron catalyst. In FT synthesis it is known that the conversion level influence the selectivity, presumably due to increasing partial pressure of water and it is therefore important to compare the catalysts at similar CO conversion level. Experiments were run at a constant conversion of CO of ca. 70%. For all start-up solvents the selectivity were also found to same, methane selectivity was ~ 8%, C_{5+} selectivity is ~ 65% and the CO_2 selectivity was ~47%. Chain growth probability was also found to be similar for all the solvents (~ 0.77). Whereas, for the cobalt catalyst selectivity was found to be differ for each solvent (Table-3) under the similar reaction conditions, The C_{5+} selectivity was found to be increasing with decreasing the molecular weight of the solvent and the methane selectivity will follow the reverse trend. However at similar CO conversion level, selectivity was found to be the same for all solvents.

The rate of CO for cobalt and iron based catalysts as a function of start-up solvent molecular weight is shown in Figure 3. The rate of CO was found to be constant for all start-up media for iron catalysts, it reveal that there is no effect with different start-up solvents. The rate of CO was found to be decrease with increasing the molecular weight of the start-up solvent. This effect might be due to the pore filling rate in the interior of the catalyst increase with increasing the start-up solvent molecular weight, so that rate of deactivation as follows the similar trend. With increasing the molecular weight of the solvent will increase the stationary part in the interior of catalyst pore.

Conclusions:

The effect of start-up solvents on the performance of cobalt based catalysts as well as potassium promoted precipitated iron catalysts were investigated during Fischer-Tropsch synthesis by using a CSTR. In this study, four different molecular weight start-up solvents were tested such as Polywax-3000, Polywax-2000, Polywax-500 and C-30 oil. The activity and selectivities (methane, C_{5+} and CO_2) were essentially the same for all the start-up solvents tested for the potassium promoted precipitated iron catalysts. Results indicated that there is no significant effect on activity and selectivity by using different startup solvents in a slurry reactor for iron catalysts. However, for the cobalt catalysts activity was found to vary with solvent media, which may be due to the particle size of the iron catalysts being very small relative to the cobalt alumina particles used. The activity of the cobalt catalysts was found to be increase with decreasing molecular weight of the solvent media. This may be as a result of pore filling rate with solvent in the interior of the catalyst, which increases with increasing molecular weight of start-up solvent. With increasing the molecular weight of the solvent will increase the stationary part in the interior of catalyst pore.

Acknowledgments

This work carried out at the CAER was supported by the Commonwealth of Kentucky.

References:

1. H. Schulz, Appl. Catal. A 186 (1999) 3.
2. C.H. Bartholomew, R.J. Farrauto, Fundamentals of Industrial Catalytic Processes, Wiley-Interscience, 2006. pp. 398.
3. H. Kolbel, M. Ralek, Catal. Rev. Sci. Eng. 21 (1980) 225.
4. S.Z. Li, S. Krishnamoorthy, A.W. Li, G.D. Meitzner, E. Iglesia, J. Catal. 206 (2002) 202.
5. H. Hayakawa, H. Tanaka, K. Fujumoto, Appl Catal A 310 (2006) 24
6. C.H. Barholomew, R.J. Farrauto, *Fundamentals of Industrial Catalytic Processes*, 2nd ed.; John Wiley & Sons, Inc.: Hoboken, NJ, 2006.
7. R.J. Gormley, M.F. Zarochak, P.W. Deffenbaugh, K.R.P.M. Rao, Appl. Catal. A: Gen., 161 (1997) 263.
8. C.N. Satterfield, H.G. Stenger, Ind. Eng. Chem. Process Des. Dev. 24 (1985) 411.
9. R.L. Espinoza, J.L. Visagie, P.J. van Berge, F.H. Bolder, U.S. Patent 5,733,839 (1998).
10. K.C. Chao, H.M. Lin, Purdue University, Final Report to DOE. Contract No. DE-AC22-84PC70024. July 1987.

Table 1: Physical properties of various start-up solvents.

Start-up solvent	Molecular weight	Average carbon number	Density, g/cc at 25 °C	Viscosity at 149 °C	Melting point (°C)
C-30 oil	437	30	0.80	1.9	-68
Polywax-500	500	35	0.93	5	88
Polywax-2000	2000	140	0.97	50	126
Polywax-3000	3000	210	0.98	130	129

Table 2: Effect of start-up solvent media on activity and selectivity for Fe based catalyst.

Startup solvent	Chain growth probability (α)	Selectivity (%)		
		CH_4 (CO2 free)	C_{5+} (CO2 free)	CO_2
C-30 oil	0.771	8.3	65.1	47.6
Polywax-500	0.767	8.2	63.7	47.7
Polywax-2000	0.765	7.9	64.7	46.4
Polywax-3000	0.775	7.9	64.8	47.7

Table 3: **Effect of start-up solvent media on activity and selectivity for Co based catalyst.**

Startup solvent	Chain growth probability (α)	Syngas Conversion (%)	Selectivity (%)		
			CH_4 (CO_2 free)	C_{5+} (CO_2 free)	CO_2
C-30 oil	0.862	61.6	8.6	81.1	1.69
Polywax-500	0.858	53.0	11.3	76.6	1.16
Polywax-2000	0.847	45.5	12.6	74.3	1.67
Polywax-3000	0.877	36.5	13.2	72.6	0.83

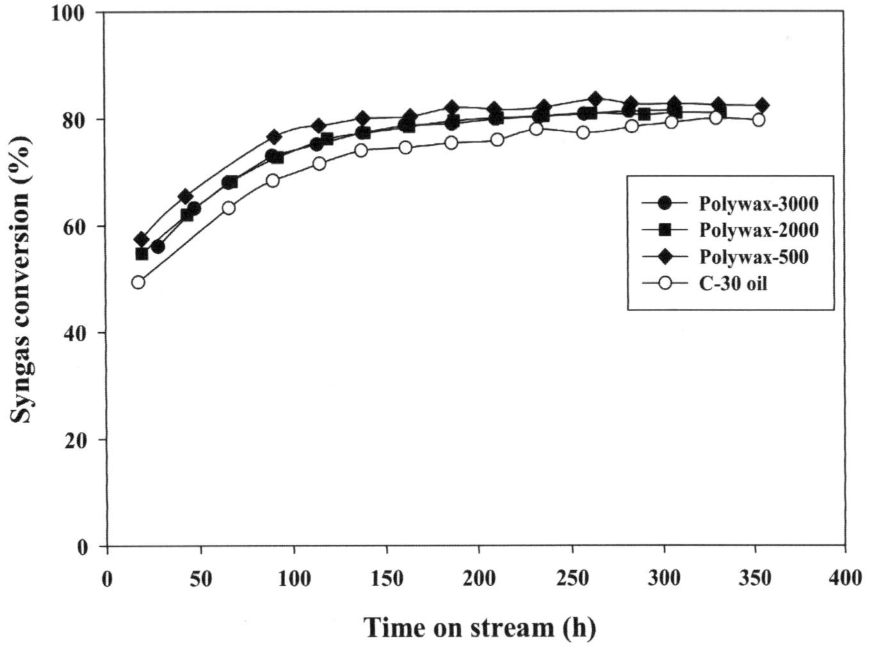

Figure 1: Effect of start-up solvent media on syngas conversion for iron based catalyst.

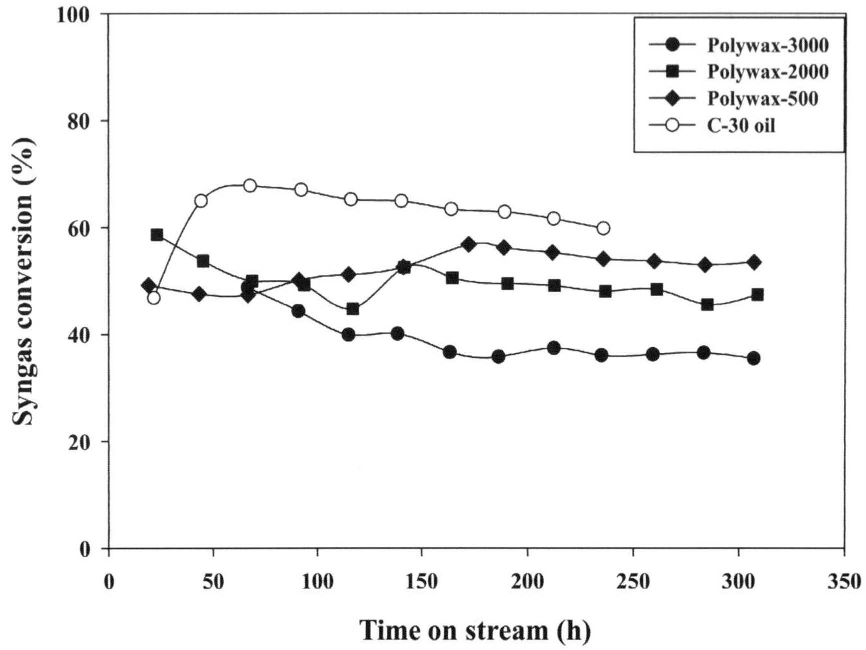

Figure 2: Effect of start-up solvent media on syngas conversion for cobalt based catalyst.

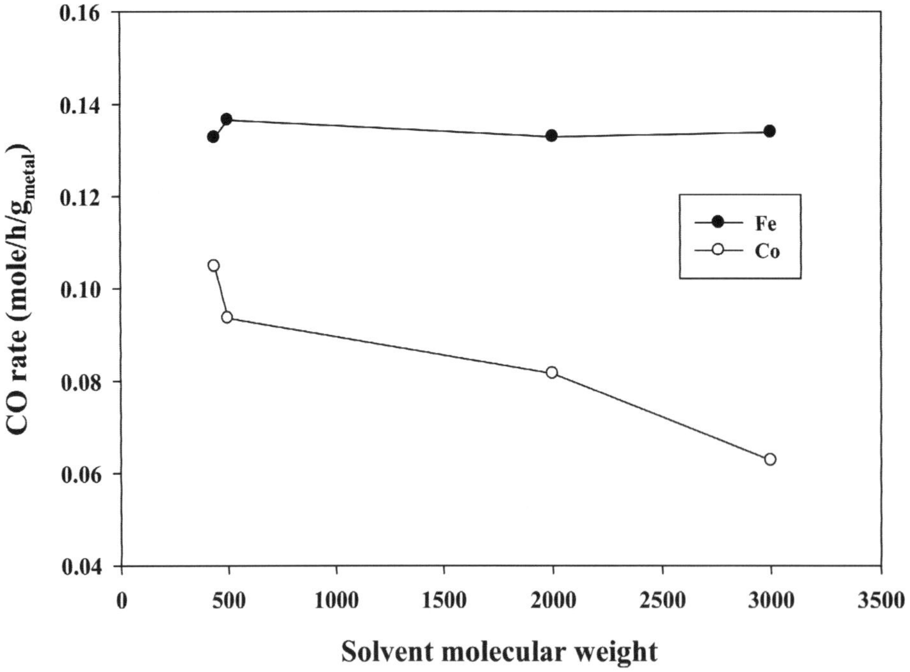

Figure 3: Effect of start-up solvent molecular weight on CO rate for iron and cobalt catalysts

(40a) Energy and Greenhouse Gas Emissions of Pyrolysis Pathways

Jeongwoo Han, Amgad Elgowainy, Ignasi Palou Rivera, Michael Wang and Jennifer M. Dunn, Energy Systems Division, Argonne National Laboratory, Argonne, IL

Pyrolysis of biomass to produce liquid fuels can play a significant role toward achieving the goal of 36 billion gallons per year of biofuels by 2022 as set by the Energy Independence and Security Act (EISA) of 2007. Argonne National Laboratory conducted life cycle analysis of various pyrolysis pathways by expanding and employing the Greenhouse Gases, Regulated Emissions, and Energy Use in Transportation Model (GREET). The fuel cycle energy use and greenhouse gases (GHG) emissions from the pyrolysis pathways were compared to the baseline gasoline pathway as well as alternative biofuel pathways such as corn and cellulosic ethanol, cellulosic biomass to Fischer Tropsch Diesel (FTD), and soybean biodiesel, renewable diesel and renewable gasoline. The life cycle results are significantly impacted by the source of hydrogen production for hydrotreatment and hydrocracking processes, and the handling of the biochar and other coproducts of the pyrolysis pathway.

See more of this Session: Advances In Alternative BioEnergy
See more of this Group/Topical: Topical 9: Waste & Biomass Valorization

(40d) Efficient Fermentation of Glucose and Xylose Mixtures Using Immobilized Saccharomyces Cerevisiae on Pine Wood Chips and Free Pichia Stipitis

Ravi Dhabhai, Chemical Engineering, Malaviya National Institute of Technology Jaipur, Jaipur, India, **S. P. Chaurasia**, Chemical Engineering, Malaviya National Institute of Technology, Jaipur, India and Ajay K. Dalai, Chemical Engineering, University of Saskatchewan, Saskatoon, SK, Canada

In this work, the ability of pine wood chips (PWC) as yeast support has been evaluated. *Saccharomyces cerevisiae* was immobilized on this biological support and mixed sugar fermentation was carried out with free culture of *Pichia stipitis* and immobilized *S. cerevisiae*. Pine wood was used in two forms; untreated (UPW) and treated (TPW). Fermentation experiments with free culture of both the yeasts were also carried out. The immobilization was carried out by co-culturing 48 h grown *S. cerevisiae* culture with PWC for 24 h with gentle shaking. The immobilization was confirmed by optical microscope and by the efficiency to ferment glucose quickly as compared to free culture. Initial experiments were carried out with 50 g/l mixture of glucose and xylose (1.5:1.0 ratio of glucose to xylose).

The efficiency of support for fermentation was evident by the fact that 63-70% of sugars were consumed in only 12 h and 84-93% fermentation efficiency was observed with almost complete utilization of both the sugars in 44 h. Doubling the sugar concentration (100 g/l) had little effects on fermentation efficiency but time required for complete sugar fermentation was increased to 78 h from 44 h. TPW worked out as a better support than UPW as the surface area of the former increased after the treatment to 1.5 times.

See more of this Session: Advances In Alternative BioEnergy
See more of this Group/Topical: Topical 9: Waste & Biomass Valorization

(40e) A Review of Biodiesel Production Via Reactive Distillation

Iman Noshadi[1], Nor Aishah Saidina Amin[1], Richard Parnas[2] and Seyed Masoom Khazraee[3], (1) Chemical Engineering, University Technology Malaysia, Johour, Malaysia, (2)Institute of Material Science, University of Connecticut, Storrs-Mansfield, CT, (3)Fars Engineering Research Center, Shiraz, Iran, Shiraz, Iran

Abstract— there are two main problems with continued use of fossil fuels, which meat about 80% of the world energy demand today. The first problem is that they are limited in amount and sooner or later will be depleted. The second problem is that fossil fuels are causing serious environmental problems. Since biodiesel is sustainable and combines high performance with environmental benefits it has become increasingly attractive as an alternative fuel. The conventional method for the production of biodiesel has several disadvantages: high alcohol demand due to shift equilibrium toward fatty esters, necessity of catalyst neutralization hence causing salt waste stream, high separation costs of fatty esters products from the reaction mixture, high production costs caused by complex processes including one or two reactors and several separation units. The reactive distillation (RD) process for the production of biodiesel overcomes these disadvantages. This paper attempts to review production of biodiesel by reactive distillation. The effect of various parameters feed ratio, type of applied catalyst, feed temperature, reboiler duty, type of column have also been summarized in this paper.

See more of this Session: Advances In Alternative BioEnergy
See more of this Group/Topical: Topical 9: Waste & Biomass Valorization

(41a) Controlled Porosity and Structure In Nanoporous Metal Oxide Monoliths by Sacrificial Vapor Phase Infiltration of Polyesters

Bo Gong, Qing Peng, Jesse Jur, Christina Devine, Kyoungmi Lee and Gregory N. Parsons, Department of Chemical and Biomolecular Engineering, North Carolina State University, Raleigh, NC

Preparation of porous metal oxide materials has attracted considerable attention because of possible use in separations, catalyst support, chemical sensors, optical and electronic devices, energy storage and solar cells. Although many methods are known to synthesize porous materials, obtaining porous materials with well defined macro-scale morphology remains an important challenge. In the present work, we demonstrate that porous metal oxide materials (including Al_2O_3 and TiO_2) with controlled pore size and distribution are produced by sacrificial infiltration of polyesters with reactive metal organic precursors and water. Results were discovered during recent experiments using atomic layer deposition (ALD) to treat and coat polymer materials. We found that exposing polyesters to Al_2O_3 and TiO_2 ALD processes, using trimethyl aluminum and titanium tetrachloride precursors, respectively, produced organic-inorganic hybrid materials with a physical macro-structure that followed the original polymer template. After post annealing, nanoporous materials with pre-defined macro-structure were prepared. Nitrogen adsorption/desorption analysis confirmed that the surface area, pore volume, and pore size distribution were tuned by the process parameters, such as temperature and vapor exposure time. The highest BET surface area of prepared porous Al_2O_3 and TiO_2 were ~440m^2/g and 210m^2/g respectively. The infiltration mechanism and chemical reactions were explored by in-situ infrared transmission analysis and quartz crystal microgravimetry. Thermal properties and air stabilities of the hybrid materials were assessed by the DSC/TGA and Ex-situ IR experiments.

See more of this Session: Advances In Aromatics and Petrochemicals Processing II
See more of this Group/Topical: Fuels and Petrochemicals Division - See Also Topicals 4, 6, and 7

(41b) Melt-Electrospinning of High Melt Flow Index Polypropylene and Polycaprolactone Utilizing An Extrusion Process

Wade S. DePolo and Eugene G. Joseph, Institute for Critical Technolgy and Applied Science, Virginia Poytechnic Institute, Blacksburg, VA

Melt-electrospinning has grown more attention relative to that of conventional solution electrospinning processes in the last 15 years due to higher production rates, removal of harmful, carcinogenic solvents, and the ability for scale-up. Therefore, melt-electrospinning is a more viable process for commercialization than that of the conventional solution electrospinning process. In this study, the effect of processing parameters, i.e. flow rate, applied voltage, and temperature, on the average fiber diameter of a high melt flow index polypropylene and polycaprolactone is investigated using an extrusion process in order to achieve fiber diameters that are less than 1 micron. Furthermore, the influence of molecular weight and, therefore, viscosity is also studied by utilizing two polycaprolactones with different melt flow indices. Co-extruded melt electrospun fibers will also be generated, which can be beneficial in the medical field, i.e. tissue scaffolding, ultra-filtration, and drug delivery systems.

See more of this Session: Advances In Aromatics and Petrochemicals Processing II
See more of this Group/Topical: Fuels and Petrochemicals Division - See Also Topicals 4, 6, and 7

Synthesis of Gemini Surfactant Disodium Dioctyl naphthalene Disulfonate

*Zhiqiang He, Jinglong Wang, Yun Fang**
School of Chemical and Material Engineering, Jiangnan University, Wuxi, Jiangsu 214122, China

Abstract In this paper, disodium dioctyl naphthalene disulfonate (DDOND) surfactant has been synthesized from cheap raw materials naphthalene, 1-octylene and chlorosulfonic acid. The alkylation of naphthalene with 1-octylene was performed in presence of acidic catalysts. It' s found that more than 90% conversion of 1-octylene and about 95% selectivity of desired dioctyl naphthalene (DON) can be obtained. The as-synthesized DON was further sulfonated by using chlorosulfonic acid as sulfonating agent in laboratory and neutralized by NaOH, and thus the target product DDOND were obtained. The purities of DDOND were determined by HPLC and the structure of DDOND was characterized by IR and ESI-MS. The surface tensions of DDOND aqueous solutions were measured with a ring method. It's found that the novel surfactants possess high capacity and efficiency of lowering both surface tension and cmc.

Keywords Gemini surfactant; disodium dioctyl naphthalene disulfonate(DDOND); alkylation

1 Introduction

Gemini-type surfactants made up of two hydrophilic heads and two hydrophobic tails in their molecular structure rather than single hydrophilic head and single hydrophobic tail in the conventional surfactants are presently attracting considerable interest in the academic and industrial communities working on surfactants [1]. The current interest is raised from three unusual properties of Gemini surfactants [2, 3, 4]. Firstly, their cmc values are generally 10-100 times lower than that of the contrastive conventional surfactants. Secondly, they lower the surface tension of water more efficiently than the contrastive conventional surfactants. The last but not the least, aqueous solutions of some quaternary ammonium Gemini surfactants with short spacers have high viscosity even at relatively low concentration, whereas the solution of the contrastive conventional surfactants remains low viscous.

Currently, a considerable number of investigations have been reported on the surface and aggregation properties of a variety of Gemini surfactants [5]. There were relatively few commercial application reports about Gemini surfactants as a result of hardly synthesized and costly. Thus, it is important to prepare the novel Gemini surfactants with cheap raw materials and simple craftwork. In this work, we describe a method to synthesize new Gemini anionic surfactants by cheap raw materials contained naphthalene, 1-octylene and chlorosulfonic acid and the surface-active properties in water of a typical member of them, disodium dioctyl naphthalene disulfonate (DDOND) are primarily investigated.

2 Materials and Methods

2.1 Materials

Cyclohexane, naphthalene, chlorosulfonic acid, anhydrous alcohol, 1,2-dichloroethane and petroleum ether are all analytical grade and purchased from Sinopharm Chemical Reagent Company (Shanghai, China). 1-octylene are purchased from BP China and used without further purification. Deionized water was obtained from a Millipore Milli-Q water purification system (Millipore, USA).

2.2 Synthesis of Disodium Dioctyl Naphthalene Disulfonate (DDOND)

2.2.1 Synthesis of Dioctyl Naphthalene (DON)

(1) One-step method:

A dry, 250-mL, four-necked and round-bottom flask was fitted with a reflux condenser, a drying tube and a thermometer, which extended well into the reaction mixture. 0.1 mol naphthalene dissolved in 20 mL cyclohexane and 5 mmol acidic catalysts were added into the flask then heated to 30 °C by a thermostated oil bath. 0.2 mol 1-octylene was added dropwise over a period of 30 min at 30 °C and continued the stirring at this temperature for 3 h. After the reaction was completed, the upper organic layer containing the octylated product was separated from the acidic catalysts in the bottom simply by decantation. The solvent was evaporated with a rotary evaporator. And the residual liquid was washed with deionized water (3 × 30 mL) and then dried over 10 g of anhydrous $MgSO_4$.

(2) Two-step method:

Preparation of mono-octylated intermediate: The reaction was carried out in a magnetically stirred glass reactor (capacity: 100 mL) fitted with a reflux condenser and a thermometer. The outlet of the reflux condenser was connected to a drying tube, and the reactor was kept in a thermostated oil bath at 80 °C. 0.1 mol naphthalene, 5 mmol acidic catalyst and 0.1 mol 1-octylene were added together into the flask then heated to 80 °C by a thermostated oil bath for 3 h. After the reaction was completed, the upper organic layer containing the mono-octylated intermediate was separated from the acidic catalysts in the bottom simply by decantation, then washed with deionized water (3 × 30 mL) and then dried over 10 g of anhydrous $MgSO_4$.

Preparation of di-octylated product: The di-octylation reaction was carried out in the same reactor as above-mentioned. 0.5 mol monooctyl naphthalene and 0.1 mol acidic catalyst were added together into the flask then heated to 20 °C. Then, 0.1 mol 1-octylene was added slowly from a dropping funnel over a period of about 30 min. The reaction was continued for 2 h. After the reaction was completed, the upper organic layer containing the di-octylated product was separated from the acidic catalysts in the bottom simply by decantation, then washed with deionized water (3 × 30 mL) and then dried over 10 g of anhydrous $MgSO_4$. The di-octylated product was simply separated from the mono-octylated intermediate by distillation under reduced pressure. The unreacted monooctyl naphthalene can be reused in next time.

All analyses were carried out with a Waters MALDI SYNAPT Q-Tof HPLC/MS equipped with Lichrospher C18 ,4.6×250 nm ,5um. The concentration of reactant and product was directly given by the system of HPLC/MS chemstation according to the area of each chromatograph peak.

2.2.2 Sulfonation of Dioctyl Naphthalene

A mixture of 7 g of DON and 15 mL of 1, 2-dichloroethane was added to a 100-mL three-necked and round-bottom flask and cooled to between –10 and –5°C. Chlorosulfonic acid (3.5 mL) was added slowly from a dropping funnel over a period of about 30 min. The temperature of the reaction increased slowly to 55°C owing to heat of the reaction and was held as near 55°C as possible. The reaction was continued for 2 h and then neutralized by adding 30% NaOH. Disodium dioctyl naphthalene disulfonate (DDOND) were deoiled followed by desalted in petholeum ether and anhydrous methanol, respectively. The final product was purified by recrystallization for three times in an ethanol/water mixture (50:50, vol/vol) and then dried for 12 h under vacuum at 65 °C. And their structures were confirmed by ESI-MS and FT-IR.

3 Result and Discussion

Alkylated naphthalenes and alkylated naphthalene sulfonates are important chemicals used in fine chemicals and pharmaceutical industries. Alkylated naphthalene sulfonates are one of the most important anionic surfactants for all kinds of industrial purposes. Their use as chemical oil displacing agent for enhanced oil recovery gave satisfactory results and the ultra low interfacial tension between oil and water could be achieved [6]. Additionally, due to their tremendous surface activity, they are widely used in the fields of cleaning, dying and printing, emulsifying, ore milling and nanotechnology. In the present work, a large number of studies focus on long chain monoalkyl naphthalene sulfonates [7] and long chain alkyl methylnaphthalene sulfonates [8]. There were relatively few reports about dialkyl naphthalene disulfonates which are Gemini anionic surfactants. In this work, a typical member of the series of disodium dialkyl naphthalene disulfonates, DDOND, has been synthesized from cheap raw materials naphthalene, 1-octylene and chlorosulfonic acid as shown in Scheme 1.

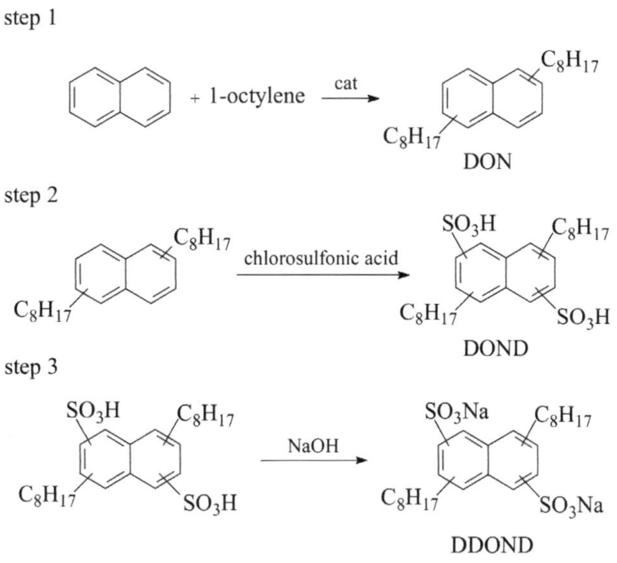

Scheme 1 The synthesis route of DDOND

3.1 Octylation of Naphthalene

Alkylated naphthalenes can be synthesized by acid catalyzed Friedel-Crafts reaction of alkylation agent. This Friedel-Crafts reaction is conventionally catalyzed by homogeneous acid catalysts like H_2SO_4, BF_3, HF, $AlCl_3$ or $FeCl_3$ but their environmental non-friendliness, troublesome product recovery and purification, and impossibility of catalyst recycling make them unsuitable for the purpose. Many efforts have been made to exploit the environmentally friendly catalytic technology for the target reaction. Zhao et al. have studied the alkylation of 1-and 2-methylnaphthalene with long chain alkenes in liquid phase using different alkyl-containing amine ionic liquid and have achieved high conversion of alkylating agent and excellent selectivity for the desired product, i.e. monoalkylated methylnaphthalene[8, 9].

In this work, ILs having different organic cations and varied acidic property were tested for alkylation reactions. Results of the alkylation of naphthalene using these ILs are outlined in Table 1. Among these ILs catalysts, IL-I and IL-II showed high activity for the alkylation reactions in low temperature near 20 °C. The high activity of these two catalysts is due to their high acidity as compared to IL-III and IL-IV. However, the alkylation products using IL-I and IL-II as catalyst are very complicated. There is nearly a third of the amount of alkylation products is undesired multialkylated, even reducing the molar ratio of 1-octylene to naphthalene. With further reduce the molar ratio the multialkylated vanished, but the conversion of naphthalene was too low to be accepted. It is hardly to recycle the unreacted naphthalene by reason of naphthalene's high melting point and easy-sublimateing. When naphthalene was reacted with 1-octylene using IL-IV as catalyst, monooctylated naphthalene was found to be the main product but dioctylated product was also observed (entry 6, 7 and 8). It was observed, however, that reducing the ratio of 1-octylene to naphthalene enhanced the conversion of 1-octylene and the selectivity to the monooctylated product. The only product of monooctylated naphthalene can be obtained when the molar ratio of 1-octylene to naphthalene is 1. Hence, we choose to synthesize the dioctyl naphthalene using the two-step method. Firstly, the monooctylated naphthalene was prepared using IL-IV as the catalyst. Then, the desired dioctyl naphthalene was synthesized by monooctylated naphthalene reacted further with 1-octylene using IL-I as the catalyst. In the second step, when the molar ratio of monooctylated naphthalene to 1-octylene was increased to 4 the selectivity of dioctyl naphthalene came up to 95 %. The unreacted monooctylated naphthalene can be recycled by distillation under reduced pressure and reused in next time.

Table 1 Activities of Lewis acidic ILs catalysts in alkylation of naphthalene with 1-octylene

Entry	Catalyst	Naphthalene/1-octylene (mol ratio)	T/°C	1-octylene conversion (%)	Product selectivity(%)		
					Mono octylated	Di octylated	Multi octylated
1	IL-I	1/2	20	> 99	35	38	27
2	IL-I	1/1	20	> 99	39	43	18
3	IL-II	1/2	20	> 99	40	44	16
4	IL-II	1/1	20	> 99	43	48	9
5	IL-III	1/1.5	50	97	59	39	2
6	IL-IV	1/1.5	80	92	86	13	0
7	IL-IV	1/1.25	80	95	93	7	0
8	IL-IV	1/1	80	96	> 99	0	0

3.2 Structural Characterization of DDOND

The ESI–MS technique was employed to measure the molecular weight of the target product DDOND in this work. The result is shown in Fig. 1. The mass-to-charge ratio (*m/z*) of $[DDOND]^-$ which lost a H^+ is 511.4, that matches well with the theoretical calculation result (511.4) of the molecule ion weight for $[DDOND]^-$. And the mass-to-charge ratio (*m/z*) of 255.2 is half of the molecule ion weight for $[DDOND]^{2-}$ which lost two H^+.

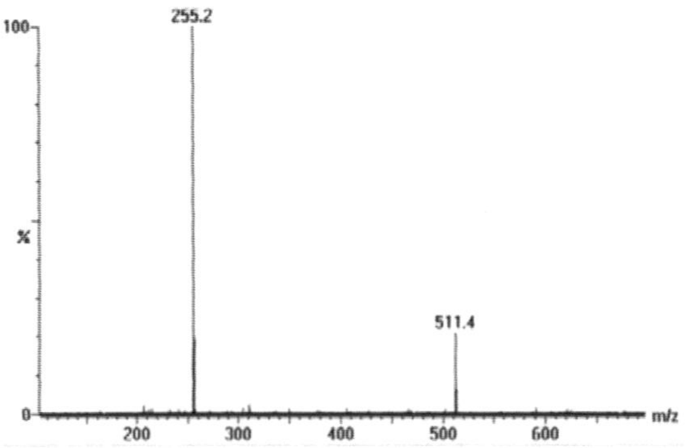

Fig. 1 The ESI-MS spectrum of DDOND at negative ion mode

The FT-IR spectrum of the DDOND is shown in Fig. 2. The absorption peaks of S=O at υ = 1202 cm^{-1} and C-S at υ =1055 cm^{-1} is feature of the spectra of sulfonate, indicating that initiation reaction has occurred.

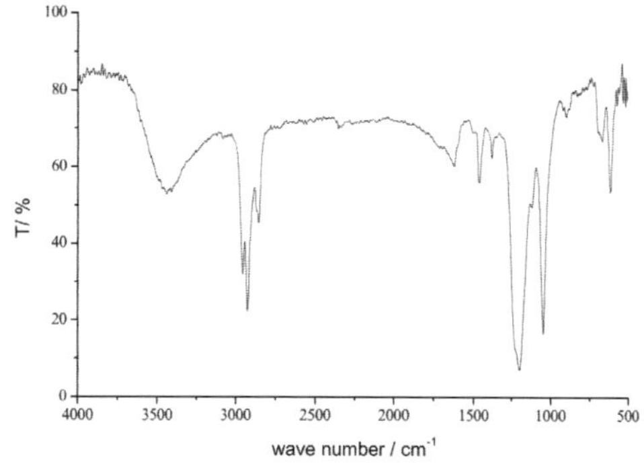

Fig. 2 FT-IR spectrum of DDOND

3.3 The Surface Properties of DDOND in Aqueous Solution

The concentration dependence of the surface tension for DDOND in aqueous solutions at 25 °C is shown in Fig. 3. The surface tension γ(mN/m) gradually decreases to a plateau region with the DDOND concentration c(mol/L) increasing. A decrease in surface tension indicates that the surfactant molecules adsorbed at the air/solution interface. The break points appear in the γ-lgc curve suggests the formation of micelles in aqueous solutions. The results of the cmc and γ_{cmc} for the surfactant were obtained from the γ-lgc curve. The value of cmc at 25 °C is 0.29 mmol/L and γ_{cmc} is 29.2 mN/m. It's found that the novel surfactant DDOND possesses high capacity and efficiency of lowering both surface tension and cmc.

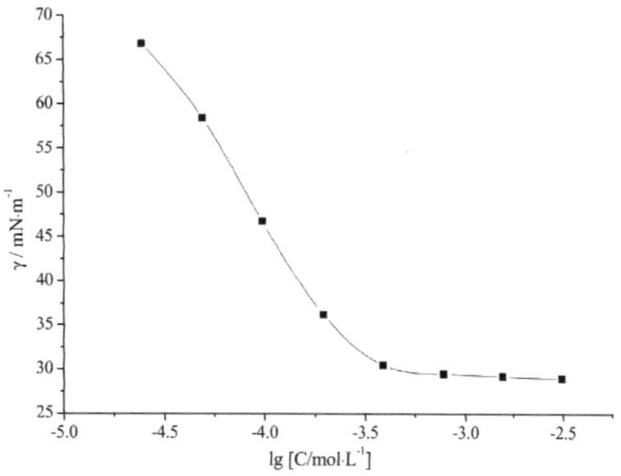

Fig. 3 The γ-lgc curve of DDOND in aqueous solutions at 25 °C

4. Conclusion

Ionic liquid catalysts, IL I to IL IV, are active Lewis acid catalysts in alkylation reactions of naphthalene with α-olefins. Among these catalysts, IL-I is highly active in low temperature, but the di-alkylated selectivity is very low. IL-IV is highly active for alkylation and good selectivity of mono-alkylated in high temperature. We combined these two catalysts to synthesis the target dialkylated dioctylnaphthalene with high selectivity. The value of cmc of disodium dioctyl naphthalene disulfonate at 25 ^{o}C is 0.29 mmol/L and γ_{cmc} is 29.2 mN/m. It's found that the novel surfactant DDOND possesses high capacity and efficiency of lowering both surface tension and cmc.

References

[1] R. Zana, Adv. Colloid Interface Sci. 97 (2002) 205.

[2] R. Zana, M. Benrraou, R. Rueff, Langmuir 7 (1991) 1072.

[3] M.J. Rosen, J.H. Mathias, L. Davenport, Langmuir 15 (1999) 7340.

[4] F. Kern, F. Lequeux, R. Zana, S.J. Candau, Langmuir 10 (1994) 1714.

[5] Xigang Du,Yao Lu, Ling Li, Jinben Wang, Zhengyu Yang. Colloids and surfaces A: Physicochem. Eng. Aspects, 290(2006)132-137.

[6] Z.K. Zhao, Y. Ba, Z.S. Li, W.H. Qiao, L.B. Cheng, Tenside Surf. Deterg. 41 (2004) 225–229

[7] Xiaoli Tan, Lu Zhang, Sui Zhao, Jiayong Yu and Jingyi An. Journal of Surfactants and Detergents, 2004, 7(2):135-139.

[8] Z. Zhao, W. Wang, W. Qiao, G. Wang, Z. Li, L. Cheng, Appl. Catal. A: Gen. 290 (2005) 133–137.

[9] Z. Zhao, W. Wang, W. Qiao, G. Wang, Z. Li, L. Cheng, Micropor. Mesopor. Mater. 93 (2006) 164–170.

(41d) Influence of Prepared Method on the Structure and Performance of VO_x-TiO_2-$(SO_4)_2$- Catalyst for the One-Step Oxidation of Methanol to Dimethoxymethane

Heqin Guo[1], **Debao Li**[1], Congbiao Chen[1] and Yuhan Sun[2], (1)Institute of Coal Chemistry, Chinese Academy of Sciences, Taiyuan, China, (2)Institute of Coal Chemistry, Chinese Academy of Sciences; Shanghai Advanced Research Institute, Chinese Academy of Sciences,, Shanghai, China

Dimethoxymethane ($CH_3OCH_2OCH_3$), as an important chemical intermediate, was widely used as diesel additive and building block in organic synthesis. Generally, DMM is produced by the two-stage synthesis. Such a flow chart could lead to a high cost, which inhibits the application of DMM. Compared with the two-step technique, the one-stage technique can synthesize DMM on the bi-founctional (employing redox sites and acidic sites) catalysts ($3CH_3OH + 1/2O_2 \rightarrow CH_3OCH_2OCH_3 + 2H_2O$). This technique is expected to avoid the drawback of the two-stage production. In this paper, the sulfated VO_x-TiO_2 (VT) catalysts were used for the one-step synthesis of DMM. The catalysts were prepared by the rapid combustion (RC) method and the impregnation (IM) method, which were marked as VTS and SVT, respectively. Herein, the effect of prepared method on the catalyst structure, acidity and reducibility as well as the catalytic performance was systematically studied. The result showed that the dispersion of vanadia and sulfate was higher on VTS catalyst than SVT catalyst, which led to the corresponding stronger reducibility and larger number of acidic sites for VTS catalyst. The catalytic test showed that the VTS catalyst exhibited higher methanol conversion and DMM selectivity due to its high dispersion of vanadia and stronger reducibility as well as larger number of acidic sites.

See more of this Session: Advances In Aromatics and Petrochemicals Processing II
See more of this Group/Topical: Fuels and Petrochemicals Division - See Also Topicals 4, 6, and 7

Rheological Characterization of Phenol-Formaldehyde (Resole) Resin

Dr. Hadeer K. Hamid, Water & Air Enterprises Ltd., Vancouver, BC. Canada

Summary

Resoles are usually made from a mixture of phenol and a molar excess of formaldehyde under basic conditions. Condensation of phenol and formaldehyde in basic medium is carried out in a series of experiments in order to serve as models in a study aimed at establishing its structure rheological characterization when dissolved in acetone. ChemCAD simulation software is adapted in this study in order to design a simulation to the real factory product. This software simulation is used for the first time in such study. The operation conditions employed in the real production site is tested on this model in order to check its applicability. The simulation showed good behavior when the modified manufacturing procedure is applied. Eight samples of Resole dissolved in acetone are used in this study. Their structure rheological characterization is carried out by plotting flow curves of different concentration and molecular mass. An important figure in this rheological study namely; the mass of network strand (M_e) is obtained. When plotted against concentration (C) i.e. Log M_e vs. Log C aslope gave rise to the scaling exponent X, which is a measure of the penetration hindrance. A scaling exponent of 0.23 is obtained in this study which is below the value expected for a network with free penetration. Therefore, partial penetration is expected.

Introduction

Resole resin is manufactured in Iraq as a substitute for imported furan. However, the main problem facing its production is the solidification of the final product in the reaction vessel or immediately after discharge. Moreover, factory manufactured resole is highly viscous and has short shelf- life at ambient temperature for subsequent fabricating. Solution to this factory persisting problem will be profitable and will tremendously reduce the cost involved in importing furan.

Resole manufactured in Iraq is used as an adhesive for wood composites such as plywood and oriented stand board. It is also used extensively in the foundry industry. Molding compounds can be formed to meet requirements of electrical, automotive, appliance and other consumer and industrial appliances.

The polymerization is normally carried out in two separate operations. The first operation involves the formation of low molecular weigh, fusible, soluble resin while the second operation involves curing reactions which leads to the cross-linked product [1-3]. Therefore it is absolutely necessary to have a fixed criterion to help stop the reaction at the desired stage, without sacrificing the final product. In order to obtain a liquid polymer required in the above mentioned applications, it is necessary to stop the reaction at the first stage before it proceeds to the cross-linking stage.

Several characterization techniques for phenol-formaldehyde Resins are cited in the literature [4], their solution properties however, are hardly studied because of their complex structure. The branching structure of long chain branched Resole Resins involves multidimensional distribution that can neither be predicted nor characterized with precision. The behavior of polymers towards solvents is characteristic and different from that of low molecular weight substances, because of the size difference between polymers and the solvents, the viscosity, the effect of texture, and molecular weight [5, 6]; therefore, special theoretical and practical treatment is required to explain the polymer solution properties.

The main objectives of this paper is to adapt the structure rheological characterization technique [7,8] using developed empirical scaling laws with molecular mass and concentration to study the behavior of Resole when dissolved in acetone at a certain specified temperature. Results of this investigation suggest the possibility of predicting the rheological behavior of polymers when its molecular structure is well-known. Such characterization technique also provides the basis of methods for using rheological data to obtain information about a structure of a polymer whose structure is not known in details. These techniques and relationships between structure and Rheology are used in this study to understand the structural behavior of Resole Resin.

Experimental

Resole Synthesis

The resin is synthesized in a pot reaction system by condensation of phenol (99.9%) with formaldehyde (37%) using NaOH (Fluka-Pure) as a catalyst since caustic soda is preferred when making liquid resins. The experimental procedure followed in this study is based on the standard reference method for making phenol-formaldehyde Resole resin by That Al-Sawari manufacturing Company in Iraq. The procedure carried out is as follows:

A calculated amount of phenol and Formaldehyde are added to the reaction vessel with continuous stirring. The catalyst, NaOH is then added slowly in three portions for the period of 15-20 min with continuous stirring. Temperature is not allowed to exceed 75 ^{0}C during addition of catalyst. After the addition is complete, the temperature is increased stepwise to the desired value and is held their for the specific reaction period. Samples are withdrawn every 45 minutes for to first few hours of the reaction then withdrawn every 25 minutes for the last hour and tested for miscibility. When the end point is detected, the final product is withdrawn and poured into a beaker then sufficient Suphuric Acid (H2SO4) is added to reduce the PH down to 6-7. The product is left to cool down, on cooling two layers are formed. The water layer is siphoned off. Product is left for 24 hours for further water separation. In this procedure removal of water is not taken to completion but is halted when the resole content of the residual aqueous solution reaches about 70%.

Rheological Characterization of Resole

A structure Rhelogical characterization is carried out experimentally in this work using a Fan VG meter Rotational Viscometer. This instrument is of the rotational coaxial – Cylinder type allowing shear rate measurements for the desired speed. Various concentrations of Resole are prepared by dissolving an appropriate amount of Resole in Acetone. The polymer is converted into a solution using a VEB MLW Labortechnik shaker type THYS 2. Shaking is applied for 5 min until a clear polymer solution with no sedimentation and no aggregation is obtained. Polymer solution is used immediately for testing and not left overnight. An Ubbelohde Capillary viscometer type from SchOH- Gerate is used in this study to determine the absolute or relative viscosity.

Results and Discussion

ChemCAD Simulation Results

ChemCad 5.2 [12] is considered as a useful tool to be used in this investigation in order to develop an ideal PC windows program for the design, development and optimization of chemical and biochemical processes. It will run on Windows platform (95, 98, 2000, NT, Windws XP). It will run quickly and with great stability.

In this work ChemCad 5.2 simulation software is used as a tool in the design of a simulation for an existing phenol formaldehyde (PF) Resole resin manufacturing plants Figure (A). The unit operations represent a simulation to those employed in the real plant. An improved production procedure adapted by the team of engineers in that Al-Sawari factory is tested in this study. Operating conditions correspond to the following:

Phenol / formaldehyde mole ratios = 0.5:1

Catalyst used is NaOH = 5 mole%

Reaction temperature = 70 ^{o}C

Formaldehyde = 37 wt%

Simulation results are presented below in the form of overall mass balance, Stream flow simulation results, Stream properties and flow sheet summary are also produced in a WinWord Window (Not included in this paper). It can be concluded from the results that follow; ChemCAD unit operations can be developed and integrated into ChemCAD system to simulate the production of phenol-formaldehyde (PF) resin. The resulting output responses generated are in line with plant data as can be indicated in the overall material balance table (A). The only difficulty facing design when polymers are applied is their varying molecular weights which might add a little confusion when introducing the components into the process flow diagram as part of the simulation design procedure.

Table (A): Overall material balance using ChemCad simulation software

Materials	**Kmol/h**		**Kg/h**	
	Input	**Output**	**Input**	**Output**
Phenol	61.700	0.000	5806.772	0.000
Formaldehyde	123.300	100.405	3702.206	3014.769
Resole	0.000	61.700	0.000	7650.800
Sodium Hydroxide	6.200	6.200	247.981	247.981
Water	2350.000	2349.634	42335.250	42328.662
Total	2541.200	2517.940	52092.210	53242.212

Figure (A): Phenol-Formaldehyde (Resole) process flow sheet constructed using ChemCAD.

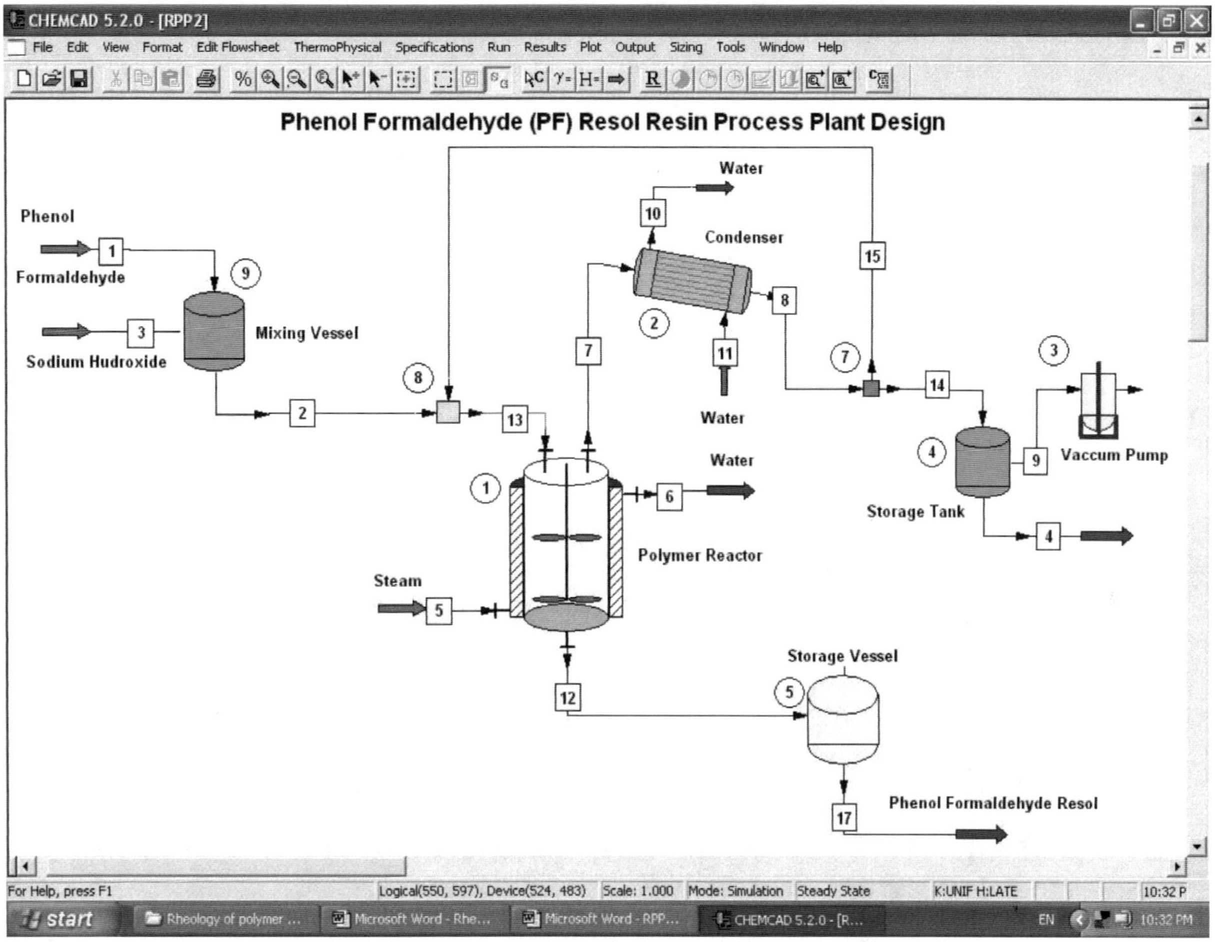

Determination of κ and α values for factory manufactured samples

Nine samples of Resol were collected from That Al- Sawari factory. Their molecular weights were measured by Osmometry technique using the vapor osmometric pressure (VOP) technique. Then the intrinsic viscosities of these samples are found viscometrically, as described in the experimental part. Numerical values of κ and α for (PF) resole dissolved in acetone have been established using the Mark-Houwink-Sakurada Equation (MHS). This relation is widely useful in polymer science. This relation is presented as follows:

$$[\eta] = \kappa M_v^{\alpha}$$

Another form of the equation is: $\quad \mathrm{Log}\ [\eta] = \mathrm{Log}\ \kappa + \alpha\ \mathrm{Log}\ M$

Table (MHS) show values of molecular weight of samples collected from the factory and the corresponding values of intrinsic viscosity [η]. Log [η] vs Log M is represented by a straight line as shown in Figure (MHS) of slope α. The intercept of this line on the viscosity axis is numerically equal to Log κ. Hence, the values of κ and α can be determined graphically.

Table (MHS): values of molecular weight of samples collected from Al-Jihad factory and the corresponding values of intrinsic viscosity [η].

Molecular weight	Log [M]	[η]	Log [η]
600	2.7782	2.4	-0.0081
670	2.8261	2.4826	0.005
710	2.8513	2.5428	0.0189
875	2.9420	2.5945	0.03
900	2.9542	2.6482	0.04
998	2.9991	2.682	0.046
1100	3.0414	2.7094	0.05
1500	3.1761	2.75	0.0713
2000	3.3010	2.7818	0.0805

Figure (MHS): Log [η] vs Log M for resole dissolved in acetone

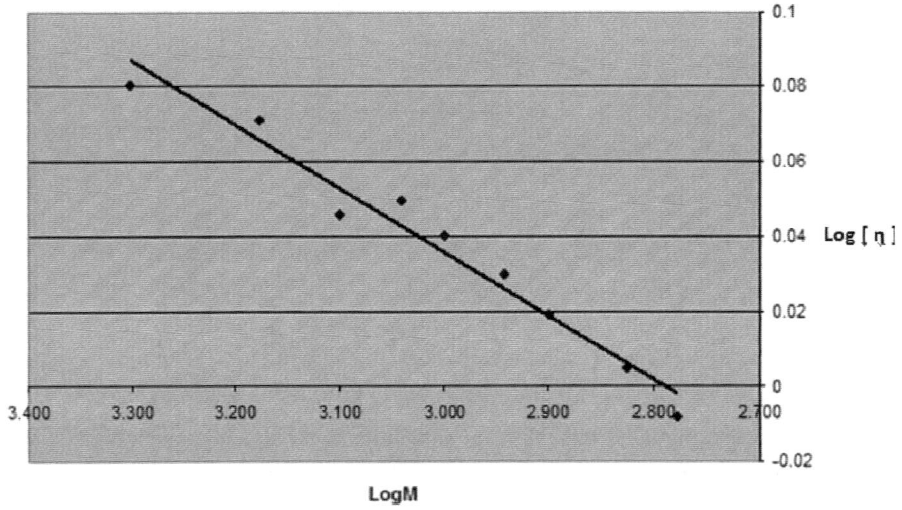

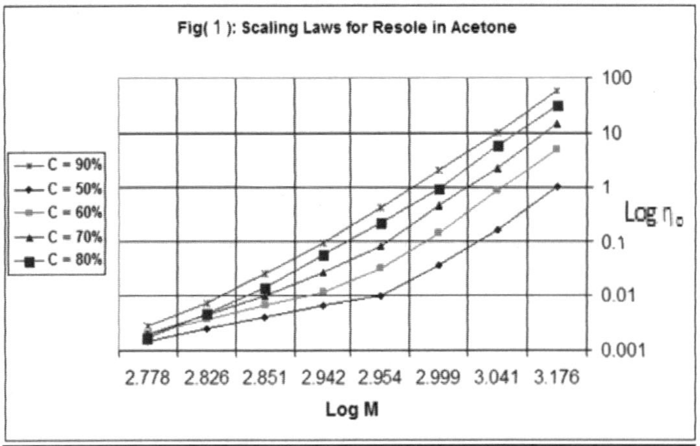

Fig(2): Values 0f (log Me) plotted against the values of (log C) for Resole dissolved in Acetone at 30 C

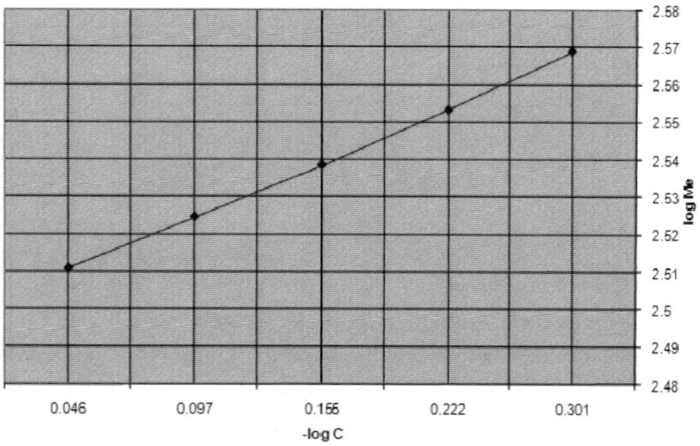

Fig(3): Apparent statistical chain element (A') as a function of concentration (C) and extrapolation to C = 0 for Resole in Aceton at 30 C

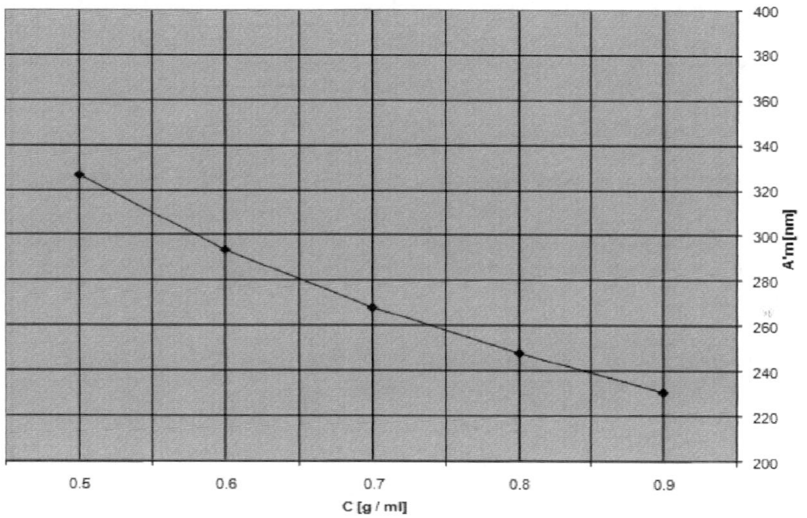

Table (1) Multiplicity of entanglement regions (X) for various concentrations (C) of Resole dissolved in acetone at 30 ^{0}C.

C [g / ml]	Me	A' [nm]	a tan □ (M> Mc)	X Multiplicity
0 (extrapolated)		350		
0.50	370.7	327	4.80	9.0
0.60	354.0	294	5.33	21.5
0.70	344.3	268	5.66	36.8
0.80	332.7	248	6.40	123
0.90	324.3	231	6.75	214

Table (2) Values of the exponent (a) for various concentrations (C) of Resole dissolved in Acetone at 30 ^{0}C.

C [g /ml]	Exponent (a)
0.5	4.8
0.6	5.33
0.7	5.66
0.8	6.4
0.9	6.75

Table (3) Values of the critical molecular weight (Mc) and the mass of network strand (Me) for various concentrations (C) of resol dissolved in acetone at 30 °C.

C [g /ml]	Mc	Me = Mc / 2	Log C	Log Me
0.50	741.3	370.7	-0.301	2.569
0.60	707.9	354.0	-0.222	2.549
0.70	688.7	344.3	-0.155	2.537
0.80	665.3	332.7	-0.097	2.522
0.90	648.6	324.3	-0.046	2.511

Table (4) Apparent statistical chain element (A') for various concentrations (C) of resol dissolved in acetone at 30 °C.

C [g/ml]	Me	A'[cm]	A'[nm]
0.9	324.3	2.31×10^{-5}	231
0.8	332.7	2.48×10^{-5}	248
0.7	344.3	2.68×10^{-5}	268
0.6	354.0	2.94×10^{-5}	294
0.5	370.7	3.27×10^{-5}	327

Results of rheological measurements are plotted in the form of flow curves, namely Log ή vs Log ɣ. The logarithm is taken since the range extends over many orders of magnitude.

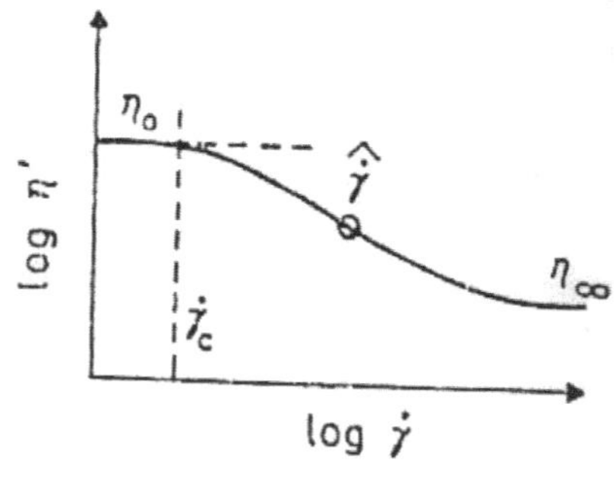

Figure (D) is a schematic representation of a complete flow curve.

The low shear rate range indicate ηo – range (first Newtonian range). Then the apparent viscosity ή is lowered in S-Shape down to the limiting viscosity η∞ (second Newtonian range) at very high shear rates (order of magnitude 106 sec -1). Values of the flow curve are ɣ c, the onset of shear dependence, and ? ɣ, the inflection point of the flow curve. These Characteristic figures are used intensively in the structure rheological characterization of Resole [7, 8].

The mass between two entanglements Me, the network strand or entanglement spacing is obtained from a plot of Log η_o with LogM, since Buech theory [9] predicts a scaling of ηo with $M^{3.5}$ for $M > M_e$ and M^1 for $M < M_e$. The intersection between the M^1 and the $M^{3.5}$ branch is Mc. Hence Me= Mc/2. Therefore a characterization of the network is possible. Figure (1) shows the scaling of Log η_o with Log M. Such diagram is referred to in the literature [8] as knick diagram. From the intersection of the two lines Mc is obtained from which M_e = Mc/2 is calculated. The break is an indication of the transition between the particle and the network regime. Values of the critical molecular weight Mc and the mass of the network strand M_e for various concentrations of Resole dissolved in acetone are presented in table (3). The well known intersection lines according to Beuche are noted with the following: (η_o Prop M^1 for M<Mc); and (η_o Prop M^a for a>= 3.5 for M>Mc). Values for the measured exponent (a) have been derived from the slopes of the knick diagram. These values are found to agree with Bueche being greater than 3.5 shown in Table (2).

Penetration Hindrance is related to the dependence of M_e on C. From a plot of Log M_e vs. Log C Figure (2), the scaling exponent (X) is obtained as a measure of the penetration hindrance to be 0.23. It is found that this value is below that expected for network with free penetration. Therefore, partial penetration is expected.

The approach introduced by Schurz [10] is adapted in order to describe the network in terms of the "Apparent Chain Element" A'. Isolated macromolecules of the statistical coil type are characterized by their mass, contour length (L) and degree of coiling, measured as statistical chain element (A). The contour length (L) is $L=p.l_o$ with $P = M / m_o$ (m_o and l_o: mass and length of the basic unit). The value of A is given as $A = h^2 /L$, whereby h^2 is the mean square end to end distance, which may also serve as a measure for the average coil diameter [6]. Now it can be assumed that at the point of transition from particle to network solution the molecular coil will just overlap, so that entanglements can be formed. Therefore, by equating these two values, a relation between M_e and the chain element (A) is arrived at. However, since the theory dealing with an equivalent network, the chain element (A) is an apparent one, and what is obtained in this way from measured M_e values is an apparent equivalent chain element A'. The relation has been worked out [11] and the following equation is arrived at:

$$\mathbf{A' = 2.16\ 3\ 10^{-16}}.\ \frac{m_0}{l_0}\ .\ \sqrt[3]{\frac{1}{Me.C^2}}$$

Where (m_o): Mass of basic unit and (l_o): Length of basic unit.

Table (4) shows a compilation A' values obtained in this way and the data in the table (Log A' vs. Log C) are plotted in figure (3) and extrapolated to C=0. Extrapolation to Zero Concentration should yield the true chain element A, which is found to be (350 nm). This value was checked against data obtained with the help of light scattering with dilute solutions and was found to agree.

Another important parameter, the multiplicity of entanglement regions have been calculated from the corresponding values of the measured exponent (a) and the molecular weight M using the relationship [8].

$\mathbf{X = M}^{\frac{a-3.5}{3.5}}$ the results obtained are inserted in table (1).

Conclusion

This work throws light on elucidating the connections between the structural parameters of phenol-formaldehyde resole resin dissolved in acetone and its rheological behavior. The solution structure of flexible chain polymers is described in terms of an equivalent entanglement network characterized by its apparent chain element (A'). Such network can exhibit free penetration of polymer segments, hindered penetration or no penetration at all. The segment interaction affects a hindrance to penetration. It can be characterized by the concentration dependence of the mass of network strand (M_e) respectively. Eight samples of Resole dissolved in acetone have been studied rheologically. A complete viscosity curve of a dissolved polymer can be characterized by several parameters: the zero viscosity; the shear rate at onset of shear thinning, the shear rate at the inflection point of the viscosity curve and the viscosity infinite shear rate. These characteristic figures are shown in a schematic viscosity curve. It has been found that these figures are connected with both concentration (C) and molecular weight (M) over scaling laws. From η_o, the mass of network strand (M_e), the apparent chain element and the penetration hindrance have been obtained. The polymer turns to form in acetone entanglement network solutions with somewhat hindered penetration. It is also important to indicate the fact that high concentration is chosen because of the low molecular weight range of Resoles synthesized. A relationship may exist between low molecular weight values of Resoles and high values calculated for the apparent chain element A' (A' = 350 nm) compared to values of other polymer solutions. ChemCAD simulation software is found to be a useful tool in the simulation design of such a complex process.

References

1. H. Yamakawa, "Modern theory of polymer solutions";(Harper and Row, New York, 1971).
2. K. Kamide, "Thermodynamics of polymer solutions"; (Elsevier Scientific, Amsterdam, 1992.
3. P. J. Flory, "Principles of polymer chemistry"; (Cornell University Press, Ithaca New York, 1953).
4. Robert A. Haupt and Terry Sellers, Jr. "Characterization of phenol-Formaldehyde Resol Resins"; Ind. Eng. Chem. Res. 1994, 33,693-697.
5. Edson Cocchieri Botelho, Natalia Scherbakoff, Mirabel Cerqueira Rezende, "Rheological Analysis of Phenolic and Furfural Resins Used in the Carbon Materials Processing", Materials Research, Vol. 3, No. 2, 19-23, 2000.
6. John M. Dealy, Ronald G. Larson, "Structure and Rheology of Molten Polymers"; (Hanser, Montreal, Quebec, Ann Arbor, Michigan, 3-446-21771-1, 2005.
7. P. J. Flory, "Statistical Mechanics of Chain molecules", (Interscience, New York, 1969.

8. Schurz, J., Marx-Figini, Muler, B and Gonzales, F.R. (1992), A StructureRheological Study of Cellulose Trinitrate in Ethyle acetate. Journal of applied polymer science, 4: 1957-1661.
9. F. Bueche, "Physical Properties of Polymers". Interscience, Newyork,1962).
10. Schurz, J. (1990), Empirical Scaling Laws in Polymer Solution Rheology. Journal of polymer Science Part C: Polymer Letters, 28: 141-145.
11. Schurz, J., Pfragner, J., Abd. Elrahim, F.M and Mustafa, A.B. (1991), A structure Rheological Study with Poly (Oxyethylene). Die Angewandte Makromolekulare Chemie, 185: 249-257.
12. http://www.chemstations.net

Ethylbenzene Dehydrogenation to Styrene Over Fresh and Used Commercial Catalysts

Morteza Baghalha*[a], Omid Ebrahimpour

Department of Chemical and Petroleum Engineering, Sharif University of Technology, Azadi St., Tehran, 11365-9465, Iran

ABSTRACT

Dehydrogenation of ethylbenzene (EB) to produce styrene (ST) monomer is commercially performed over potassium-promoted iron oxide catalysts. Catalyst deactivation due to coke deposition and loss of potassium is a major operation and economic drawback in this process. To better evaluate the performance change of these catalysts over their life-cycle, dehydrogenation of EB to ST over the fresh and used commercial catalysts was experimentally investigated at temperatures from 590 to 639°C under atmospheric pressure. The fresh and the used catalysts were obtained from a styrene plant. The used catalyst had been continuously operated for two years under severe plant conditions. The catalyst activity tests were performed in a lab-scale continuous fixed bed reactor, maintained at constant temperature. For the fresh catalyst, the yield of styrene increased with a decrease in LHSV and reached its highest value of 79% at 0.48 h^{-1} and T=627 °C. At the same experimental conditions, it was found that the selectivity of styrene for the used catalyst was higher than that for the fresh catalyst. However, the reaction rates and conversion in the used catalyst were considerably lower than that in the fresh catalyst. The transient effect of coke deposition on the EB conversion of the fresh catalyst was also investigated. Due to rapid coke deposition in the first 6 hour of these experiments, the fresh catalyst lost about 25% of its initial activity before reaching pseudo steady state conditions.

Keywords: ethylbenzene, styrene, dehydrogenation, potassium-promoted iron oxide catalyst, used catalyst, conversion, selectivity, catalyst deactivation

Introduction

Styrene is one of the most important starting compounds for the synthesis of many polymers. 90% of styrene is commercially produced by catalytic dehydrogenation of ethylbenzene [1]. The commercial styrene catalyst contains oxides of Fe and K, as well as one or more promoters [2]. Recently, our group investigated the structural changes of internal areas of the styrene catalyst during deactivation by comparing the fresh and used catalyst [3]. Through this research, it was found that the surface catalytic activity for ethylbenzene dehydrogenation follows the order: $KFeO_2$>Fe_2O_3/Fe_3O_4>carbon at temperatures less than 600 °C. We also found that these catalysts are of macroporous nature with rather large pores and low internal surface areas of 1 to 2 m^2/g.

Most of the industrial styrene plants carry out the dehydrogenation reaction adiabatically in multiple reactors operating in series. The dehydrogenation reaction is usually carried out at temperatures above 600 °C with a combined feed of steam and

[a] baghalha@sharif.edu, Tel: (+98-21) 6616-6423, Fax: (+98-21) 6602-2853

ethylbenzene vapor (generally, a mass ratio of 1 to 2.5 steam-to-ethylbenzene). Steam dilution serves three purposes; supplying the heat, lowering the partial pressure of the reactants to shift the equilibrium to styrene, and inhibiting the formation of carbonaceous deposits to preserve the catalyst activity [4].
Many attempts have been carried out to gain understanding on the kinetics of the dehydrogenation reactions [5,6,7], the role of promoters [8,9,10], the nature of the active phase in the catalyst [11,12], and factors affecting the deactivation of catalyst under reaction conditions [13,14].

Because of the thermodynamic limitation, formation of styrene is favored by low partial pressure of EB and hydrogen. Thus, to provide the heat for reaction and to reduce coking of the catalyst, the dehydrogenation process is usually carried out in the presence of overheated steam at ambient or lower pressure and at temperature as high as 823–973 °K.

The scrutiny of the open literature reveals three main reasons for the iron oxide catalyst deactivation: (i) formation of carbonaceous deposit, (ii) change in oxidation state of iron, and (iii) loss of potassium promoter. It has been shown that all three causes are mutually related. Although the role of coke is still controversial, it is generally accepted that its presence is harmful for the catalyst performance. Historically, addition of potassium to iron oxide catalysts was believed to assist in coke removal through its gasification with the co-fed steam.

Many studies have evidenced that at the reaction conditions, the active phase on the Fe_2O_3-K_2O catalyst is potassium ferrite, $KFeO_2$. The role of other promoters in the Fe–K–O system is to uphold and support the $KFeO_2$ phase formation and to stabilize it under reaction conditions at a H_2O/EB ratio as low as possible.

In the present work, the influence of changes of the operating conditions such as temperature, steam to ethylbenzene ratio, and LHSV on the catalytic activity of both fresh and used catalysts is studied.

Experimental

A fresh and used commercial catalyst was obtained from the styrene monomer plant. The used catalyst was downloaded from an industrial reactor which had been used continuously under severe condition (LHSV = 1 h^{-1}, T = 610 °C, mass ratio of steam to ethylbenzene = 1.2, P = 1.2 atm). The catalyst activity tests were performed in a lab-scale continuous tubular fixed bed reactor. For each catalytic test, the catalyst bed in the reactor was maintained at a desired constant temperature using an electrically heated furnace and a proportional-integral-derivative (PID) controller. The reactor feed and liquid products were analyzed with an Agilent 6890 N gas chromatograph with an FID detector. More detailed information on the reactor system may be found elsewhere [3].

Results and discussion

In order to separate the effects of catalytic and thermal EB dehydrogenation mechanisms, a set of EB dehydrogenation reactions were performed in the absence of any catalyst. For this purpose, inert silica particles were used to fill the reactor to produce the same LHSV as the catalytic reactions. The pyrolysis tests revealed that the thermal conversion of ethylbenzene to styrene and other products is only 9.7% and 1.5% at the temperatures of 639 and 591 °C, respectively.

The effect of reaction temperature on EB conversion

Ethylbenzene conversion as a function of reaction temperature is shown in Fig.1. As shown, the catalyst activity increases with an increase in temperature.

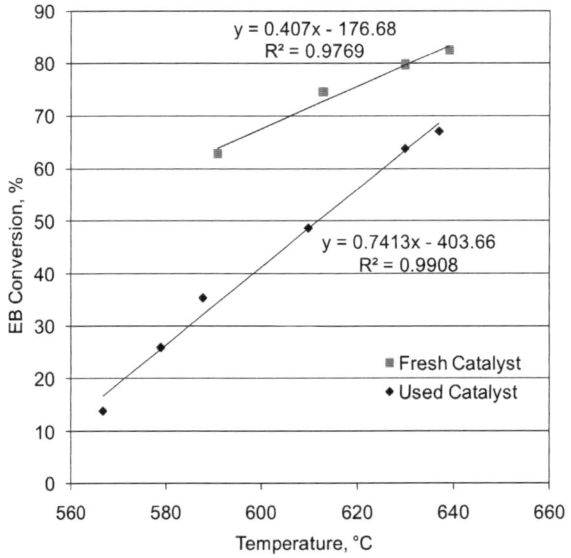

Fig. 1. The effect of reaction temperature on the conversion of ethylbenzene to styrene. Test conditions: LHSV = 0.525 h^{-1}, H_2O/EB = 2.8 (mL/mL).

For fresh catalyst ethylbenzene conversion increased from 62.9 at 591 °C to 82.5 at 639 °C; while for the spent catalyst, the conversion of ethylbenzene at the reaction temperature of 591 °C is blow 36%, and increased to approximately 67.1 at the temperature of 639 °C. The considerably lower conversion of ethylbenzene on the used catalyst at all temperatures indicates severe deactivation of the used catalyst during the two years of the plant operation.

As seen in Fig. 1, the effect of temperature on EB conversion is more pronounced in the case of the used catalyst. This indicates that the activation energy of EB reaction to ST for the used catalyst is higher than that of the fresh catalyst. Therefore, in deactivation modeling works, it is reasonable to assume a gradual increase of activation energy of EB dehydrogenation with increase in catalyst life.

Another important observation from Fig. 1 is that the rate of increase of the conversion of ethylbenzene at high temperatures of 630 to 639 °C, is lower than that at the low temperatures of 591 to 613 °C. These can be interpreted by limitation of equilibrium conversion of EB to ST.

For the fresh catalyst, the styrene selectivity was 94.8% at the reaction temperature of 591 °C. It decreased to 88.7% when the reaction temperature increased to 639 °C. The noticeable decrease in styrene selectivity at 639 °C is assumed to be due to enhanced thermal cracking of ethylbenzene into benzene and toluene. However, the styrene selectivity on the used catalyst (where EB to ST reaction may occur partly on the carbon surface) was found to be 1 to 3 % higher than that on the fresh catalyst (where the active phase is mostly $KFeO_2$ surface).

The effect of LHSV on EB conversion

The effect of LHSV (i.e., inverse of contact time) on ethylbenzene dehydrogenation conversion for the fresh and the used catalysts are shown in Fig. 2. In both cases, with an increase in contact time (i.e., decresse in LHSV), ethylbenzene conversion increased. One major observation in Fig. 2 for both catalysts is that the effect of change in LHSV on EB conversion is rather small at low LHSV values. This is due to the fact that at lower LHSV's, the residence time is high enough for the reactions to approach their equilibriums. However, as evident from Fig. 2, the equilibrium conversions of EB to ST in the case of the fresh and used catalysts are appreciably different. This observation, for the 1st time in the context of ethylbenzene dehydrogenation reactions, shows that the EB to ST reaction and its equilibrium constant is shifted as the catalyst deactivation proceeds. This can be explained by the possibility that the forward and the backward directions of the EB dehydrogenation reaction are affect differently as deactivation continues.

The range of selectivity values for the fresh catalyst and for the used catalyst are 89-93% and 90-96%, respectively. In both cases, the fractions of by-products (toluene and benzene) increases with increasing contact time. Hence, at longer residence times, as the conversion increases, the selectivity to styrene decreases. This is due to the fact that at longer residence times, the main reaction to styrene slows down as it approaches equilibrium while the by-product reactions to toluene and benzene keep moving forward through the catalytic and thermal mechanisms.

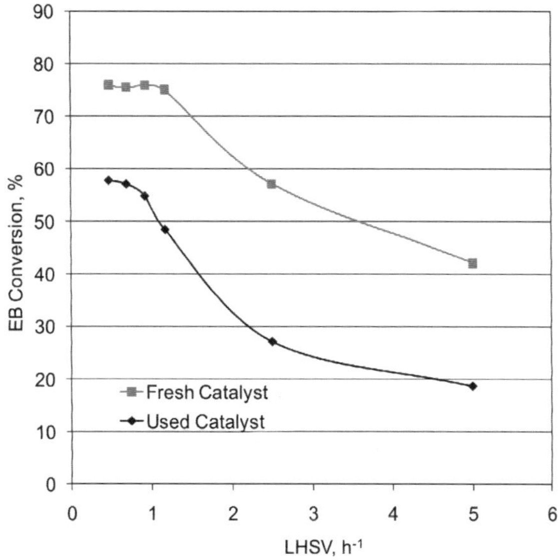

Fig. 2. The effect of LHSV on the conversion of EB to ST.
Test conditions: Temperature = 627 °C, H_2O/EB = 2.0 (mL/mL).

The effect of steam/ethylbenzene ratio in the feed on EB conversion

The effect of steam/ethylbenzene ratio in the feed on the conversion of ethylbenzene and selectivity of styrene was also studied. The results for EB conversion for both the fresh and the used catalysts are shown in Fig. 3. As shown, when the steam to ethyllbenzene ratio increases, the conversion of both the fresh and the used catalysts increases rather linearly. Fig. 3 also shows that there is a rather large ~32% difference in the conversion of EB to styrene for the fresh and the used catalysts, at the same H_2O/EB ratio.

Increasing the EB/H_2O ratio also increases the selectivity in the cases of both catalysts. This can be attributed to the effect of steam on the equilibrium conversion of EB to styrene. By increasing steam content of the reactor feed, the equilibrium conversion to styrene is shifted to higher values; hence, the actual reactor conversion becomes further apart from the equilibrium value. As explained before for the effect of LHSV on the styrene selectivity, the further the styrene reaction from the equilibrium state, the higher the styrene selectivity.

Water vapor does not participate in the reaction itself, i.e. not involved in the rate determining step of the reaction mechanism, but acts purely as an inert dilution agent. The dilution of EB with superheated steam is believed [4] to have several desired advantages for the styrene synthesis:

- Shift of equilibrium towards higher EB conversion, (reduce products partial pressures).
- Supplies part of the heat needed for the endothermic reaction.
- Decreases carbonaceous deposits by steam reforming reaction.
- Avoids catalyst over reduction and deactivation due to the presence of large amounts of hydrogen in the reaction gas mixture. Adding steam effectively controls the valence state of iron; i.e. prevents its over reduction to metallic state. Under the reaction operating conditions, the reduction of Fe_2O_3 is normally limited to Fe_3O_4.

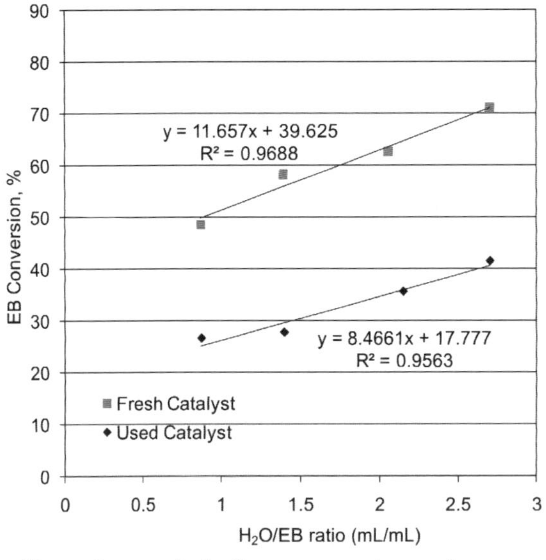

Fig. 3. The effect of steam/ethylbenzene ratio on the conversion of EB.
Test conditions: LHSV = 0.925 h^{-1}, Temperature = 615 °C

The effect of adding styrene and toluene to the feed on EB conversion

In a typical industrial styrene monomer plant, fresh ethylbenzene is mixed with the recycled ethylbenzene and steam. After heating to reach the desire temperature, the feed enters the 1st reactor. The partially reacted and cooled effluent from the 1st reactor is then heated to re-gain the desired temperature before entering the second reactor. The conversion of ethylbenzee in the first reactor is normally about 35-40% and the yield of toluene is about 2-5%. To investigate the role of styrene and toluene in the feed to the 2nd reactor (and also in the feed to the 3rd reactor) on ethylebenzene conversion and selectivity, six additional reactor tests were performed, as listed in Tables 1 and 2, for the fresh and the used catalysts, respectively.

Table 1: The effect of styrene and toluene in the feed on EB conversion and ST selectivity for the fresh catalyst

Fresh Catalyst	Temp. °C	LHSV h^{-1}	H_2O/EB mL/mL	EB Conversion %	ST Selectivity %
EB	627	0.925	0.88	64.5	90.4
EB(62%),ST(38%)	627	0.925	0.88	55.9	93.2
EB(80%),TO(20%)	627	0.925	0.88	64.1	69.1

Table 2: The effect of styrene and toluene in the feed on EB conversion and ST selectivity for the used catalyst

Used Catalyst	Temp. °C	LHSV h^{-1}	H_2O/EB mL/mL	EB Conversion %	ST Selectivity %
EB	627	0.925	0.88	42.0	90.1
EB(64%),ST(36%)	627	0.925	0.88	14.1	92.6
EB(90%),TO(10%)	627	0.925	0.88	31.8	63.7

As seen from Tables 1 and 2 for both types of catalysts, adding styrene to the ethylbenzene feed reduces the EB conversion, as the backward direction of the main EB/ST equilibrium reaction is now more dominant. However, contrary to the expectation of lower ST selectivity when the main reaction is closer to the equilibrium state, Tables 1 and 2 show that by adding styrene to the EB feed, ST selectivity increases. It seems that another factor also contribute in determining the ST selectivity. When large amounts of ST are present in the EB stream, there is a competition between EB and ST in adsorption on the catalyst active sites. In this case, decreased formation rates of benzene and toluene (and hence increased ST selectivity) indicates that styrene adsorption on the active catalyst sites is more dominant than that of the ethylbenzene.

The results of addition of toluene to the EB feed are more puzzling. While the effect of toluene addition to the EB feed on the EB conversion for the fresh catalyst is negligible, it seriously reduces the EB conversion in the case of used catalyst. The effect of toluene addition on the ST selectivity for both fresh and used catalysts is dramatic, as seen in Tables 1 and 2. The complicated behavior of toluene in this system is postulated to be due to the competition between EB, ST, and TO adsorption on the catalyst active sites. A detailed intrinsic kinetic modeling work should be implemented to deconvolute the complicated adsorption phenomena in this system.

Initial transient coke deposition and catalyst deactivation

When the fresh catalyst is used for EB reaction to ST, during a short period of time from the start of operation, a rapid transient coke deposition on the catalyst occurs which leads to a fast initial catalyst deactivation. This phenomenon was investigated in one test under rather normal conditions. The experimental conditions and the results for EB conversion for this test are shown in Fig. 4. As shown in this figure, it would take about 6 hours under normal operating conditions for the fresh catalyst to reach a pseudo steady state conditions, with an approximately 25% drop in catalyst activity and EB conversion. A small variation in EB conversion under pseudo steady state conditions is observed in Fig. 4. This variation is mostly due to small changes in reactor temperature that would deviate ±3 °C around the set point. A power law fit of the transient conversion data points is also shown in Fig. 4. We had previously [3] shown that the internal surface area of a used catalyst with 1.6% coke deposition had decreased sharply. It was also shown [3] that the decrease in surface area of the used styrene catalyst was due to the deposition of coke in the micropores of the catalyst. Such deposition is the major cause of the catalyst deactivation during early stages of operation. FT-IR spectra also indicated that styrene is the major precursor for coke formation.

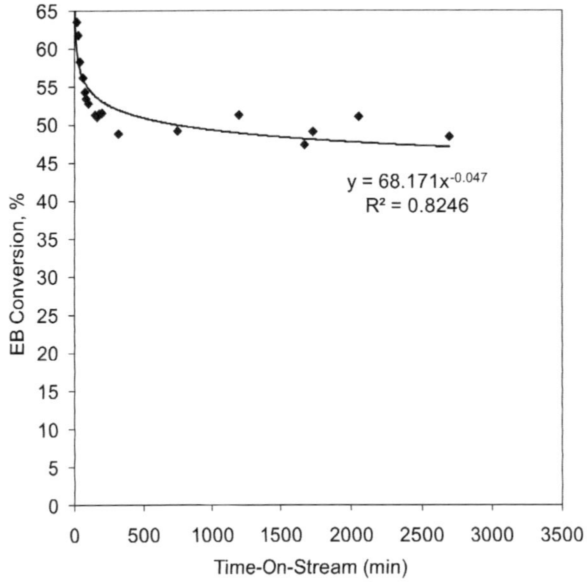

Fig. 4. Effect of transient coke deposition on the EB conversion of the fresh catalyst. Test conditions: LHSV = 0.575 h^{-1}, H_2O/EB = 1.8 (mL/mL), Temperature = 606 °C

Conclusions

The effects of changes of operating conditions on the activity of the fresh and the used catalysts were studied. Under the same experimental conditions, EB conversion in the fresh catalyst was much higher than that of the used catalyst. In contrary, the styrene selectivity in the used catalyst was higher than the fresh catalyst. This is due to the fact that EB to ST reaction in the used catalyst is further away from the equilibrium, hence, this reaction is faster and can better compete with the side reactions that produce benzene and toluene. For fresh catalyst the yield of styrene increased with a decrease in *LHSV* and reached its highest value of 79.2% at 0.48 h^{-1} and T=627 °C. The yield of styrene increased with increasing reaction temperature, but the selectivity to styrene slightly decreased with an increase in reaction temperature from 591 to 639 °C. The selectivity to styrene increased a little with an increase in the H_2O/ethylbenzene ratio. It was also found that the main cause of the catalyst deactivation at the initial stage of the catalyst life is the coke deposition which would lead to a ~25% reduction of the initial catalyst activity after 6 h of operation under normal operating conditions. This deactivation of the dehydrogenation catalyst in the early stages of operation is mostly caused by coke formation and plugging of the microporous structure of the catalyst.

Acknowledgment
The financial support of Deputy of Research of Sharif University of Technology is greatly appreciated.

References

1. Shekhah, O., Ranke, W., Schlogl, R., J. Catal. 225, 56-68 (2004).
2. Denis, H.J., Castor, W.M., Styrene, in Ullmann's Encyclopedia of Industrial Chemistry, vol. A25, Ed.: B. Elvers, S. Hawkin and W. Russey, John Wiley, New York, 1992, p.325.
3. Baghalha, M., Ebrahimpour, O., "Structural changes and surface activities of ethylbenzene dehydrogenation catalysts during deactivation", Applied Catalysis A: General, 326, 143–151 (2007).
4. Matsui, J., Sodesawa, T., Nozaki, F., Appl. Catal. 67, 179-188 (1991).
5. Wenner, R.R., Dybdal, E.C., Chemical Engineering Progress 44, 275 (1948).
6. Sheel, J.G.P., Crowe, C.M., Canadian Journal of Chemical Engineering 47, 183-187 (1969).
7. Abdalla, B.K., Elnashaie, S.S.E.H., Alkhowaiter, S., Elshishini, S.S., Appl. Catal. A: General 113, 89-102 (1994).
8. Stobbe, D.E., van Buren, F.R., van Dillen, A.J., Geus, J.W., J. Catal. 135, 533-547 (1992).
9. Hirano, T., Appl. Catal. 28, 119-132 (1986).
10. Cavani, F., Trifiro, F., Appl. Catal. A: General 133, 219-239 (1995).
11. Hirano, T., Appl. Catal. 26, 65-79 (1986).
12. Muhler, M., Schutze, J., Wesemann, M., Rayment, T., Dent, A., Schlogl, R., Ertl, G., J. Catal. 126, 339-360 (1990).
13. Matsui, J., Sodesawa, T., Nozaki, F., Appl. Catal. 51, 203-211 (1989).
14. Meima, G.R., Menon, P.G., Appl. Catal. A: General 212, 239-245 (2001).

(42a) Reactor Design for Supercritical Fischer Tropsch

Ed Durham, Department of Chemical Engineering, Auburn University, Auburn, AL, Mario R. Eden, Department of Chemical Engineering, Auburn University, Auburn University, AL and Christopher B. Roberts, Auburn University, Auburn University, AL

Fischer Tropsch Synthesis (FTS) is a mature process for the production of fuels and chemicals from syngas. In low temperature mode (LTFT) it is used primarily for the production of diesel fuel and wax. The FTS reaction is strongly exothermic, and managing this heat is a major factor in LTFT reactor design. The traditional fixed bed design utilizes a high gas velocity in a high number of narrow tubes. This reactor suffers from poor economies of scale and high pressure drop. Fixed bed reactors have been utilized by Sasol and Shell for LTFT. To address problems associated with fixed-bed LTFT, Sasol developed a slurry reactor. This reactor offers superior economies of scale, pressure drop, and thermal uniformity, but also increased development cost and the technical challenge of separating produced wax from the catalyst.

In an attempt to realize the advantages of each reactor design, Kaoru Fujimoto's research group pioneered the use of a supercritical fluid as the reaction media for LTFT (SC-FTS) in a fixed-bed reactor. This has been shown to give suppress methane and CO_2 formation, to enhance the selectivity of the reaction to potentially valuable primary products, and to improve activity maintenance. Two issues preventing the industrial utilization of SC-FTS are higher capital cost of the SC-FTS reactor (relative to the Sasol ARGE reactor) due to the higher pressure and higher operating cost due to compression of the media and syngas.

We will present work that we have done to decrease the capital cost of the SC-FTS reactor to improve its potential for industrial utilization.

See more of this Session: Coal, Biomass and Natural Gas to Liquids I
See more of this Group/Topical: Topical 6: 11th Topical Conference on Gas Utilization

(42b) Bench Scale Biomass to Liquid (BTL) Facility

Xinying Liu[1], Xiaodi He[2], Yinsuo Jia[2], Diane Hildebrandt[1] and David Glasser[1], (1)Center of Material and Process Synthesis, University of the Witwatersrand, Johannesburg, Johannesburg, South Africa, (2) Institute of Genetics and Physiology, Hebei Academy of Agricultural and Forestry, Shijiazhuang, China

A bench scale biomass to liquid facility is constructed and commissioned. A down draft fixed bed biomass gasification facility is used to convert the biomass into wood gas. The wood gas is then compressed into the gas cylinder after initial gas clean up. The gas is to be used in the Fischer Tropsch Reaction. A single tube fixed bed reactor with 50 mm inner diameter and 1 meter in length is used as the Fischer Tropsch reactor. The reactor is heated with oil bath at the controlled temperature. Wood gas from the cylinder is cleaned up in a special designed gas clean up facility to lower the total sulphur content to less than 10 ppbv. The reactor has the capacity of producing 100 g liquid products a day. This facility is to demonstrate COMPS's new Once-Through Fischer Tropsch Process and will lead to the design of a BTL pilot plant.

See more of this Session: Coal, Biomass and Natural Gas to Liquids I
See more of this Group/Topical: Topical 6: 11th Topical Conference on Gas Utilization

(42c) Compact Heat Exchange Reactor In FT Applications

Zhijun Jia, Chart Energy & Chemicals, La Crosse, WI, Dennis E. Sparks, Center for Applied Energy Research, University of Kentucky, Lexington, KY and Burtron H. Davis, Center for applied energy research, University of Kentucky, Lexington, KY

Fixed bed reactors are being considered for use for FT process at a small scale. Since the FT reaction is highly exothermic, the reaction typically is controlled by heat transfer and not the kinetics of the reaction. The removal of the heat of the reaction is a demanding challenge for a fixed bed reactor to operate to obtain a high productivity per catalyst unit. A compact heat exchange reactor (CHER) provides a new solution for this application. It is capable of removing the heat of the reaction along with reactor length; therefore the reaction becomes essentially isothermal. As a result, the reactor productivity is improved significantly. This paper describes the test results for using the CHER in FT application and demonstrates the capability of CHER for this purpose.

See more of this Session: Coal, Biomass and Natural Gas to Liquids I
See more of this Group/Topical: Topical 6: 11th Topical Conference on Gas Utilization

Catalytic upgrading biosyngas from a downdraft gasifier to liquid hydrocarbons

Qiangu Yan, Fei Yu
Department of Agricultural & Biological Engineering, Box 9632, Mississippi State University, MS 39762

Introduction

There are two strategies to convert syngas to a wide range of fuels and chemicals. Fischer–Tropsch (FT) synthesis is a major part of gas-to-liquids (GTL) technology [1,2], which converts syngas into liquid fuels with a wide range of liquid hydrocarbon fuels and high-value added chemicals [3,4].However, the FT products are controlled by the Anderson–Schulz–Flory (ASF) polymerization kinetics, resulting in a nonselective formation of any hydrocarbons [5,6].Another approach is first to convert syngas to methanol over a methanol synthesis catalyst and subsequently polymerize methanol to hydrocarbons over ZSM-5 [7,8]. Syngas is first converted to methanol over a copper-based hydrogenation catalyst. In the second stage, methanol is converted to gasoline over a ZSM-5 catalyst. During the recent years, many investigations have demonstrated the advantages of the one-stage processes by using bifunctional catalysts compared with the two-stage and three-stage processes of synthesis gas conversion to gasoline [9, 10]. The use of a bifunctional catalyst allows for simultaneously carrying out the synthesis of methanol from syngas over the metallic function and the transformation of methanol into hydrocarbons over the acidic function. The fulfillment of both steps in the same reaction medium promotes the displacement of the thermodynamic equilibrium of methanol synthesis. Also, the shape selectivity of the acidic function provides a high selectivity that cannot be reached in the Fischer–Tropsch synthesis. Moreover, previous syngas to gasoline technologies are mostly based on using pure syngas or low nitrogen syngas [1-10], which mainly derive from natural gas or coal, there are limit publications using nitrogen-rich syngas to produce hydrocarbons [11-13]. The existing downdraft gasifier is producing syngas from biomass. Currently, the syngas from the gasifier contains about 20 % hydrogen, 19 % CO, 12 % CO_2, 2% CH_4 and 49% N_2 [14]. The nitrogen content is too high for the hydrocarbon synthesis if we use existing technologies including catalysts. Developing high activity and high stability catalysts is the key for a better overall performance when using the bio-syngas from the gasifier.

Existing syngas to gasoline technologies are mostly based on using pure syngas or low level nitrogen syngas, while the biomass derived syngas from the gasifier contains N_2 up to 50% to 60% (vol). Most of the catalysts for the syngas to gasoline process have low performance under such high nitrogen content. Moreover, the operation cost of the syngas compression press is a large portion of the overall invests. A new catalyst with high activity and high stability is developed for a single stage hydrocarbon mixture production process from biomass-derived nitrogen rich syngas when using the bio-syngas from the gasifier.

Experimental

Catalyst preparation

A typical F-T catalyst was used in the present work. First at least one of the transition metals is deposited or ion-exchanged onto ZSM-5 (SiO_2/Al_2O_3 = 50) obtained from Zeolyst International.the zeolite by one of the methods as the conventional impregnation. After the transitional metal is deposited on the zeolite, the mixture is calcined at 400 °C for 3 h. The transition metal deposited zeolite is ready for use.

Test Syngas

Four different compositions of commercial compressed syngas (Airgas, Inc.) and a wood syngas produced from a down-draft gasifier (Community Power Corporation Bio-Max, Littleton CO)were used in this test : 1) 50% H_2, 50% CO; 2.) 20.1%H_2, 19.9%CO, balance N_2; 3) 40% H_2, 20% CO, 12% CO_2, 2% CH_4, 26 % N_2, 4) 19% H_2, 20% CO, 12% CO_2, 2% CH_4, 47 % N_2. Unlike the clean syngas, gasifier syngas mix contained certain impurities. This syngas was made up of a near 1:1 mole ratio of H_2 to CO (19% H_2, 20% CO, 12% CO_2, 2% CH_4, 46 % N_2, and trace of contaminants).

Cleaning and Compression of Wood Syngas

To prepare the wood gasifier syngas, a feedstock of woodchips was converted to syngas with a down-draft gasifier (Community Power Corporation Bio-Max, Littleton CO) [14].The wood chips were gasified, and the syngas was cooled with an ice bath. The syngas was then compressed to approximately 50 psig. After compression, a water scrubber and gas drying apparatus were assembled to scrub the wood syngas from the gasifier. Two 25-gallon 304 stainless steel (SS) wide mouth tanks were connected for the water scrubber. The first tank was outfitted with 13 SS exhaust mufflers with a filtration rating of 50 microns. These mufflers were used to create the maximum amount of syngas bubbles so that the utmost possible surface area of syngas could come in contact with the water and contaminates (i.e. ammonia) could be scrubbed by dissolving in the water [14]. After the water scrubber process, the gasifier syngas was flowed through another 25-gallon tank which contained 80 lbs of silica gel desiccant to remove moisture in the wood syngas. Finally, the gasifier syngas was compressed to approximately 2000 psig using a 2-stage pneumatic air pump obtained from Hydraulics International (Chatsworth, CA).

Catalytic reaction

The synthesis gas conversion reaction was carried out in a home-built continuous flow fixed-bed reactor system. Three (3) grams of the catalyst was loaded to the reactor. The system was first purged by a helium flow for 30 min, followed by prereducing with a syngas mixture at 400 °C for 8 h, then syngas was fed in until reaching the desired pressure with slow adjustment of the system to the desired temperature. The reaction was operated under the following conditions: 250- 400°C, a gas volume hourly space velocity

(GHSV) of 500 to 5,000 h^{-1} and a pressure of 500-1500psi. Liquid products were collected using a condenser kept at 270 K, and the effluent gas from the condenser was analyzed with an on-line gas chromatograph (GC, HP 6980) equipped with thermal conductive detector (TCD) and flame ionization detector (FID). A packed Molecular Seive 5A column and a HP-1 capillary column were employed for separation of inorganic gases and light hydrocarbons. Liquid products, collected from the condenser, were separated into oil phase and water phase, and analyzed with GC–mass spectrometer (Agilent) equipped with DB-Wax capillary column for oxygenated compounds and HP-5ms capillary column for hydrocarbons [15].

Results and discussion

Effect of temperature

The temperatures studied are: 290, 300, 310, 320, 330 and 350 °C, with the remaining conditions maintained at the values previously indicated. Figure 1a shows the variation of CO conversion, $C_1 \sim C_5^+$ selectivity with temperature, the other variables remaining constant. Figure 1a shows that the influence of temperature on CO conversion is very important. CO conversion increases from 55% at 290°C to 83% at 350°C. Figure 1b clearly shows the hydrocarbon distribution changes with temperature, the light hydrocarbons being favored at higher temperatures. Higher hydrocarbons are more sensitive to temperature, undergoing cracking and dealkylation reactions that become increasingly important with increasing temperature.

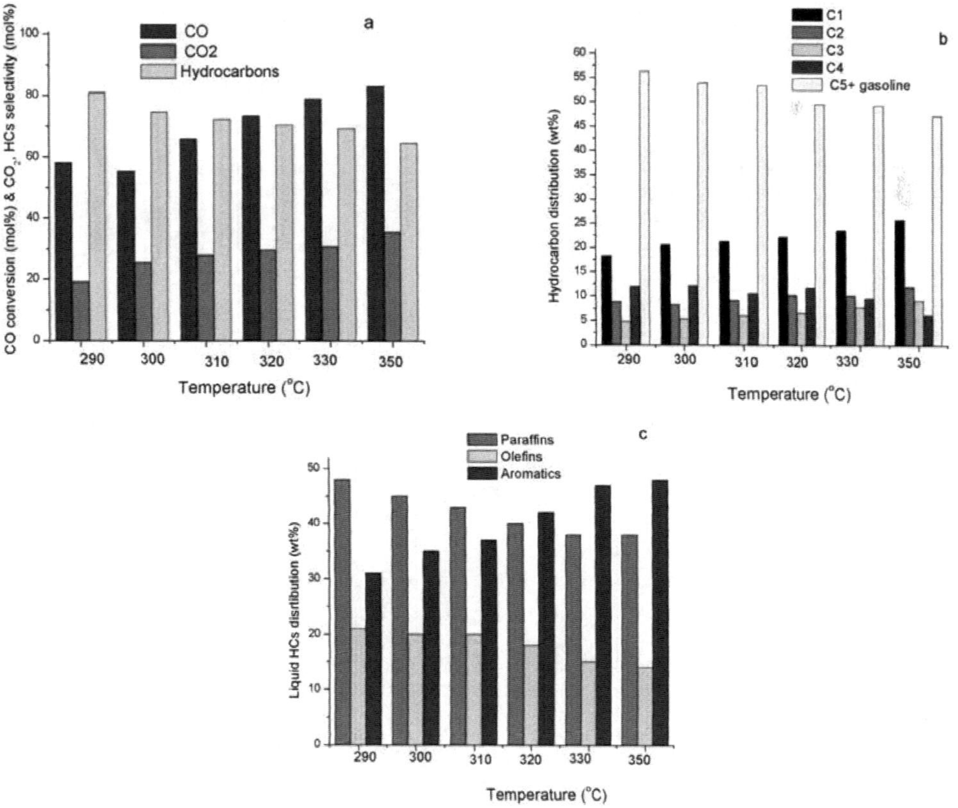

Figure 1 Effect of temperature on (a) CO conversion, CO_2 and hydrocarbon selectivity, (b) hydrocarbon distribution, and (c) liquid hydrocarbon distribution at 1000psig, with syngas of 19.0%H_2, 20.0%CO, 12.0%CO_2, 1.921%CH_4, balance N_2, GHSV of 2000 h^{-1}; 3 g of the catalyst was used in the reaction. Time on stream of 48-100 h.

Influence of pressure on the catalyst performances

An increase in total pressure will increase the equilibrium towards the product side, and increase the conversion. The change in Gibbs free energy will always be negative for a system where the change in number of moles upon reaction is negative and when pressure increases. Furthermore, a change in total pressure will not affect the water gas shift reaction but will change the equilibrium product distribution. Experiments under pressures of 500, 750, 1000, and 1250 psig have been carried out. The other operating conditions are: temperature, 310°C; gas space velocity 2000 h^{-1}, with syngas of 19.0%H_2, 20.0%CO, 12.0%CO_2, 1.921%CH_4, balance N_2.The effect of pressure on both CO conversion and process selectivity shows that when pressure increases, CO conversion increases (Figure 2). Within the C_5^+ fraction, an important effect of pressure on selectivity was observed. The increase in pressure clearly favors the formation of C_5^+, while C_1~C_3 gaseous hydrocarbons clearly decrease with increasing of pressure.

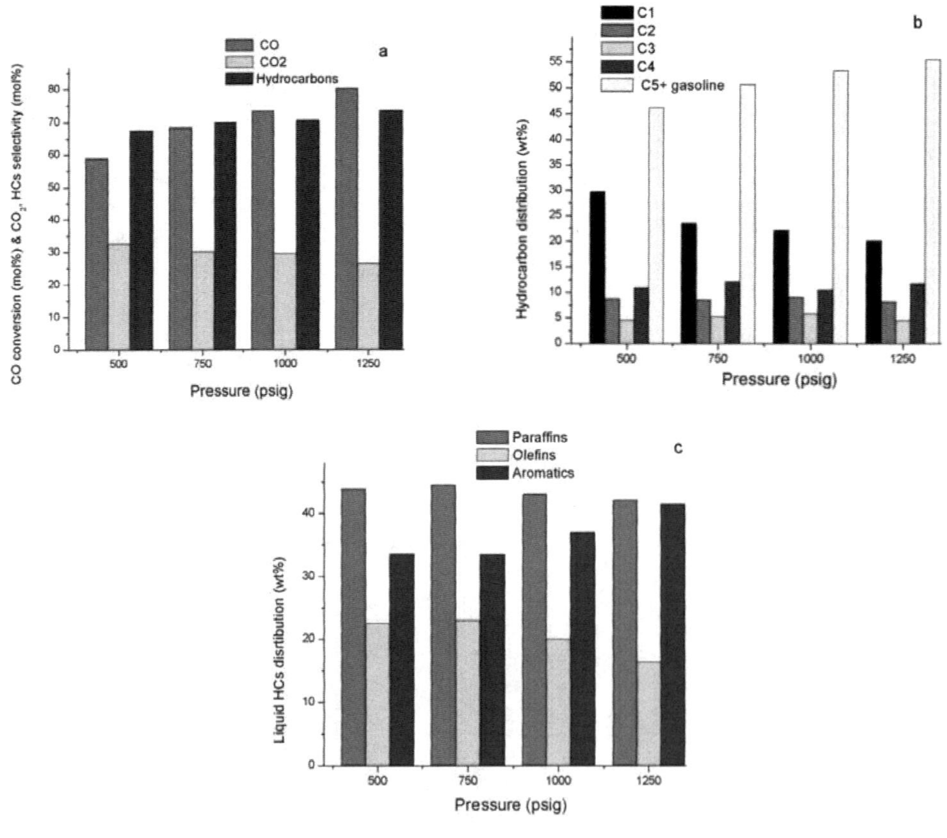

Figure 2 Effect of pressure on (a) CO conversion, CO_2 and hydrocarbon selectivity, (b) hydrocarbon distribution, and (c) liquid hydrocarbon distribution at 310 °C, with syngas of 19.0%H_2, 20.0%CO, 12.0%CO_2, 1.921%CH_4, balance N_2, GHSV of 2000 h^{-1}; 3 g of the catalyst was used in the reaction. Time on stream 48-100 h.

Impact of space velocity on the catalyst performances

Experiments were completed with different GHSV, 500, 1000, 2000, 2500 and 3000 h^{-1}; the remaining operating conditions have been maintained at 310°C, 1000 psig, CO/H_2 = 1. The effect of the space velocity on CO conversion and hydrocarbon distribution shows that when the space velocity increases, CO conversion decreases as expected (Figure 3). At the same time, C_5^+ selectivity also decreases. It can be concluded that CO conversion decreases with the increasing of space velocity. The effect of space time on product distribution is significant. It was observed that the yields of hydrocarbons decreased with increasing of space velocity. The gasoline fraction (C_5^+) continuously decreases with increasing space velocity.

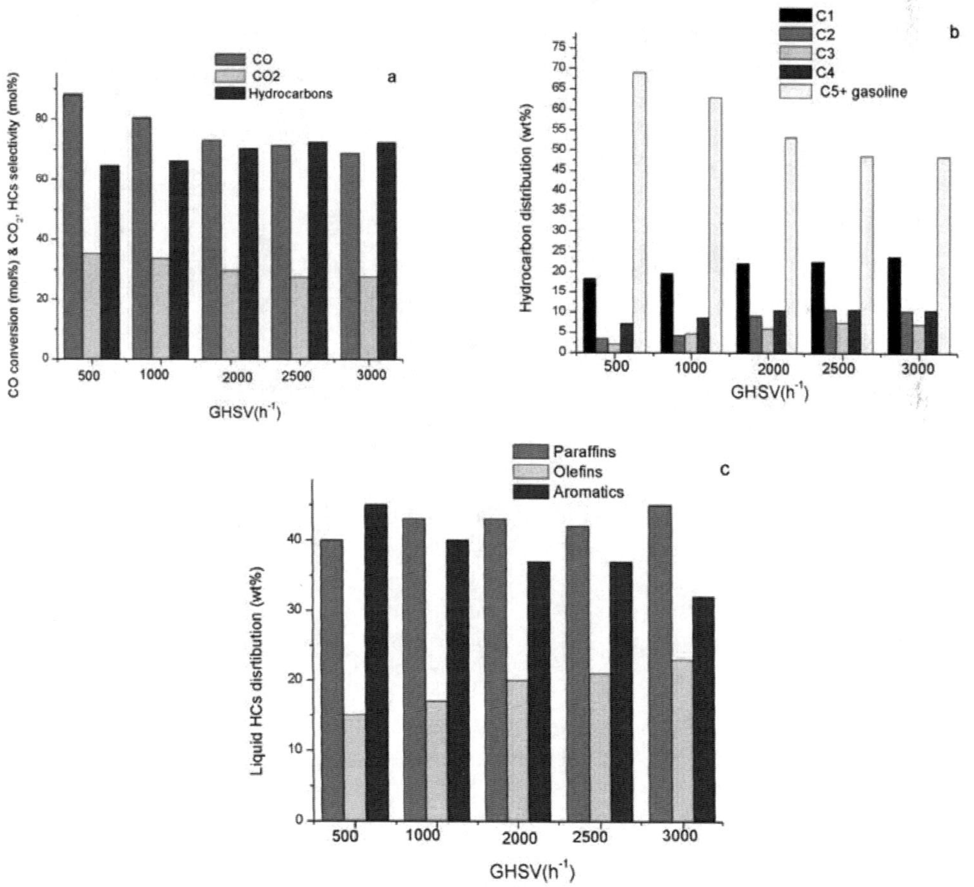

Figure 3 Effect of gas hourly space velocity (GHSV) on (a) CO conversion, CO2 and hydrocarbon selectivity, (b) hydrocarbon distribution, and (c) liquid hydrocarbon

distribution at 310 °C, 1000psi, with syngas of 19.0%H_2, 20.0%CO, 12.0%CO_2, 1.921%CH_4, balance N_2; 3 g of the catalyst was used in the reaction.

Role of syngas composition on the catalyst performances

Synthesis gas composition is an important parameter in higher hydrocarbon synthesis. It can affect both reaction rates and activity. It was found that the catalyst has relatively high activity and selectivity in producing liquid hydrocarbons when running with nitrogen-rich syngas.

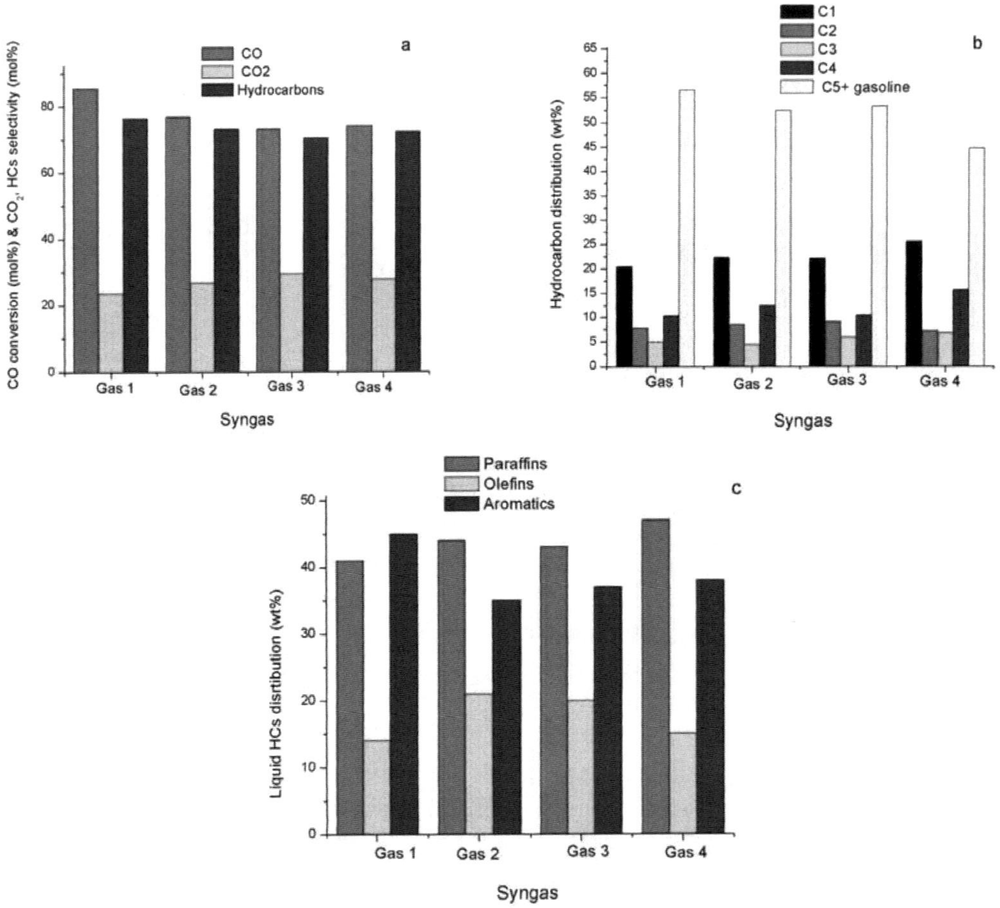

Figure 4 Effect of syngas composition on (a) CO conversion, CO2 and hydrocarbon selectivity, (b) hydrocarbon distribution, and (c) liquid hydrocarbon distribution at 310 °C, 1000psig, GHSV of 2000 h^{-1}; 3 g of new generation catalyst was used in the reaction. (Gas1: 49.64% H_2, 50.36%CO; Gas 2: 20.1%H_2, 19.9%CO, balance N_2; Gas 3: 19.0%H_2, 20.0%CO, 12.0%CO_2, 1.921%CH_4, balance N_2; Gas 4: 40.054%H_2, 19.98%CO, 11.99% CO_2, 2.006%CH_4, balance N_2.)

Time-on-Stream Performance

The effect of time on stream on the performance of the catalyst was studied at 310 °C over a period of 120 h. The evolution of CO conversion and hydrocarbon distribution with time on stream are shown in Figure for the new catalyst. The activity of the catalyst reached a steady state after a period of approximately 20 h under operating conditions. Time-on-stream results showed the catalyst was stable. CO conversion reached 74% after 20 hrs run at 310°C, and kept constant after 120 hrs run.

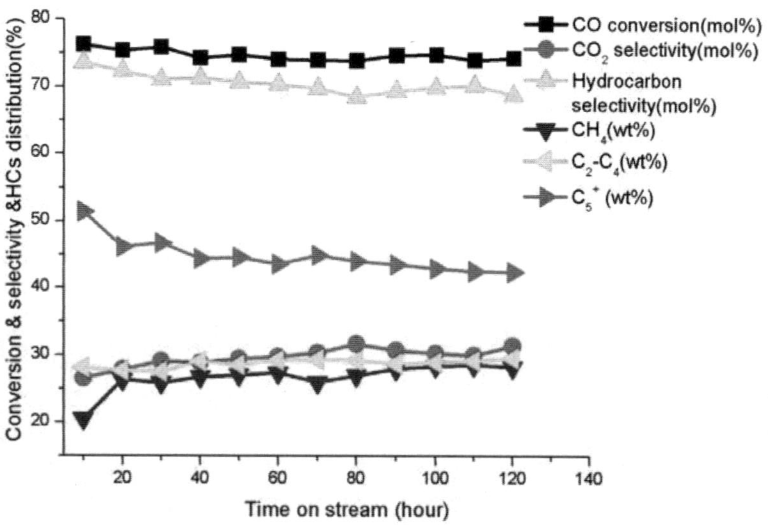

Figure 5 Time on stream of CO conversion, hydrocarbon selectivity and distribution on the new generation catalyst at 310°C, 1000 psig, 2000 h^{-1} with syngas of 19.0%H_2, 20.0%CO, 12.0%CO_2, 1.921%CH_4, balance N_2.

Wood syngas activity

The compressed wood syngas was feed to F-T synthesis reactor for liquid hydrocarbon production. The wood syngas activity and selectivity was similar to that of the simulator syngas (19.0%H_2, 19.0%CO, 12.0%CO_2, 1.921%CH_4, balance N_2) (not list here). Liquid product samples were collected and analyzed. The composition of the gasoline phase was also similar to the simulated syngas except there were more aromatics and less olefins in the wood syngas product.

Conclusion

A bi-functional catalyst has been prepared and evaluated for catalytic converting nitrogen-rich syngas to gasoline range hydrocarbons. It was found that the catalyst has relatively high activity and selectivity in producing liquid hydrocarbons when running with nitrogen-rich syngas. The CO conversion, hydrocarbon selectivity and hydrocarbon distribution as a function of temperature, pressure, GHSV, composition of the feed, and reaction time were examined. The increase of temperature and pressure favors carbon monoxide conversion and formation of more hydrocarbons. The conversion of carbon monoxide in the syngas was 73% under a reaction temperature of 310°C, and a pressure of 1000 psig and 58% under a reaction pressure of 500 psig when using a simulated gasifier-derived syngas. This catalyst gave higher conversion efficiency under a lower $CO-H_2$ pressure. The conversion into hydrocarbons reached about 70 mol% while the conversion into carbon dioxide was about 30 mol%. The resultant hydrocarbons were hydrocarbons of the gasoline boiling point range, which contained 40 to 50 wt% of light hydrocarbons (C_1-C_4), it is significantly lower than in the case of using the Mo/HZSM-5 catalyst which producing more than 70% light hydrocarbons. According to detailed analysis, the composition of gasoline fraction contained up to 40 wt% aromatics, 15wt % of olefins and 40wt % of paraffins, while there were more than 90% of the liquid hydrocarbons were aromatics in the case of the Mo/HZSM-5 catalyst. The compositions of liquid hydrocarbon mixture products can be adjusted through the operating conditions. Time-on-stream results showed the catalyst was stable. CO conversion reached 74% after 20 hrs run at 310°C, and kept constant after 120 hrs run. Bio-syngas was tested for F-T synthesis reaction. The wood syngas activity and selectivity was similar to that of the simulator syngas (19.0%H_2, 19.0%CO, 12.0%CO_2, 1.921%CH_4, balance N_2). The composition of the gasoline phase was also similar to the simulator syngas except there were more aromatics and less olefins in the wood syngas product.

Acknowledgement

Funding for this work was provided by the U.S. Department of Energy under Grant DOE-DE-FG36-06GO86025 to the MSU Sustainable Energy Research Center.

References

1. Haag, Werner O., and Huang, Tracy J., United States Patent 4279830.
2. Woo, H. C.; Nam, I. S.; Lee, J. S.; Chung, J. S.; Kim, Y. G. J. Catal. 1993 142 672.
3. Epling, W. S.; Hoflund, G. B.; Hart, W. M.; Minahan, D. M. J. Catal. 1997 169 438.
4. Girardon, J.-S.; Quinet, E.; Griboval-Constant, A.; Chernavskii, P. A.
5. Gengembre, L.; Khodakov, A. Y. J. Catal. 2007, 248, 143.
6. Gormley, R. J.; Rao, V. U. S.; Anderson, R. R.; Schehl, R. R.; Chi, R. D. H. J. Catal. 1988 113 193.
7. Shi, B.; Davis, B. H. Appl. Catal. A: General 2004, 277
8. Chang,, C. D.Lang,, W. H.; Silvestri,, A., U.S. Patent 4 096 163, 1978.
9. Qiangu Yan, et al., J. Phys. Chem. C, 2008, 112 (31), pp 11847–11858
10. Mysov, V. M.; Reshetnikov, S. I.; Stepanov, V. G.; Ione, K. G. Chem. Eng. J. 2005, 107, 63.
11. Comelli, R. A.; Fìgoli, N. S. React. Kinet. Catal. Lett. 1994, 52, 139.
12. Xu, D., H. Duan, et al., Energy Fuels, 2006, 20(3): 955-958
13. Jess, A., R. Popp, et al. (1999), Applied Catalysis A: General, 186(1-2): 321-342.
14. Jason Tyler Street, Master thesis,"GASOLINE-RANGE HYDROCARBONS PRODUCED FROM THREE TYPES OF SYNTHESIS GAS USING A MO/HZSM-5 CATALYST", Mississippi State University, December, 2010.
15. Shetian Liu, Amit C Gujar, Peter Thomas, Hossein Toghiani; Mark G White, Applied Catalysis, A: General, 2009, 357, 18-25.

(42e) A Study of Low Temperature Fischer-Tropsch Synthesis by Switching Between $CO_2/H_2/N_2$ and $CO/H_2/N_2$ Syngases Over a Cobalt Based Catalyst

Yali Yao, Centre of Material and Process Synthesis, University of the Witwatersrand, Johannesburg, South Africa, **Diane Hildebrandt**, Center of Material and Process Synthesis, University of the Witwatersrand, Johannesburg, South Africa, Johannesburg, South Africa, David Glasser, Centre of Material and Process Synthesis, University of the Witwatersrand, Johannesburg, South Africa, Johannesburg, South Africa and Xinying Liu, Center of Material and Process Synthesis, University of the Witwatersrand, Johannesburg, Johannesbueg, South Africa

Abstract:

The Fischer-Tropsch synthesis (FTS) provides a well-established commercial technology for conversion of syngas to clean transportation fuels and chemicals. The raw synthesis gas or syngas derived from coal, natural gas or biomass is a mixture of H_2, CO, CO_2 and CH_4. Because the syngas composition is dependent on many factors such as gasifier type, operating conditions, gasifying agents, the composition of CO_2 in the raw syngas varies from around 1% to 30%. Although the need for CO_2 separation before using the syngas in FTS is mentioned in the patent literature, for some cases, recent process development studies discuss a potential cost advantage if CO_2 is not removed before the synthesis step [1]. In addition, fixation of CO_2 into hydrocarbons through FTS, in an attempt to reduce CO_2 emission, has become of greater interest in recent years. It is therefore interesting to investigate the effect of CO_2 on a cobalt catalyst under low-temperature FTS conditions.

In the present work, 10 wt % of Co/TiO_2 catalyst used in this study was prepared by impregnation of TiO_2 with a cobalt nitrate solution. The experiment, called the cobalt-based catalyst stability testing during CO and CO_2 hydrogenation, was conducted by repeatedly switching between the two feed gases, CO_2 feed ($CO:H_2:N_2$=30%:60%:10%) and CO feed ($CO_2/H_2/N_2$ = 22.5%:67.5%:10%), into a micro plug flow reactor at 180-220°C, 20bar and 30ml(NTP)/(min·gcat).

For instance the CO or CO_2 reaction rate as a function of time on stream for low temperature FTS is shown in Figure 1. The data shows that both of the CO and CO_2 are readily hydrogenated over cobalt based catalyst and with an increase in temperature both the CO and CO_2 reaction rates are increased. At the lower temperature of 180∞Ê, the catalyst reactivity for CO_2 reaction is close to that of CO. However, when we increase the reaction temperature from 200 to 220°C, a lower reactivity of CO_2 to CO is obtained. It is however quite interesting that with the reaction temperature at 180°C: (1) when the CO_2 feed mixture is first introduced into the fixed bed reactor, the CO_2 reaction rate achieves its highest rate at that temperature; (2) after this, when the feed gas is switched from CO_2 feed to CO feed and then switched back to CO_2 feed, the CO_2 reaction rate is two times lower than the first time; (3) with subsequent repeat switching between the two feed gases, both of the CO_2 and CO reaction rates remain constant and do not change for CO_2 and CO reaction respectively, which indicates that the catalyst is not de-activated.

It should be noted that as the feeds are changed back and forth between H_2/CO_2 and H_2/CO there is no apparent catalyst deactivation. In this paper we further discuss the catalyst activity, product selectivity, product distribution and paraffin to olefin ratio (P/O) during switching between the two feed gases at different reaction temperatures. The results provide some hints on how to design FTS processes and FT catalysts to improve the product selectivity.

THE GRAPHIC ON THIS PAGE WAS NOT AVAILABLE
ON THE E-MEDIA VERSION OF THIS DOCUMENT

Figure 1: CO or CO_2 reaction rate as a function of time on-stream for a Co/TiO_2 catalyst at 180-220°C, 20bar and 30ml (NTP)/(min·gcat).

References : [1] T. Riedel, G. Schaub *Topics in Catalysis,* 2003, **26**: 145-156.

See more of this Session: Coal, Biomass and Natural Gas to Liquids I
See more of this Group/Topical: Topical 6: 11th Topical Conference on Gas Utilization

Reaching Higher Productivity In a Fixed-Bed Fischer-Tropsch Reactor at Laboratory Scale and Pilot Scale

Vladimir Z. Mordkovich[1, 2], *Vadim S. Ermolaev*[1], *Ilia S. Ermolaev*[1], *Eduard B. Mitberg*[1], *Lilia V. Sineva*[1, 2] *and Igor G. Solomonik*[1]

[1]*INFRA Technologies Ltd., Troitsk, Russia*
[2]*TechnologicaL Institute for Superhard and Novel Carbon Materials., Troitsk, Russia*

It is commonly accepted that the productivity of Fischer-Tropsch reaction is one of the keys to economic efficiency of the whole XTL process. The expression for calculation of productivity is shown below in equation (1) as mass of liquid product per volume of catalytic bed in period of time (kg/(m^3×hr).

$$P_{FT} = \frac{M_{liq}}{V_{bed} \times t} \tag{1}$$

where P_{FT} is productivity of the bed; M_{liq} is weight of liquid Fischer-Tropsch product released from the reactor bed (usually accounted as sum of hydrocarbon from pentane and heavier or C_{5+}); V_{bed} is the volume of the catalytic bed, and t stands for the period of time necessary for the release of M_{liq}.

Although there are many other types of productivity definition in literature, the one shown in equation (1) is the best for understanding relation between catalyst properties and reactor performance parameters. The productivity is closely related with the catalyst activity but not equal to it. Due to severe diffusion limitations and heat transfer complications, which are typical for Fisher-Tropsch reaction, catalysts with very high activity may show quite low productivity if placed into a catalytic bed. The above-mentioned limitations and complications can be relatively easily removed if one uses a catalyst in the form of small micrometric grains diluted with thermally conductive neutral agent such as quartz sand or silicon carbide powder. The active centers of the catalyst would release a large amount of product under such conditions. However, the volume of such diluted bed V_{bed} would be too large and hence P_{FT} would not reach any significant level as is obvious from equation (1). So this work does not consider experiments with diluted catalytic beds.

Higher productivity means smaller reactor size and, hence, lower capital investment in XTL plant. Also higher productivity means that smaller amount of the catalyst needs to be exchanged for fresh one during regular reloading.

This work reports the results of the study devoted to the development of a fixed bed with significantly increased productivity 240 to 400 kg/(m^3×hr). The research includes:

a) The mathematical modeling of the bed;
b) Formulation of pre-conditions for higher productivity, including requirements to the catalyst, requirements to the reactor and requirements to the temperature and pressure of the process;
c) The results of developing and laboratory testing pelletized catalyst capable of manifesting productivity higher than 240 kg/(m^3×hr) in a fixed bed;
d) The results of designing, building and starting up a scaled-up pilot GTL unit ($1/8$ bbld capacity) with a highly productive Fischer-Tropsch reactor.

The mathematical model describes stationary process occurring in a fixed pelletized bed of a Fischer-Tropsch reactor. The model integrates the following contributing parts:

- Chemical kinetics of paraffine and olefine synthesis from CO and H_2 at the surface of a Co-based catalyst;
- Diffusion model of heat and mass transfer inside quasi-spherical catalytic particles including capillary condensation of water inside pores of the particles;
- Heat and mass transfer model of inter-particle medium of the bed;
- The pressure drop correlation of vapor-liquid flow inside porous medium of the trickle bed;
- Model for mass transfer of vapor phase components through the liquid film, which is formed at the surface of the catalytic particles and at the points of interconnection between particles during the process;
- Modern equation of state (PC-SAFT) for estimation of thermodynamic properties of vapor and liquid phases in three-phase systems (gas phase-water-rich phase - hydrocarbon-rich phase).

The model allows estimation of thermal stability of a spherical catalytic particle with internal heat generation and diffusion resistance of gaseous component.

This model was integrated into a dialogue computer software, which allows calculation and visualization of temperature profile along and across the reactor as well as pressure gradient, velocity vector field and the map of concentration distribution of components in the reactor.

All numerical results of mathematical modeling were compared with experimental data. The experimental data were obtained on two types of chemical reactors with fixed catalytic bed: a laboratory scale single tube reactor 0.25 meter long and a demonstration scale reactor 1.5 meter long. The data include variation of different technological parameters such as pressure and temperature of the process, syngas composition and gas hour space velocity. The comparison showed good agreement with experimental data in the examined range of parameters. For example, the influence of both temperature T and gas hour space velocity u on carbon monoxide overall conversion K_{CO} is shown in Fig. 1 beneath.

It was found that the most thermally unstable zone of the catalytic bed is located in the front area of the bed. Thermal stability of this zone most strongly depends on the size of catalytic particles and the reactor tube diameter. Decreasing the size of catalytic particles leads to increasing the productivity and, as a consequence, to stronger specific heat release. Larger diameter of the tubes leads to increasing specific heat release per unit length.

The analysis of the calculations showed that a highly productive catalytic bed should include a catalyst with high thermal conductivity and advanced pore system. It was also shown that a chosen highly productive catalyst can manifest high productivity if aspect ratio of a fixed bed is high enough. The value of aspect ratio determines an optimal value of the diameter of a fixed bed. Another important result of the calculations was a conclusion that high productivity can be reached at gas hour space velocity (GHSV) exceeding certain limit, which most usually varies between 2000 hr^{-1} and 4000 hr^{-1}.

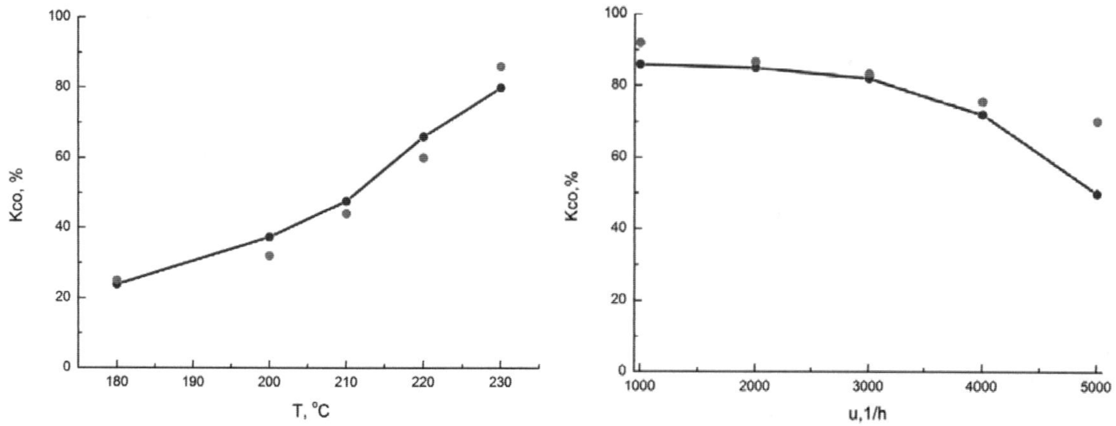

a b

Fig. 1. Carbon monoxide overall conversion: a – influence of temperature, b – influence of gas hour space velocity of syngas. Blue lines – simulation results. Red circles – experimental data. Experimental data presented were obtained for laboratory scale reactor. Parameters of the process were 21 bar and 3000 h^{-1} for the left figure and 240 °C for the right one.

In addition to steady state theoretical analysis of the reactor the parametric analysis of thermal stability of catalytic particles was performed. Spherical porous catalyst particles are considered in this study, with due regard to both diffusion resistance of the components of syngas within the pores of particles, and thermal effects of the process of synthesis, and heat and mass transfer between the surface of particles and a fluid flowing in the catalyst layer. A catalytic particle may lose its thermal stability; this event is associated with the predominance of heat release in the volume of particle over the intensity of the extraction of heat to the outside. In this case, uncontrolled heating of particles will result in the loss of thermal stability by the entire reactor.

Note that the processes of diffusion resistance in the pores of particles contribute to the reduction of the rate of chemical reactions and intensity of heat release. But, besides the stabilization, diffusion resistance inside a porous particle can cause oscillation regime with significant amplitude of both temperature and concentration variations.

It is established that there are four main regimes from the point of view of thermal stability:

- Stable stationary values of temperature and concentrations of all intra-particle components;
- Oscillation of temperature and concentrations in a particle with stable stationary values;
- Regime with very large amplitude of oscillations;
- Unstable stationary conditions accompanied by oscillation.

The stable stationary regime is a target regime of the reactor operation. Meanwhile, the oscillation regime is the most interesting and it is illustrated as such in Fig. 2.

The analysis allows formulating the limitations on particle size and heating conductivity for different process conditions.

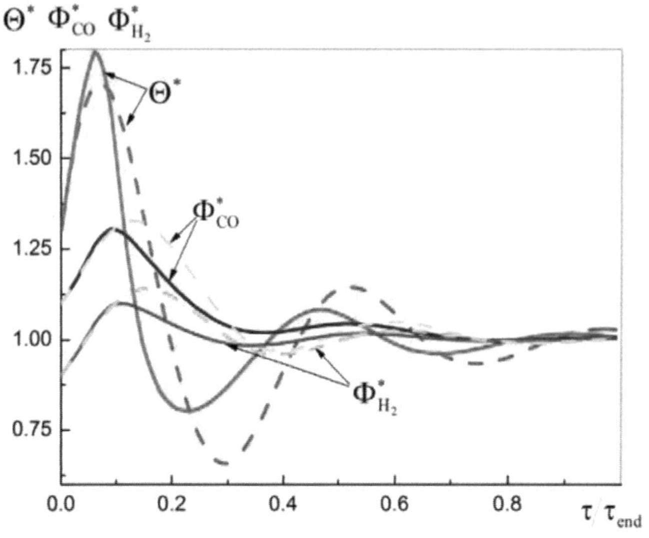

a

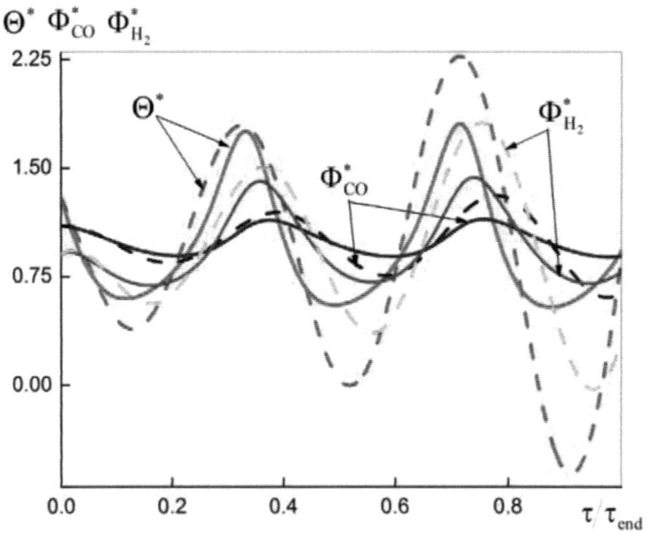

b

Fig. 2. Oscillation regimes of thermal stability. a - developments of oscillation of temperature and concentrations in a particle with stable stationary values. b - unstable stationary conditions accompanied by oscillation. Dimensionless parameters: Θ - temperature, Φ - concentration.

We developed a catalyst for a highly productive Fischer-Tropsch process in compliance with recommendations of the mathematical models. This is a cobalt-based catalyst, which is manufactured by impregnation of a pelletized composite thermally conductive support with cobalt salts. The catalyst composition and manufacturing procedure are described in more detail in the following patents and patent applications: PCT/RU2010/000323 (Rus. Pat. 2405625 of 10.12.2010), PCT/RU2010/000429 and PCT/RU2010/000618.

The catalyst was tested at a laboratory test rig in a fixed pelletized bed (pellet diameter 2.5 mm, pellet length 2.5 - 3 mm) at the pressure of 21 bar. The testing procedure included activation by hydrogen at 400°C and ambient pressure and consequent measurement of catalytic properties in a stream of synthesis gas at the temperatures 150 to 250°C and GHSV from 1000 hr^{-1} to 5000 hr^{-1}. An extract from experimental results is shown in Table 1.

Table 1. Extract from experimental records of testing the highly productive palletized catalyst.
(Conditions: H_2/CO ratio in feedstock 1.98, P = 21 bar)

No	GHSV (hr^{-1})	T (°C)	Conversion, %	Productivity (g/(m^3×hr))
1	1000	230	86	112
2	2000	233	85	244
3	3000	233	82	335
4	4000	235	72	361
5	5000	237	50	276

It can be seen from the Table 1 that the catalyst is indeed capable of producing more than 240 kg/(m^3×hr) of liquid Fischer-Tropsch product in a fixed bed. The data of Table 1 suggest that optimal conditions for operation of Fischer-Tropsch process with this catalyst are as follows:

- GHSV 3000 hr^{-1}
- Temperature of flow 233°C.

The colorless clear liquid produced at the above-mentioned parameters is further referred to as synthetic oil. This synthetic oil was analyzed in our laboratory by a number of techniques, namely:

- Liquid analysis by gas liquid chromatography with a capillary column DB Petro);
- Liquid analysis by simulated distillation (SimDist) ASTM D 2887;
- Liquid analysis (selected gasoline samples) by P.I.A.N.O. (ASTM D 5134);
- Liquid analysis by distillation into fractions;
- Computation of material balance using the analysis of gaseous flow compositions;
- Computation of molecular weight distribution, group and fraction composition of liquids basing on experimental chromatography data.

The analysis showed that the synthetic oil can be characterized as a mixture of normal paraffins with significant amounts of iso-paraffins and alpha-olefins. No traceable amounts of aromatics or oxygenates have been detected. The boiling point curve data suggest that the synthetic oil comprises mostly gasoline and diesel fractions. The properties are controllable in certain limits as shown in Table 2.

An example of a typical synthetic oil composition and properties is shown in Table 3. The Table 3 illustrates that this product is suitable for pumping into oil pipelines because the product complies with pipeline standards.

Table 2. Controllable limits (in wt %) of group composition in synthetic oil product

n-Paraffins	Iso-paraffins	Olefins	Total	Extra info
35–60	20–35	10–30	100	Oxygenates — < 200 ppm Aromatics — traces Fractions: C_5–C_{10} — 50–80% C_{11}–C_{18} — 20–47% C_{19}–C_{30} — <3%

Table 3. Group composition, fractions and properties of a typical sample of synthetic oil product

	n-Paraffines	Iso-paraffines	Olefines	Total	
C_5–C_{10}	35	14	17	66	C_5–C_{10}
C_{11}–C_{18}	18	11	2	31	C_{11}–C_{18}
C_{19}–C_{30}	2	1	0	3	C_{19}–C_{30}
Total	55	26	19	100	Total

Specific weight — 663 kg/m3 (pipeline standard requires: below 830 kg/m3)
Saturation pressure — 0.14 atm (pipeline standard requires: below 0.66 atm)
Viscosity — 0.623 mPa•s (pipeline standard requires: below 5 mPa•s)

The results of testing the developed catalyst allowed us to design, build and start up a scaled-up pilot unit for modeling a complete XTL process. The pilot unit with capacity of ¼ bbld models a full cycle of XTL process with natural gas as a feedstock. The unit includes a reactor of desulfurization of feedstock; a reactor for steam-CO_2 reforming, heating-cooling station, compressor station and two parallel highly productive Fischer-Tropsch reactors. The unit in the form of a truck-transportable block was complete and commissioned in October 2010 – see photograph in Fig. 3.

Fig. 3. A photograph of the running pilot unit

(56b) The Effect of Accumulated Products In the Fischer-Tropsch Reactor on the Product Distribution

Cornelius Mduduzi Masuku, Center for Applied Energy Research, University of the Kentucky, Lexington, KY, Wilson D. Shafer, Center for Applied Energy Research, University of Kentucky, Lexington, KY and Burtron H. Davis, Center for applied energy research, University of Kentucky, Lexington, KY

The phrases ''product accumulation'' or ''accumulated products'' or ''product holdup'' have appeared in the literature during the past several decades to qualitatively explain the experimental results for Fischer-Tropsch Synthesis (FTS). However, since the amount of accumulated products was not determined experimentally, the seriousness of accumulated products in FTS was underestimated by many researchers. The apparent products of the FTS reaction is a mixture of freshly produced FTS products and the products left in the reactor. In order to obtain the correct product distribution and the paraffin to olefin ratio of a FTS reaction, it is necessary to find a way to evaluate or eliminate the contribution of the accumulated products in the product distribution which requires conducting FTS research in a different way. This study develops a method to evaluate product accumulation inside the FT reactor.

See more of this Session: Production of Fuels and Chemicals From Fischer-Tropsch Hydrocarbons II
See more of this Group/Topical: Topical 6: 11th Topical Conference on Gas Utilization

(56c) Characterization and High Temperature Water Gas Shift Activity Testing of Cu-Doped Transition Metal Doped Ferrites

Krishna Reddy[1], Punit Boolchand[2] and Panagiotis Smirniotis[1], (1)Chemical & Materials Engineering, University of Cincinnati, Cincinnati, OH, (2)Electrical and Computer science Engneering, University of Cincinnati, Cincinnati, OH

In the present study, the role of copper on the shift activity of various transition metal doped catalysts is being investigated. The idea was to stimulate the ferrite formation via doping with copper and to promote the Fe^{3+} « Fe^{2+} redox couple during the WGS reaction. For this purpose, various copper doped modified ferrite catalysts of type Fe: M: Cu (M= Cr, Ce, Ni, Co, Mn, and Zn) with the chemical composition 10: 1: 0.25 were prepared by ammonia assisted coprecipitation method. The prepared and activated catalysts were characterized by using X-ray diffraction (XRD), Temperature programmed reduction (TPR), Raman, Mössbauer, and X-ray photo electron spectroscopic (XPS) techniques. The WGS feed condition were chosen to mimic conditions found in a membrane reactor. The WGS reaction conducted in the temperature region 400-550 ^{o}C and at a Steam/CO ratio 3.5 which the conditions used in our membrane reactor. Interestingly, our shift activity findings reveal that copper act as a promoter for the transition metal doped ferrites expect for the Fe/Ce catalyst. While, Cu acts as an inhibitor for the Fe/Ce catalyst. XRD and TPR results suggest that addition of Cu to Fe/Ce catalyst leads to the formation of wustite phase during the activation. Mössbauer spectroscopy studies indicate that copper selectively enters in to the octahedral sites of the magnetite cubic lattice during the activation of the catalysts and promotes the WGS activity. O1s XPS results show that copper suppress the carbonate formation during the activation of modified ferrites. Cu 2p XPS spectra show that, Cu is not accessible to the surface in the Fe/Ce/Cu catalyst. It is completely incorporated in to the Fe/Ce lattice.

See more of this Session: Production of Fuels and Chemicals From Fischer-Tropsch Hydrocarbons II
See more of this Group/Topical: Topical 6: 11th Topical Conference on Gas Utilization

(56d) Production of Middle Distillate Range Transportation Fuels From Synthesis Gas Using Fischer Tropsch Synthesis Technology Under Supercritical Phase Conditions

Sihe Zhang[1], Rui Xu[2], Ed Durham[1] and Christopher B. Roberts[3], (1)Department of Chemical Engineering, Auburn University, Auburn, AL, (2)Chemical Engineering, Auburn University, Auburn, AL, (3)Department of Chemical Engineering, Auburn University, Auburn University, AL

Fischer-Tropsch (FT) synthesis involves surface catalyzed polymerization reactions that inherently yield a high selectivity towards methane and a broad product distribution of gaseous light products, middle distillate hydrocarbons (olefins + paraffins), and heavy wax, as well as CO_2 and water. Subsequent product upgrading via catalytic oligomerization and hydrocracking/isomerization can be employed to convert light olefin FT hydrocarbons into middle distillate range liquids, while also converting the long-chain hydrocarbons of heavy wax into fuel range products. This results in a reduction in olefin selectivity and an enhancement of cold-flow properties.

In an effort to further direct the FT product distribution towards middle distillate transportation fuel range products, we have undertaken a series of experiments where we have sequentially incorporated catalytic oligomerization and hydrocracking/isomerization stages subsequent to FT in single pass operation. Moreover, we have utilized the enhanced heat transfer properties of supercritical fluid media in the FT and subsequent upgrading stages to reduce the selectivity towards CH_4 and CO_2. Moreover, the use the supercritical solvent media provides prolonged catalyst life and activity maintenance. In particular, this study involves a systematic investigation into the influence of the supercritical fluid reaction medium on the overall product distribution and selectivity towards fuel range products.

See more of this Session: Production of Fuels and Chemicals From Fischer-Tropsch Hydrocarbons II
See more of this Group/Topical: Topical 6: 11th Topical Conference on Gas Utilization

(62a) Setting Biorefinery Manufacturing Fundamentals to Produce a Portfolio of Commodity and Fine Chemicals

Alain A. Vertés[1], Masayuki Inui[1] and Hideaki Yukawa[2], (1)Molecular Microbiology and Biotechnology Group, Research Institute of Innovative Technology for the Earth (RITE), Kizugawa, Japan, (2)Molecular Microbiology and Biotechnology Group, RITE, Kizugawa, Japan

The chemurgy concept, that is, manufacturing industrial commodity and chemical products from organic raw materials such as agricultural biomass and lignocellulosic waste, predates the petrochemical industry. However, petrochemicals prevailed over fermentation chemicals given critical industrialization factors, comprising a low cost of the petroleum feedstock and a high process economic efficiency, as exemplified by the full utilization of products and by-products from petrochemical plants for the production of gasoline. These economies of scale and scope have played an important role in the booming industrial development at the beginning of the XXth century and were realized on a very large scale in World War II and the following decades[1].

These extremely efficient cost structures that have been derived in the petrochemical industry over a century are now both a blessing and a curse. The blessing is that these economies of scale and scope have made possible the manufacturing of cheap fuels and materials. The resulting abundant supply of chemicals for industry and transportation has enabled modern life and thus now appear to be indispensable. The curse is that efforts to displace and replace the petroleum industry by the chemurgy industry for alleviating global warming and fossil fuel supply threats meet a very high economic and resistance barrier[2]. This has resulted over the years in a detrimental delay in renewable chemistry technology investment, development, and deployment[3].

The biorefinery concept aims at recreating for agricultural feedstock cost structures that are similarly enabling as the ones at play in the petrochemical industry[4, 5]. This change is of course fundamental and will have far reaching consequences, as it integrates the agricultural value chain into the petrochemical value chain, thus colliding the transportation, polymer, and chemical value chains with other established value chains and notably of the food and feed industry[3]. A key difference nevertheless remains in that the higher oxygen content of biomass-based compounds makes possible to produce totally novel materials and chemicals with novels properties[6]. An enabling factor of the biorefinery vision is of course the availability of versatile, cost-effective, and robust industrial processes to manufacture an array of chemicals to serve various sectors including the fuel, chemical, industrial polymer, feed, food, cosmetic, and pharmaceutical industries. Notably, it is important for the biorefinery plant manager to be able to produce in addition to fuels, such as ethanol or isobutanol, higher value compounds for the chemical industry, such as ethyl-acetate or isobutanol.

We report here on a growth-arrested biotechnological process to manufacture ethanol, organic acids, and amino acids from mixed sugars, based on *Corynebacterium glutamicum*, an organism that has a long history of use to produce amino acids as feed and food supplements or cosmetic ingredients. The intrinsic design of the process is to decouple in a separate fermentor and a reactor biological catalyst-production and product-production phases[7] with cell recycling and using very high cell concentrations; it builds upon the intrinsic properties of this organism. *C. glutamicum* is a facultative anaerobic bacterium that grows in the absence of oxygen only in media containing a terminal electron acceptor such as nitrate[8]. *C. glutamicum* does not sporulate but remains metabolically active when placed under growth-arrested conditions[9, 10] and is not inhibited by common fermentation inhibitors present in typical lignocellulosic hydrolysates[11, 12]; moreover, it can catabolize a range of sugars in parallel with glucose[7, 13, 14]. A complete toolbox of molecular tools enables one to modulate at will its plastic genome and general metabolism[10, 15]. The RITE process presents the entire set of fundamental attributes necessary for reaching cost-effectiveness and enabling a rich product portfolio: simplicity, industrial robustness[16], raw material versatility (glucose, fructose, glucuronic acid, glucosamine, maltose, mannose, α-methyl-glucoside, ribose, sucrose, trehalose, arbutin, salicin, cellobiose, arabinose, xylose)[13, 14, 17, 18], product versatility (ethanol, organic acids, amino acids)[9, 16, 19, 20], reduced utilities requirement (reduced fermentation heat[12], no oxygenation requirement[9]), high conversion rates of carbohydrate feedstocks-to-product given cell recycling and the absence of growth requirements (e.g., 110 gl^{-1} or 40 $gl^{-1}h^{-1}$ D-lactate)[21], possibility to conduct continuous fermentations for long periods of time, dramatically reduced risks of contamination in the reaction vessel given the use of very high cellular concentrations, secreted product, no expensive reagent, portfolio of processes that enable the practitioner to manage the biorefinery plant in swiftly interchangeable processes or to capture synergies by implementing multiplex productions[22].

The stage is now set for this core technology to be translated into a biorefinery demonstration plant, as a first step towards full-scale commercialization.

References

1. Benninga, H. A history of lactic acid making. (Kluwer Academic Publishers, Dordrecht, The Netherlands; 1990).
2. Vertès, A.A., Inui, M. & Yukawa, H. Implementing biofuels on a global scale. Nat Biotechnol 24, 761-764 (2006).
3. Vertès, A.A., Qureshi, N., Blaschek, H.P. & Yukawa, H. (eds.) Biomass to biofuels: strategies for global industries. (Wiley, Chichester, UK; 2010).
4. Cherubini, F. The biorefinery concept: using biomass instead of oil for producing energy and chemicals. Energy Conversion Manag 51, 1412-1421 (2010).
5. Octave, S. & Thomas, D. Biorefinery: toward an industrial metabolism. Biochimie 91, 659-664 (2009).
6. Marquardt, W. et al. The biorenewables opportunity: toward next generation process and product systems. AIChE J 56, 2228-2235 (2010).
7. Vertès, A.A., Inui, M. & Yukawa, H. in 14th European Biomass Conference 1068-1071, Paris,

France; 2005).
8. Nishimura, T., Vertès, A.A., Shinoda, Y., Inui, M. & Yukawa, H. Anaerobic growth of Corynebacterium glutamicum using nitrate as a terminal electron acceptor. Appl Microbiol Biotechnol 75, 889-897 (2007).
9. Inui, M., Kawaguchi, H., Murakami, S., Vertès, A.A. & Yukawa, H. Metabolic engineering of Corynebacterium glutamicum for fuel ethanol production under oxygen-deprivation conditions. J Mol Microbiol Biotechnol 8, 243-254 (2004).
10. Yukawa, H., Inui, M. & Vertès, A.A. in Amino acid biosynthesis, Vol. 5. (ed. V.F. Wendisch) (Springer Verlag, Berlin, Germany; 2006).
11. Sakai, S. et al. Effect of lignocellulose-derived inhibitors on growth of and ethanol production by growth-arrested Corynebacterium glutamicum R. Appl Environ Microbiol 73, 2349-2353 (2007).
12. Inui, M. et al. in 28th Symposium on Biotechnology for Fuels and Chemicals, Nashville, TN, USA; 2006).
13. Kawaguchi, H., Sasaki, M., Vertes, A.A., Inui, M. & Yukawa, H. Engineering of an L-arabinose metabolic pathway in Corynebacterium glutamicum. Appl Microbiol Biotechnol 77, 1053-1062 (2008).
14. Sasaki, M., Jojima, T., Inui, M. & Yukawa, H. Simultaneous utilization of D-cellobiose, D-glucose, and D-xylose by recombinant Corynebacterium glutamicum under oxygen-deprived conditions. Appl Microbiol Biotechnol 81, 691-699 (2008).
15. Vertès, A.A., Inui, M. & Yukawa, H. Manipulating corynebacteria, from individual genes to chromosomes. Appl Environ Microbiol 71, 7633-7642 (2005).
16. Vertès, A.A., Inui, M. & Yukawa, H. Technological options for biolofical fuel ethanol. J Mol Microbiol Biotechnol 15, 16-30 (2008).
17. Yukawa, H. et al. Comparative analysis of the Corynebacterium glutamicum group and complete genome sequence of strain R. Microbiology 153, 1042-1058 (2007).
18. Kawaguchi, H., Vertès, A.A., Okino, S., Inui, M. & Yukawa, H. Engineering of a xylose metabolic pathway in Corynebacterium glutamicum. Appl Environ Microbiol 72, 3418-3428 (2006).
19. Inui, M. et al. Metabolic analysis of Corynebacterium glutamicum during lactate and succinate productions under oxygen deprivation conditions. J Mol Microbiol Biotechnol 7, 182-196 (2004).
20. Okino, S. et al. An efficient succinic acid production process in a metabolically engineered Corynebacterium glutamicum strain. Appl Microbiol Biotechnol 81, 459-464 (2008).
21. Okino, S., Suda, M., Fujikura, K., Inui, M. & Yukawa, H. Production of D-lactic acid by Corynebacterium glutamicum under oxygen deprivation. Appl Microbiol Biotechnol 78, 449-454 (2008).
22. Vertès, A.A., Inui, M. & Yukawa, H. Alternative technologies for biotechnological fuel ethanol manufacturing. J Chem Technol Biotechnol 82, 693-697 (2007).

See more of this Session: Developments in Biomass to Fuels
See more of this Group/Topical: Topical 9: Waste & Biomass Valorization

(62b) Evaluation of Pretreatment Methods In the Production of Ethanol From Cattail Leaves

Elijah Wreh, Biological Sciences, St. Cloud State Graduate School, Champlin, MN

Abstract

Previous research in this lab indicated that cattails are a potential source of biomass for the production of cellulosic ethanol since their carbohydrate composition is comparable to that of other plants being considered for biofuel production. To further test their viability, we tested various pretreatment methods on dried cattail leaves. Before polysaccharides in plants can be enzymatically hydrolyzed to fermentable sugars, the plant material must be pretreated to render the polysaccharides accessible to the enzymes. The purpose of this project has been to compare the efficiency of sulfuric acid and ammonia pretreatment methods in preparing cattail biomass for ethanol production. In this project, dried, powdered cattail leaves were pretreated either by autoclaving them with 2% sulfuric acid for one hour or by incubating them overnight at 40 C° in 15 % aqueous ammonia. Samples of the dried, pretreated solid were treated with cellulase and -glucosidase for 48 hours. To compare the efficiency of these pretreatment methods, glucose liberated in these samples was measured by a glucose oxidase assay. It was found that more glucose was recovered in the enzymatic hydrolysis (step two) than in the pretreatment step (step one.) In step two, more glucose was liberated from biomass pretreated with ammonia than from biomass pretreated with sulfuric acid. However, more glucose was recovered from in step one by sulfuric acid pretreatment. Overall 27.8 % of the starting biomass was recovered as glucose with ammonia pretreatment compared to 11.7 % for sulfuric acid pretreatment. Interestingly, 22.3% of the starting biomass was recovered as glucose when no pretreatment was used.

See more of this Session: Developments in Biomass to Fuels
See more of this Group/Topical: Topical 9: Waste & Biomass Valorization

HEMICELLULOSE AND LIGNIN RECOVERY FOR AMMONIUM HYDROXIDE PRETREATMENT PROCESSES

*STEVEN R. SHERMAN, JACOB A. MORRIS, JOHN H. GOODELL**
Savannah River National Laboratory, Aiken, SC, USA
**Adrian College, Adrian, MI, USA*

Introduction

Pretreatment is a step performed in fermentation-based processes that convert lignocellulosic materials into fuels and chemicals. During pretreatment, size-reduced (granulated, chopped, ground, etc.) biomass is subject to physical or chemical steps to make it more amenable to hydrolysis into sugars [1]. While physical pretreatment methods operate by creating more biomass surface area on a microscopic level, chemical pretreatment operates by removing or disrupting the lignin structure that hinders access to cellulosic fibers in the biomass.

Since 2008, Savannah River National Laboratory (SRNL) and Clemson University have been funded by the U.S. Department of Energy, Office of Energy Efficiency and Renewable Energy (DOE-EERE), to develop a fermentation-based process for converting switchgrass, sorghum, and wood into ethanol and other biofuels. The process backbone consists of the following steps: size reduction, pre-treatment, hydrolysis, fermentation, and product purification. The Soaking in Aqueous Ammonia (SAA) pretreatment method [2–4] was initially selected for this work due to prior experience with the method, and it has since proven effective at pretreating these feedstock materials. Application of this method dissolves between 40-60% of lignin, < 20% of hemicellulose, and < 5% of the cellulose from the pretreated biomass. Although the amount of lignin dissolved by this method may be less than can be achieved using other methods, this amount of lignin dissolution has proven sufficient to allow for greater than 80% hydrolysis of pretreated biomass into into sugars when commercially available cellulase mixtures are used to perform the hydrolysis.

As this work proceeds, we have begun to appreciate that the economics of the fermentation-based process may be improved by generating multiple products from the same feed material. During the pretreatment step, lignin is removed from the biomass. This lignin may be recovered from the pretreatment fluid and used as a fuel or a feedstock material higher-value chemicals. Following pretreatment, cellulose is hydrolyzed into glucose (a C-6 sugar), and hemicellulose is hydrolyzed into xylose (a C-5 sugar) and other sugars. In the fermentor, C5-sugars such as xylose are not easily metabolized by natural yeast. As a result, process throughput is reduced and the break-even price of the ethanol product is increased. Although C5-sugar utilization could be improved by using a genetically modified yeast or alternative organism, ethanol has a relatively low market price, and the C5-sugars might be better used to make products that have a higher market value than ethanol.

SRNL recently developed a process for separating and isolating lignin from ammonium hydroxide pretreatment fluids [5], which allows for the generation of a lignin side stream. Some hemicellulose is also extracted during ammonium hydroxide pretreatment, but it is not recovered by the lignin recovery process. Hemicellulose is better extracted using a separate process step, and hemicellulose extraction methods are described in the literature that might be used to recover hemicellulose as a distinct product [4, 6]. When performed before the hydrolysis step, both extraction steps are categorized as pretreatment.

In this work, the performance of the lignin extraction step and the hemicellulose extraction step are examined separately, and then in combination. When both extractions are performed in sequence, the order of the steps is alternated and compared, so that a recommended sequence can be discovered. The lignin extraction is performed using the SRNL method cited previously. The hemicellulose extraction is performed using a method adapted from [6] that uses $Ca(OH)_2$ solution as an extracting agent. Due to lack of chemical analytical data, the chemical composition of the lignin and hemicellulose samples collected during the course of this work is unknown. Instead, sample color and appearance are used as qualitative indicators of product purity in the tests.

Description of Lignin Recovery Method

Lignin is recovered from the ammonium hydroxide pretreatment fluid by the following procedure, and is described more fully in [5].

- L1: The pretreatment fluid is separated from the pretreated biomass material using filtration.
- L2: The ammonium hydroxide pretreatment solution containing dissolved lignin and other materials is heated to the boiling point in order to evaporate ammonia from solution. Ammonia-rich vapor is allowed to evaporate from the solution until the boiling point of the solution reaches 96-98°C at 101.1 kPa.
- L3: While hot, the solution is titrated with 10% sulfuric acid solution until the color of the solution changes from black to brown and the pH of the solution falls to between 2 and 3. The color change indicates the precipitation of a colloid of lignin particles.
- L4: The titrated solution is heated again to the boiling point, and water vapor is allowed to evaporate from the titrated solution until the volume of liquid has fallen to approximately half the starting volume and a "skin" of brown solid begins to form on the sides and top of the liquid. The formation of the "skin" indicates coagulation of colloidal lignin particles into larger particles that are separable from solution by gravity or filtration.
- L5: Allow the mixture to cool, and separate the lignin particles from the liquid by high-speed centrifugation or filtration using a coarse (>25 μm) filter and then a fine (>4 μm) filter.
- L6: Dry the recovered solids in the open air, or in a vacuum oven at a temperature less than 65°C to avoid conglomeration and glazing of the recovered solid.

Description of Hemicellulose Recovery Method

A hemicellulose extraction method adapted from [6] was used to extract and isolate hemicellulose from untreated and pretreated biomass in this work. The extraction steps are provided below. This method was selected for this work because it was well described in the literature, was free to use because the applicable patents had expired, and could be performed using the same equipment as was used for performing lignin extractions.

- H1: Size-reduced biomass (untreated or pretreated) is combined with 0.25 wt% $Ca(OH)_2$ solution in a 10:1 (volume-to-mass) ratio, and the slurry is mixed and heated for 2 hours at 80°C.

- H2: The slurry is allowed to cool, and then the liquid is separated from the solid material by filtration.
- H3: The recovered filtrate is titrated using 10-20% sulfuric acid until the pH of the filtrate is between 2 and 3. The solution may appear yellow-tinted with a fine suspension of white material after titrating it with sulfuric acid.
- H4: The titrated fluid is mixed with methanol in a volumetric ratio of 1.25:1 (volume-to-volume). This causes hemicellulose to precipitate out of solution. The appearance of the solution changes from pale yellow to milky white.
- H5: The precipitated solid may be allowed to settle and the liquid decanted, or the solid-liquid separation process may be accelerated by high-speed centrifugation or filtration.
- H6: The recovered solid, which is white, is allowed to dry in the open air or in a drying oven at a low temperature to avoid melting or caramelizing the recovered hemicellulose.

Results and Discussion

All tests were performed using 50 g samples of biomass as the starting amount of material. The amount of pretreatment solutions used to extract lignin and hemicellulose from the biomass materials were then adjusted to match the mass of the starting material.

Figures 1 and 2 show the percentage of original lignin recovered from previously untreated biomass. Four different feedstock materials were tested: switchgrass, sweet sorghum, sweetgum, and loblolly pine. The percentage of original lignin recovered is defined as the mass of lignin recovered as a solid lignin product divided by the estimated amount of lignin originally present in the biomass, based on characteristic compositional data provided in [8]. According to Figure 1, the amount of lignin recovered varies with the type of biomass when the pretreatment ratio (volume:mass) is 8:1. At an 8:1 pretreatment ratio, the lignin extraction method worked best on switchgrass, and was the least effective on loblolly pine. When the pretreatment ratio was increased from 8:1 to 16:1, the performance of the lignin extraction method on sweet sorghum and loblolly pine was improved, and, for loblolly pine, became statistically indistinguishable from the results obtained for switchgrass (see Figure 2). The convergence of results when the pretreatment ratio was increased from 8:1 to 16:1 suggests that ammonium hydroxide pretreatment, rather than the lignin recovery process, is responsible for the poorer lignin recovery from sweet sorghum, sweetgum, and loblolly pine. The ammonium hydroxide pretreatment process may be affected by a lignin solubility limit, the composition of the biomass, or, more likely, mass transfer limitations that are alleviated at higher liquid-to-solid ratios. The solid material recovered from all of the biomass samples using the lignin recovery procedure was dark brown and powdery, and matched the appearance and consistency of samples that were chemically analyzed and identified as lignin in previous work [5].

The fraction of original hemicellulose recovered from previously untreated samples of switchgrass and sweet gum is shown in Figure 3. In the figure, only about 20% of the original hemicellulose content in the switchgrass samples was recovered as a solid product, while more than 40% of the original hemicellulose content was recovered from samples of sweet gum. Since the same extraction and hemicellulose recovery procedure was used for both materials, it is likely that the differences in the results are due to mass transfer limitations during the extraction. No additional work was performed to

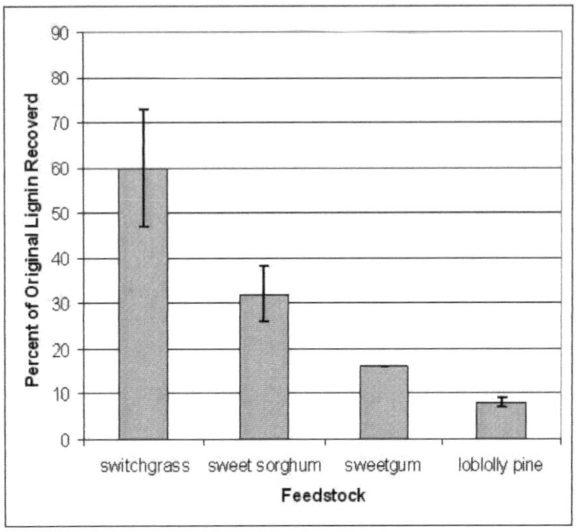

Figure 1: Lignin recovery varies with feedstock at an 8:1 pretreatment ratio.

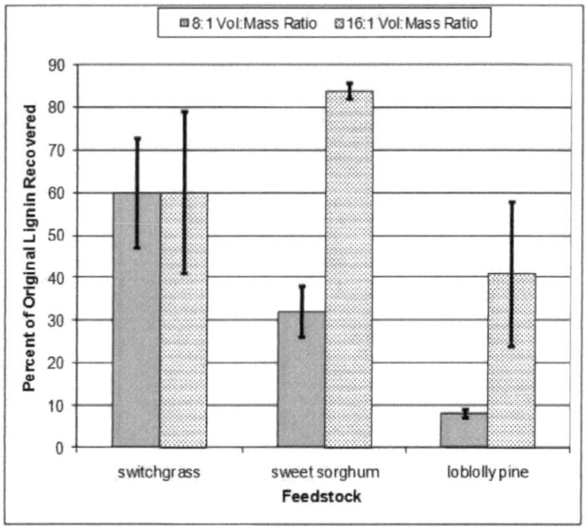

Figure 2: Lignin recovery is improved at a 16:1 pretreatment ratio.

study this question, and more work remains to be done if this hemicellulose extraction method is to be improved and optimized for the biomass feed materials of interest. The hemicellulose recovered from switchgrass had a dull, white, semi-crystalline appearance, while the hemicellulose recovered from sweet gum was bright white and powdery. Both types of hemicellulose turned light brown in color when heated in a drying oven.

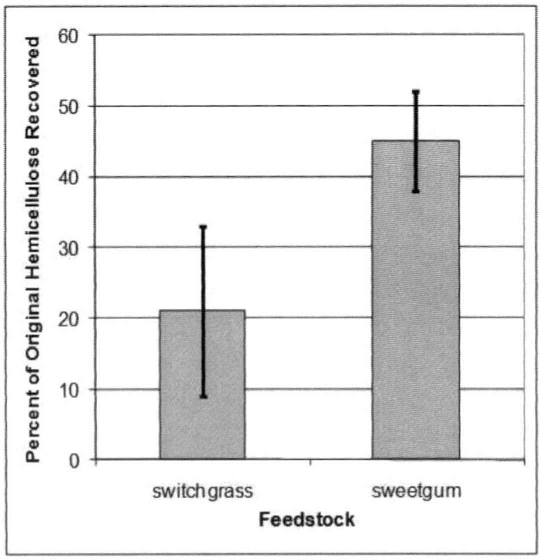

Figure 3: Hemicellulose recovery varies with feedstock.

The switchgrass and sweet gum samples from which hemicellulose had been extracted were then subjected to ammonium hydroxide pretreatment, and lignin was recovered from the ammonium hydroxide pretreatment fluid. The percent of original lignin recovered for these samples is shown in Figure 4. The percentage of original lignin recovered was lower for switchgrass from which hemicellulose had been extracted than for previously untreated switchgrass. For sweetgum, there was no distinguishable difference in the percentage of original lignin recovered, though it is difficult to assign a clear trend for sweetgum when the error bars on the data are so wide.

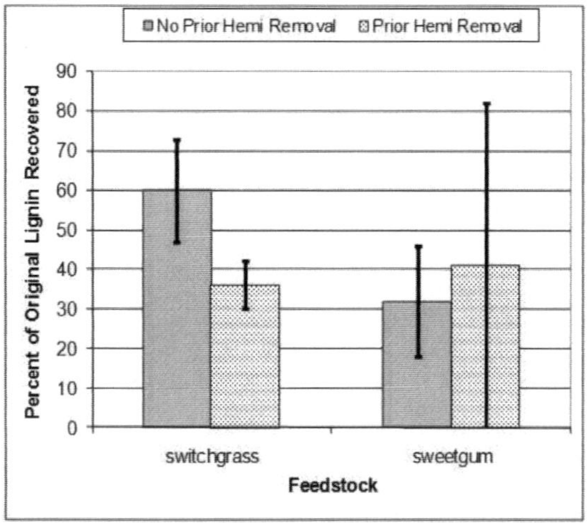

Figure 4: Variation in lignin recovery after some hemicellulose has been removed from the biomass.

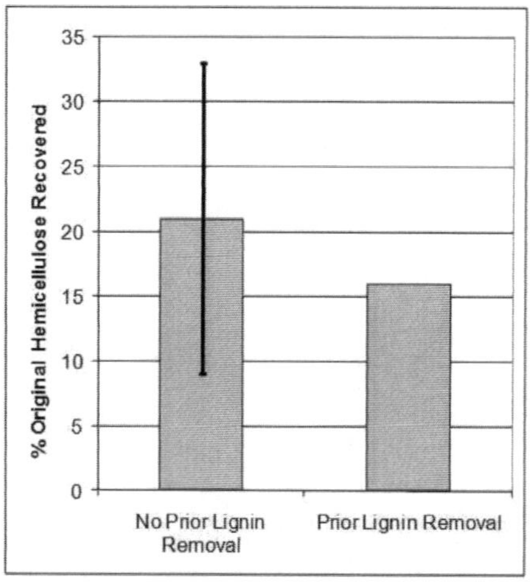

Figure 5: Comparison of hemicellulose extraction results on untreated and NH_4OH pretreated switchgrass.

Figure 5 shows a comparison of the percentage of hemicellulose recovered from untreated switchgrass, and switchgrass that had been pretreated with ammonium hydroxide. Although the chart shows a lower percentage of hemicellulose recovered from switchgrass that had been pretreated, statistically there is no difference between the two cases, and it appear that hemicellulose removal is not affected by ammonium hydroxide pretreatment, at least for the experimental conditions tested here.

Conclusions

The lignin extraction procedure and hemicellulose extraction procedure were shown to be capable of recovering lignin and hemicellulose as separate products from a range of biomass materials (i.e., switchgrass, sweet sorghum, sweetgum, and loblolly pine). The amount of lignin recovered from each type of biomass varied, and it appears that the variation in the amount of lignin recovered was due to mass transfer problems during pretreatment, rather than due to the lignin recovery procedure. The percentage of original hemicellulose recovered from sweetgum was higher than was recovered from switchgrass, which shows that the hemicellulose extraction procedure is affected by biomass type and perhaps mass transfer limitations. The amount of hemicellulose recovered from switchgrass that had been pretreated with ammonium hydroxide was lower than the amount recovered from untreated switchgrass, though this trend was not seen with sweetgum. The amount of lignin recovered from switchgrass that had previously been pretreated with $Ca(OH)_2$ solution was statistically similar to the amount of lignin recovered from untreated switchgrass.

With no clear trends in the data indicating an ideal sequence of extraction steps, it appears that the order in which the hemicellulose and lignin extraction steps is not important. Although there may be some amount of hemicellulose that is extracted along with lignin during ammonium hydroxide pretreatment, apparently the amount of hemicellulose lost from the biomass during the lignin extraction step is small enough that it does not significantly affect the performance of the subsequent hemicellulose extraction step. The reverse sequence, performance of the hemicellulose extraction step before extracting lignin with ammonium hydroxide, shows no effect on the amount of lignin recovered from switchgrass, and so is also a valid extraction sequence.

The variation in the data are wide. More experiments with repeated trials are needed to generate better statistical information before strong conclusions can be drawn on the proper sequencing of the extraction steps. Also, it would be helpful to collect chemical compositional data on the pretreated biomass and recovered products to discern the effects of varying the sequence of extractions on biomass composition and product purity.

References

[1] Y. Sun, J. Cheng, "Hydrolysis of lignocellulosic material for ethanol production: a review", *Bioresource Technology* (2002), 83:1-11.

[2] T.H. Kim, F. Taylor, K.B. Hicks, "Bioethanol production from barley hull using SAA (soaking in aqueous ammonia) pretreatment", *Bioresource Technology* (2008), 99:5694-5702.

[3] T.H. Kim, Y.Y. Lee, "Pretreatment of corn stover by soaking in aqueous ammonia at moderate temperatures", *Applied Biochemistry and Biotechnology* (2007), 137-140:81-92.

[4] T.H. Kim, Y.Y. Lee, "Fractionation of corn stover by hot-water and aqueous ammonia treatment", *Bioresource Technology* (2006), 97:224-232.

[5] S.R. Sherman, J.J. Goodell, C.E. Milliken, J.A. Morris, M.B. Gorensek, "A New Process Developed for Separation of Lignin from Ammonium Hydroxide Pretreatment Fluids", accepted for publication in *Environmental Progress and Sustainable Energy*, December 2010.

[6] M.W. Rutenberg, W. Herbst, "Process for Extraction of Hemicellulose", U.S. Patent 2,801,955, Patented August 6, 1957.

[7] W. Shu-Lai Mok, M.J. Antel Jr., "Uncatalyzed Solvolysis of Whole Biomass Hemicellulose by Hot Compressed Liquid Water", *Industrial Engineering Chemistry Research* (1992), 31:1157-1161.

[8] W.J. Frederick Jr., S.J. Lien, C.E. Courchene, N.A. DeMartini, A.J. Ragauskas, K. Iisa, "Production of ethanol from carbohydrates from loblolly pine: A technical and economic assessment", *Bioresource Technology* (2008), 99:5051-5057.

(62d) Accelerated Hydrolysis of Starch Using Cationic Polyacrylamide for Production of Biofuels

Kendra E. Maxwell, School of Chemical & Biomolecular Engineering Georgia Tech, Georgia Institute of Technology, Atlanta, GA and Sujit Banerjee, School of Chemical & Biomolecular Engineering, Georgia Institute of Technology, Atlanta, GA

Improving the efficiency of the hydrolysis of starch to glucose is key to reducing energy use during the conversion of starch to biofuels. We have found that the addition of positively charged polyacrylamide increases the conversion rate of starch to glucose. The aim of this study is to determine the major factors which affect the mechanism so that optimal production rates are achieved in the presence of the cationic polyelectrolytes. A combination of analytical methods including dynamical light scattering, optical imaging, and uv-vis spectroscopy was used to study polyelectrolyte, starch, and enzyme interaction as a function of process conditions crucial to starch hydrolysis such as composition and temperature. The results indicate that interactions caused by electrostatic forces largely drive the increase in hydrolysis rate. These and related results will be discussed in this presentation.

See more of this Session: Developments in Biomass to Fuels
See more of this Group/Topical: Topical 9: Waste & Biomass Valorization

(62e) Production of Bio-Fuels From Giant Miscanthus

Sanjeev K. Gajjela, Brain Mitchell, Qi Li, El barbary M. Hassan and Philip H. Steele, Forest Products, Mississippi State University, Mississippi State, MS

The energy crop giant miscanthus was pyrolyzed in a laboratory-scale auger reactor at Mississippi State University (MSU). The influence of pyrolysis temperature and biomass pyrolysis time on bio-oil yields is investigated. Pyrolysis temperature of giant miscanthus was varied between 375-450 °C. The highest bio-oil yield of 60 wt% was obtained at 425 °C. Best yield was also a function of biomass feed rate during pyrolysis. We found that maximum yield was attained for an approximate doubling of feedstock feed rate for giant miscanthus as compared to that for pine wood. The reason for this is likely due to differences in particle thickness with wood particles having approximately double or more the thickness of pine wood particles. This apparently allowed the giant miscanthus feedstock particles to pyrolyzed fully at a higher rate. When fed at a the same rate as for pine particles over pyrolysis of the miscanthus occurred as evidenced by lower yields and higher water content of the resulting bio-oil.

Further, raw bio-oil from giant miscanthus in the presence of hydrotreating/hydrocracking catalyst with high temperature and high hydrogen pressure yields hydrocarbons and water. MSU has developed effective hydrotreating (HT) and hydrocracking (HC) catalysts that have successfully produced a hydrocarbon mixture from pine-based bio-oil. These catalysts were applied to the hydroprocessing of giant miscanthus bio-oil. Giant miscanthus bio-oil demonstrated some differences in response to the first-stage HT and second-stage HC treatments as compared to pine bio-oil. These differences will be described in our paper. The physical and chemical properties of giant miscanthus bio-oil and upgraded bio-oil are determined. The fuel-value characteristics of the hydrocarbons produced by two-stage HT and HC will be fully described by analysis by GC/MS and FTIR. Simulated distillation will be performed on the hydrocarbons to describe the boiling range distribution of the hydrocarbon weights produced with comparison of these weights to those of petroleum fuels in the same molecular weight classes.

See more of this Session: Developments in Biomass to Fuels
See more of this Group/Topical: Topical 9: Waste & Biomass Valorization

DEVELOPMENT OF A DETAILED REACTION KINETIC MODEL FOR CELLULOSE FAST PYROLYSIS

Vinu Ravikrishnan and Linda J. Broadbelt, Chemical and Biological Engineering, Northwestern University, Evanston, IL

Fast pyrolysis is one of the potential strategies to convert whole biomass to sustainable transportation fuels. Fast pyrolysis involves heating the biomass in an inert atmosphere to moderate temperatures (400 – 500 ^{o}C) in a very short period of time (< 1 s). This leads to a complex network of competing reactions, which results in the formation of bio-oil, non-condensable gaseous species, and solid char. Bio-oil is a mixture of anhydro sugars, phenolic compounds, furan derivatives and low molecular weight oxygenated compounds. The development of a detailed reaction kinetic model to predict the yields of the various pyrolysis products is of utmost importance to understand the relative contributions of the various reaction pathways, and the effect of process parameters like temperature, pressure, and mineral content of biomass.

The existing kinetic models of biomass pyrolysis are global, as they lump the various product species into different fractions like char, condensable and non-condensable gaseous species, thus ignoring the individual identity of the reaction steps that lead to the various products. We have developed a comprehensive mechanistic model for the fast pyrolysis of cellulose, which incorporates the individual steps for the formation of important products like levoglucosan, glycolaldehyde, 5-hydroxymethyl furfural, furfural and formic acid. The underlying mechanism involves the scission of the cellulose chain both at random mid-position and at the chain ends. The heterolytic glucosidic bond scission and the retro-Diels-Alder reactions occur at the mid-chain, while the retro-aldol reaction, 1,2-dehydration, 1,3-dehydration, hydrolysis, and cyclic Grob fragmentation reactions occur at the chain end. The kinetic rate coefficients for most of the above reaction steps were obtained from the literature, based on either experimentally determined Arrhenius parameters for cellulose and model compounds, or theoretically computed values of activation energies using quantum calculations. A computational framework based on continuous distribution kinetics was constructed to solve the kinetic model and predict the time evolution of the product yields. The predictions are compared with experimental data from different sources reported in the literature. The promising results of this study and the applicability of the model for complex biomass systems will be addressed in this presentation.

(68a) Contaminant Tolerant Water-Gas-Shift Catalyst for Coal-Biomass-to-Liquids Process

Gökhan Alptekin, **Ambalavanan Jayaraman** and Sarah Devoss, TDA Research Inc., Wheat Ridge, CO

Gasification technologies convert coal and biomass into synthesis gas feed streams that can be used as a chemical feedstock to Coal-Biomass-to-Liquids (CBTL) processes. The CBTL process can produce transportation fuels and value added chemicals. It uses domestic feedstock, has a greenhouse gas footprint better than conventional coal or petroleum fuels, and is projected to be economical at world oil price significantly below the existing level of $82 per barrel (IEC, November 2009).

The commercialization of the CBTL technologies requires the development of clean-up and purification processes that can virtually eliminate the contaminants from the synthesis gas feed stock. While the gas clean-up is vital to protect the catalysts, it will also increase the cost of the CBTL process, and the more stringent the removal requirements are, the more the gas clean-up will cost. Therefore, it is desirable to develop catalysts that can resist poisoning to reduce the extent and severity of the gas clean-up process.

The water-gas-shift catalyst is tolerant to the contaminants in the synthesis gas streams generated by coal and biomass co-gasification. We will present the results from our contaminant tolerant catalyst development efforts in the meeting.

See more of this Session: Coal, Biomass and Natural Gas to Liquids II
See more of this Group/Topical: Topical 6: 11th Topical Conference on Gas Utilization

The U.S. Air Force's Assured Aerospace Fuels Research Facility Fischer-Tropsch Liquids Upgrading Reactor

Heinz J. Robota, University of Dayton Research Institute, Dayton, Ohio;*
Russell K. Smith, Battelle Memorial Institute, Columbus, Ohio;
Robert W. Morris, Jr., Air Force Research Laboratory AFRL/RZPF, WPAFB, Ohio

Abstract

As part of a strategic initiative to prepare for the broader use of synthetic aviation fuels, the United States Air Force is constructing a facility which will allow it to produce synthetic aviation fuels and fuel components in quantities sufficient for small scale engine testing. When completed, it will be possible to produce synthesis gas, convert synthesis gas to liquids, and further process these liquids to fuel and fuel components which can be separated by distillation. Here we describe the operation of a catalytic hydrocracker for the conversion of normal alkane feedstocks to jet fuel as the first phase of this effort. The unit is designed to produce in the range of 10-15 gallons of jet fuel per day. A commercially available cut of a Fischer-Tropsch wax with carbon numbers ranging from about C_{20} to C_{60} and a distribution peak in the C_{30}-C_{35} range, serves as the initial test feed. This broad distribution of paraffins is isomerized and hydrocracked over a commercially manufactured, unsulfided, Co/MoO_3/Silica-Alumina catalyst to yield naphtha and fuel fractions which are subsequently isolated in a fractionator and vacuum distillation column, respectively. The insufficiently converted distribution left in the vacuum bottoms is then recycled to the entrance of the catalytic cracking unit where it is continuously combined with fresh feed. We describe the initial operating characteristics, cracking, and product recovery results.

Introduction

The U.S. Air Force's current target for 2016 is to: "Cost competitively acquire 50% of the Air Force's domestic aviation fuel requirement via alternative fuel blends in which the alternative component is derived from domestic sources produced in a manner that is greener than fuels produced from conventional petroleum."[1] Achieving this target is being enabled through the certification of 50/50 JP-8/Synthetic Paraffinic Kerosene (SPK) blends based on either Fischer-Tropsch (FT) or Hydrotreated Renewable Jet (HRJ) synthetic components. In these blends, the petroleum-derived fraction provides molecular components which fulfill a variety of performance requirements that pure SPK cannot. Further, current fuel specifications derive from a long, successful history of using petroleum-derived fuels. In order to move beyond the 50/50 blend to synthetic-rich formulations, a better understanding of how specific molecular components contribute to performance characteristics is valuable. Similarly, distinguishing petroleum-derived empirical specifications from performance driven requirements based on molecular composition in synthetics will be necessary to allow for the potentially wide variety of synthetic molecular compositions.

The Assured Aerospace Fuels Research Facility (AAFRF) has been commissioned to deliver a flexible variety of purely synthetic fuels and fuel components in quantities large enough to enable intensive laboratory evaluation and where appropriate, full scale engine testing. AAFRF consists of two components. The first is a laboratory-scale facility in which the synthetic steps enabling the production of targeted fuel compositions are identified and verified. The second is a practical scale sample production reactor system capable of producing 10+ gallons per day of the targeted composition. We present here the results from the commissioning of the practical scale sample plant in which a Fischer-Tropsch wax fraction was converted to a full range SPK blendstock composition.

Experimental

A schematic diagram of the AAFRF sample plant is shown in Figure 1. The unit can be considered in three sections: 1) Feed introduction; 2) Hydrocracking reactors; and 3) Product recovery. Hydrogen is recycled back to the reactor loop and the liquid stream can be operated in either once-through or recycle modes. The reactors were charged with a commercially obtained $Co/MoO_3/SiO_2$-Al_2O_3 catalyst which was used in an unsulfided form. The catalyst was activated and pre-conditioned according to the manufacturer's recommendations. For the purpose of verifying the sample plant's capabilities, a commercial Fischer-Tropsch wax fraction, IGI-1339A, was used as feed material. This is in keeping with other examples where fractions of a full spectrum wax were used as feed for middle distillate production.[2] GC analysis showed the wax to be >98% normal paraffins. Temperatures within the catalytic reactors were monitored using thermocouples within a central thermowell at 6 points along the axis of each reactor. Gas samples from the reactor effluent stream could be directed to a refinery gas analyzer for monitoring hydrogen and hydrocarbon content in the recycled gas stream. Reaction progress and product recovery behavior could be monitored by collecting and analyzing liquid samples by GC from various points throughout the process loop.

When operating in production mode, the liquid effluent from the catalytic reactors was directed to a naphtha fractionator operating between 15 and 60 psia. The light ends of the product distribution, typically $<C_9$, were recovered at the top of this distillation column and directed to a naphtha recovery tank. The remaining stream was then directed to a vacuum distillation column for recovery of the SPK fraction, typically C_9-C_{16}. The remaining heavier fraction could then be recycled to the front of the unit and combined with fresh feed to maintain a constant reactor feed rate. Final product analysis was done on the material collected in the product recovery tank.

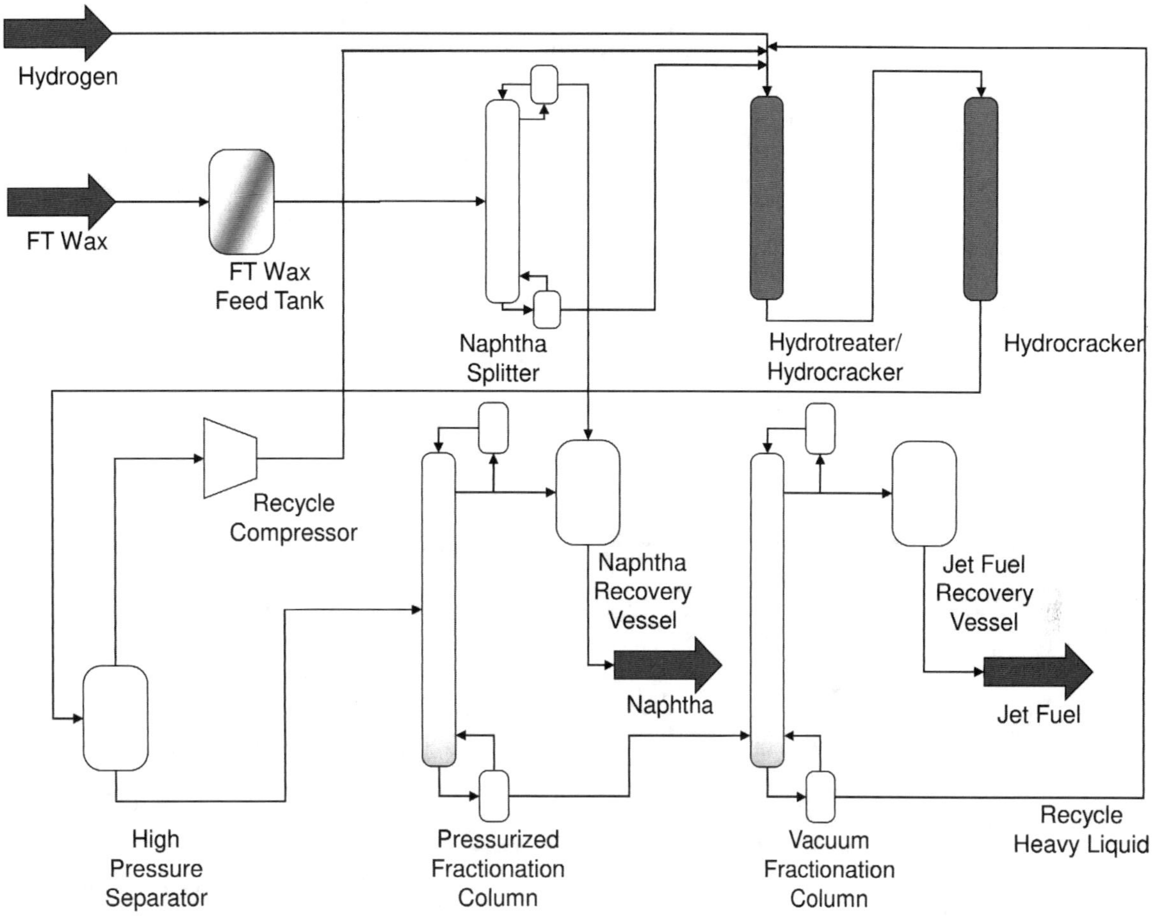

Figure 1. A schematic depiction of the AAFRF hydrocarbon upgrading reactor system.

Results and Discussion

In order to verify the performance of the hydrocracking catalyst, the unit was initially operated without product recovery in a single pass mode. 2.7 kg/hr of feed wax were directed to the head of the first catalytic reactor while 20 L/hr of fresh hydrogen were introduced. For this operation, each of the reactors was loaded with 1.7 L of upgrading catalyst, resulting in a LHSV=1/hr. Gases exiting the reactor were directed back to the compressor section where they were combined with fresh H_2. The hydrocracking section was operated at 880 psig and the return gas stream was monitored with the on-line GC to assure an H_2 content of at least 85 vol%. The recycled gas flow rate was kept high enough to maintain a H_2/hydrocabon mole ratio of at least 10 within the hydrocracking reactors. In the event that the H_2 fraction fell below 85 vol%, the rate at which fresh H_2 was fed was increased.

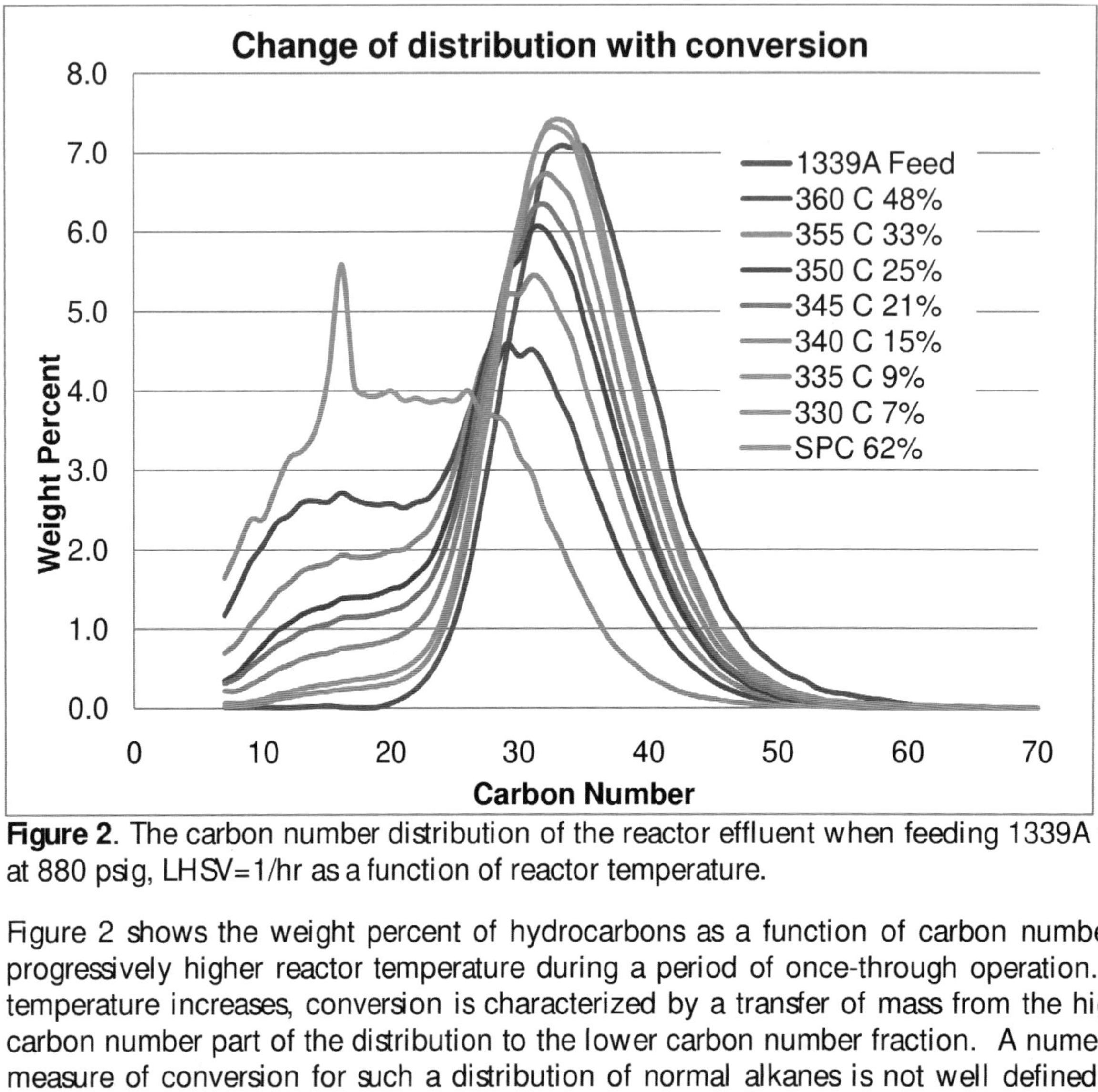

Figure 2. The carbon number distribution of the reactor effluent when feeding 1339A wax at 880 psig, LHSV=1/hr as a function of reactor temperature.

Figure 2 shows the weight percent of hydrocarbons as a function of carbon number at progressively higher reactor temperature during a period of once-through operation. As temperature increases, conversion is characterized by a transfer of mass from the higher carbon number part of the distribution to the lower carbon number fraction. A numerical measure of conversion for such a distribution of normal alkanes is not well defined. In single component studies of hydrocracking, conversion typically includes any catalyst mediated transformation of the feed molecules. Thus, even in the absence of cracking, high degrees of isomerization constitute high conversions. However, when cracking ensues, longer chain feed molecules yield cracked fractions which coincide with the feed distribution and give rise to a normal alkane fraction as well. In order to provide some numerical gauge of conversion, we have calculated a minimum molar conversion by calculating the change in the number of moles of a particular carbon number between the feed and the product distributions. Summing only those carbon numbers which show a net decline in number of moles and normalizing to the number of moles in the fresh feed, we calculate the lower bound on molar conversion shown as the percentage in the legend of the graph. At the highest temperature, the net molar cracking conversion is no less than about half of the feed molecules. Also shown for comparison is the distribution of product observed in single pass operation of a laboratory scale reactor (designated SPC) at the same temperature, LHSV, but at a pressure of 400 psig. These reactions exhibit a negative kinetic reaction order with the partial pressure of H_2 such that the difference in pressure has the same effect on reaction rate that roughly a 10°-15° C temperature increase would have[3]. With that caveat, we believe the correspondence between the lab scale reactor operating with 3 mL of catalyst and the sample plant scale reactor operating with 3.4 L of catalyst is quite good. (The peak at C_{16} in the lab scale result comes from the slow washout of n-C_{16} feed used prior to feeding the 1339-A wax.)

For the steady state production of SPK, the product recovery section of the sample plant must be operated and the fraction of heavier than target hydrocarbons recycled for further conversion. In order to meet the boiling range and flash point requirements of synthetic fuel blendstock, first the low boiling components in the reactor effluent are removed in the naphtha fractionator without removing too much of the targeted C_9-C_{16} range molecules. Next, the targeted C_9-C_{16} fraction is recovered in the vacuum column while minimizing the proportion of heavier components collected in order to meet both the upper boiling range and freeze point specifications. At a given throughput and cracking severity, meeting these targets required the adjustment of: column pressure; reboiler temperature and power; column temperature profile; and reflux rate. This was accomplished using an iterative process, while holding total liquid feed (recycle plus fresh) to the reactors constant with a constant reactor temperature profile. At each set of conditions, it was important to wait sufficiently long so that the liquid samples collected were representative of a steady state at the selected conditions. Figure 3 shows how the composition of the hydrocarbon fluid evolves in traversing the system. At steady state, the feed to the reactors is composed of a blend of fresh 1339-A wax and the vacuum column recycle stream. The distribution begins abruptly at C_{15} due to partitioning between the recovered and recycled streams. The fresh feed contribution is readily apparent with its peak near C_{33}. Upon traversing the cracking reactors, the heaviest components react with greatest facility, essentially transferring the fresh feed distribution into the much lighter distribution between C_5 and C_{17}. In contrast to once through operation, it appears that the fresh feed portion of the stream undergoes complete cracking conversion with some secondary cracking likely. After the reactors, the naphtha fractionator removes the C_5-C_8 fraction followed by the vacuum column recovering most of the C_9-C_{16} fraction. Once a steady state is established, rates and compositions remained stable over the product collection period of at least 24 hours.

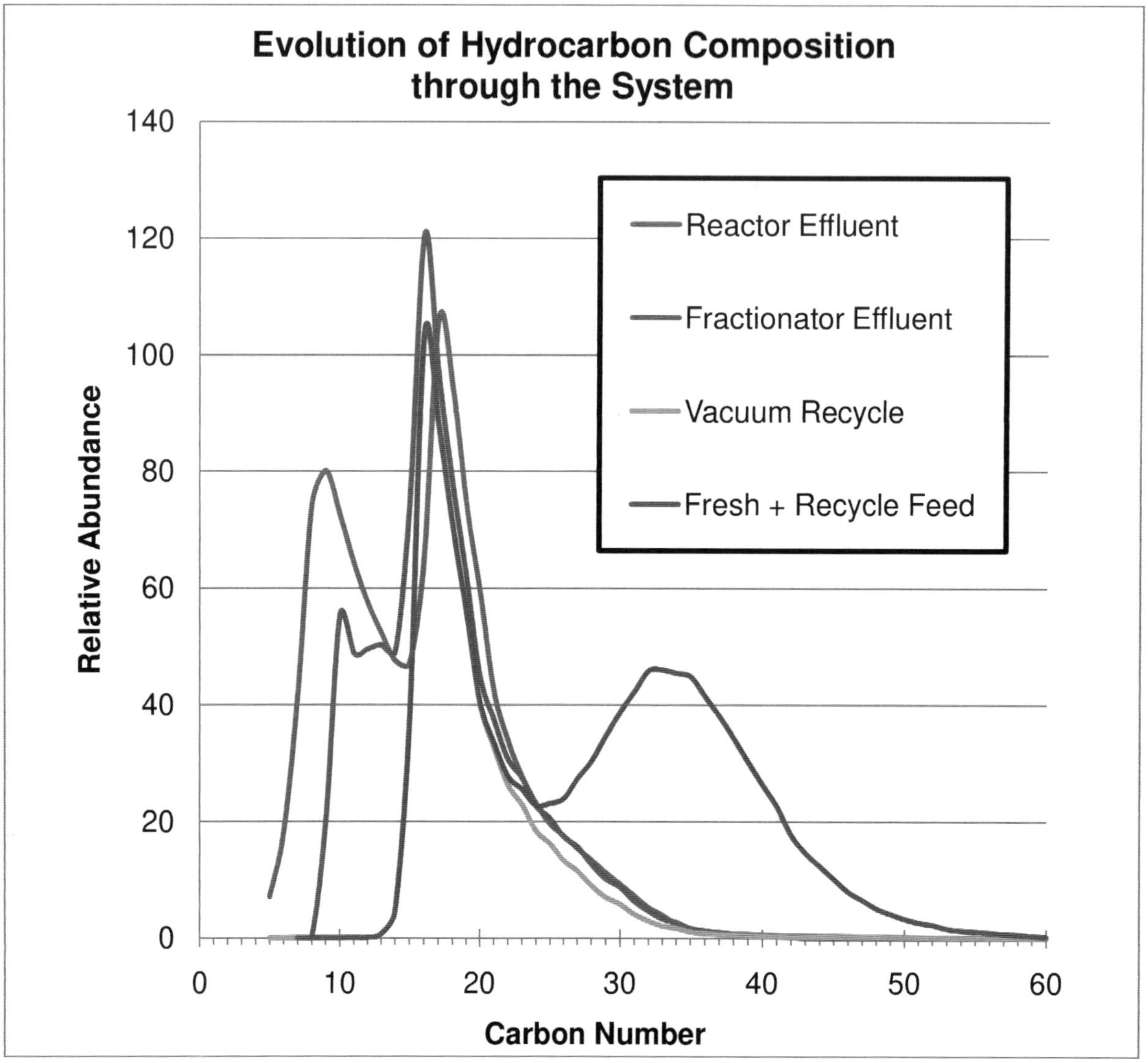

Figure 3. Changes in hydrocarbon composition at different points throughout the system.

The most sensitive target to meet proved to be the freeze point of -47° C. The freeze point of such a complex mixture of isomers of different chain lengths depends on the relative proportions of each carbon number, the degree of isomerization, the fraction remaining as normal paraffin, and the proportion of C_{17}+ hydrocarbons remaining in the recovered SPK. From among these parameters, the most readily adjustable one was the C_{17}+ fraction in the SPK. Through iterative adjustments, a maximum value of 1.3 wt% was set as the operating target with values less than 1% typically achieved. Figure 4 shows the chromatogram of the AAFRF SPK compared with several other SPKs from commercial manufacturers. The similarity in composition among the SPKs is apparent, with a pattern of multibranched isomers followed by monomethyl isomers and concluding with the normal alkane for each carbon number component.

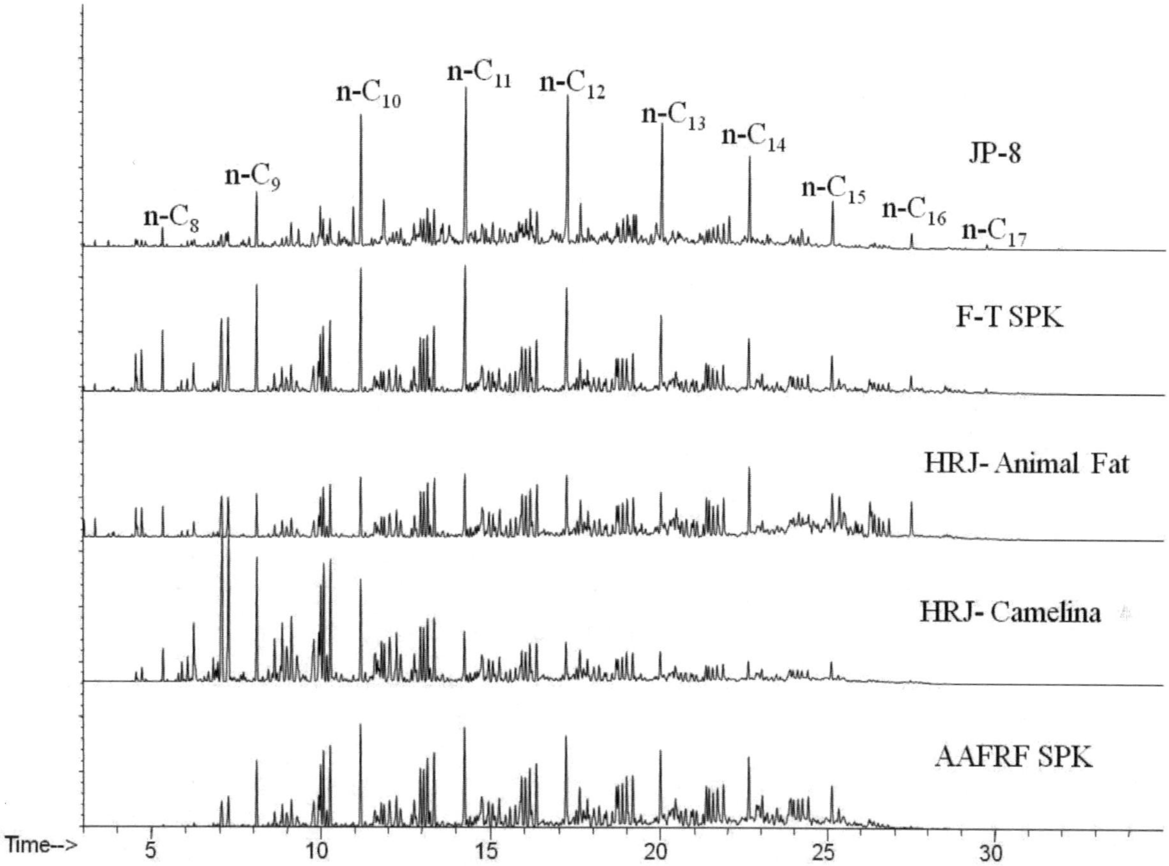

Figure 4. A comparison of the AAFRF SPK with petroleum-derived JP-8 and SPKs from a variety of commercial sources.

Once satisfactory operations had been established with a 2.7 kg/hr combined feed rate to the reactors, the rate of SPK production was found to be below the design target of 10+ gallons/day. In order to achieve the targeted production rate, the combined feed rate was doubled to 5.4 kg/hr and the following system performance criteria established:

- Reactor C_9-C_{16} output >40 wt% of the stream
- Fractionator C_9-C_{16} overhead retention <5% net
- Fractionator C_4-C_8 bottoms retention <1wt%
- Vacuum column C_9-C_{16} yield >70%
- Vacuum column product C_{17}+ content <1.3 wt%.

At the higher feed rate, reaction rates needed to be increased as well. This was accomplished by first lowering the operating pressure from 880 psig to 600 psig followed by a reactor temperature increase to 370° C. Due to a limited operating time window, we elected to accept a slightly higher fractionator C_9-C_{16} overhead retention of 13% in exchange for a SPK C_{17}+ content <0.3 wt%. Even with this compromise, a product delivery rate of just over 10 gallons/day was achieved. With time for some further tuning, at least another gallon/day could have been recovered. Assuming such further process tuning, 2.7 kg/hr of fresh wax feed was converted to 1.3 kg/hr of specification SPK.

The objective of this commissioning was to verify the unit's ability to convert a Fischer-Tropsch wax to a SPK suitable for blending with petroleum-derived JP-8. To that end, the collected product was evaluated against the MIL-DTL-83133G specification. All values were within the specification range except for those requiring additives for conductivity and water ice suppression.

Conclusions

The Air Force's Assured Aerospace Fuels Research Facility hydrocracking unit has been demonstrated to successfully convert Fischer-Tropsch wax to specification synthetic paraffinic kerosene suitable for use in preparing 50/50 synthetic/petroleum blended JP-8 aviation fuel. This was accomplished under "recycle to extinction" conditions using a commercially obtained unsulfided $Co/MoO_3/SiO_2$-Al_2O_3 hydrocracking catalyst acting on an IGI-1339A wax. The desired product was recovered by first stripping light components in a naphtha fractionator followed by vacuum distillation of the SPK fraction. Evaluation of the recovered SPK verified its compliance with the MIL-DTL-83133G specification. The 10 gallon/day production rate achieved the targeted 10+ gallon/day design goal.

Acknowledgements

This research was supported, in part, by the U. S. Air Force Cooperative Grant Numbers F33615-03-2-2347 and FA8650-10-2-2934,Mr. Robert W. Morris Jr., Air Force Grant Monitor. The research was also sponsored by the State of Ohio Subrecipient Award No. COEUS # 005909 to the University of Dayton (Dr. Dilip Ballal as the Grant Monitor) under the "Center for Intelligent Propulsion and Advanced Life Management," program with the University of Cincinnati (Prime Award NO. TECH 09-022). The authors gratefully acknowledge this grant support.

References

1. Braun, Jeff, "AF Synthetic Fuel Certification Program 3 Aug 10", http://www.sebioenergy.org/2010/PDF_10/August_3/Ballroom_A/3.30-5.00/Jeffrey%20Braun-UofGA_Brief_27Jul10.pdf
2. Leckel, Dieter and Liwanga-Ehumbu, Maurice, "Diesel-Selective Hydrocracking of an Iron-Based Fischer-Tropsch Wax Fraction (C_{15}-C_{45}) using a MoO_3-Modified Noble Metal Catalyst', Energy & Fuels **2006**, 20, 2330-2336; Leckel, Dieter, "Hydrocracking of Iron-Catalyzed Fischer-Tropsch Waxes", Energy & Fuels **2005**, 12, 1795-1803; Lee, Joongwon, Hwang, Sunhwan, Lee, Sang-Bong and Song, In Kyu, "Production of middle distillate through hydrocracking of paraffin wax over $NiMo/TiO_2$-SiO_2 catalysts", Korean J. Chem. Eng. 27(6), 1755-1759 (**2010**)
3. Steijns, Matt, and Froment, Gilbert F., "Hydroisomerization and Hydrocracking. 3. Kinetic Analysis of Rate Data for n-Decane and n-Dodecane", Ind. Eng. Chem. Prod.Res. Dev. **1981**, 20, 660-668

(68c) Supported Iron-Based Catalysts for the Fischer-Tropsch Synthesis of Lower Olefins

Hirsa M. Torres Galvis[1], Johannes H. Bitter[1], Krijn P. de Jong[1] and **Ard C. J. Koeken**[2], (1)Inorganic Chemistry and Catalysis. Debye Institute for NanoMaterials Science, Utrecht University, Utrecht, Netherlands, (2)Chemistry - Inorganic Chemistry and Catalysis, Utrecht University, Utrecht, Netherlands

Lower olefins can be produced from synthesis gas via methanol to olefins process (MTO), via SDTO process (syngas via dimethylether to olefins) or by cracking of Fischer-Tropsch products [1]. These processes involve a large number of steps which are energy and time consuming. One interesting approach is the modification of the Fischer-Tropsch synthesis to directly produce lower olefins.

The Fischer-Tropsch to olefins (FTO) process has to be carried out at high temperatures in order to drive the selectivity towards short hydrocarbon chains. Under these conditions, undesired side reactions as methanation and the Boudouard reaction (carbon deposition) are also thermodynamically favored [2]. Fe-based catalysts are preferred to perform this type of process due to their higher selectivity to olefins and lower methane selectivity at high temperatures compared with cobalt catalysts. However, carbon buildup on large iron particles may block the active sites in the catalyst surface, leading to deactivation. Moreover, carbon deposits can induce fragmentation of the carbide particles resulting in catalyst attrition [3].

A new type of catalyst has been developed in order to overcome activity and stability problems encountered when using bulk iron catalysts under FTO process conditions. These novel catalysts consist on Fe_2O_3 particles dispersed on an inert support. Here we have used several supports with a restricted reactivity towards iron oxide. More specifically, carbon nanofibers, a-Al_2O_3 and b-SiC have been loaded with 2-10%wt Fe. The support improves the mechanical stability of the catalyst, restricting the mobility of the particles and avoiding aggregation. Smaller Fe_2O_3 particles are less prone to carbon deposition which not only preserves catalytic activity but also avoids catalyst attrition.

Supported iron catalysts showed 3 to 7 times increase in catalytic activity after 15 h of reaction at 1 bar, 350°C and $H_2/CO = 0.5$ when compared with an unpromoted bulk iron catalyst. Additionally, C_2-C_4 olefins selectivity was increased from 11 to 52-58%wt and a reduction of methane selectivity from 85% wt to 11-30%wt was observed. The dispersion of Fe_2O_3 particles on an inert support results in improved stability, activity and selectivity. These results were confirmed with high pressure measurements performed at 20 bar.

[1] Janardanarao, M., Ind. Eng. Chem. Res. 29 (1990) 1735.

[2] Wang, C. et al., J. Nat. Gas Chem. 12 (2003) 10.

[3] Shroff, M. D. et al., J. Catal. 156 (1995) 185.

See more of this Session: Coal, Biomass and Natural Gas to Liquids II
See more of this Group/Topical: Topical 6: 11th Topical Conference on Gas Utilization

FISCHER-TROPSCHSYNTHESIS: INFLUENCE OF CO CONVERSION ON SELECTIVITIES, H_2/CO USAGE RATIOS, AND CATALYST STABILITY FOR A Ru PROMOTED Co/Al_2O_3 CATALYST USING A SLURRY PHASE REACTOR

Wenping Ma[a], Gary Jacobs[a], Yaying Ji[a], Tejas Bhatelia[b], Dragomir Bukur[b], Syed Khalid[c] and Burtron H. Davis[a]*

[a] *Center for Applied Energy Research, University of Kentucky, 2540 Research Park Drive, Lexington, KY, 40511*
[b] *Department of Chemical Engineering, Texas A&M University at Qatar. PO Box 23874, Doha, Qatar.*
[c]*Brookhaven National Laboratory, Brookhaven Ave., Upton, NY, 11973*

Introduction

Development of active Co-based Fischer-Tropsch synthesis (FTS) catalysts has been the subject of a larger number of studies in the past decades. The primary objective has been focused on maximizing the production of waxy hydrocarbons and improving catalyst stability through understanding support structure effects [1-5], the role of promoters - such as noble metals (Pt, Ru, Pd, Re, Rh) [6-17], Group 11 metals (Cu, Ag, and Au) [18], alkali promoters (Li, Na, K, Cs) [19,20], and rare earth (La, Ce, Th) [21,22] as well as other oxides (Zr, Mo, B, and Mn) [21-25], and Co particle size effects [25-30]. The strong interaction between Co and the Al_2O_3 support results in a better Co dispersion than moderate and weak interactions between Co and TiO_2, and Co and SiO_2, respectively. However, for a standard H_2 reduction step (e.g., 350°C for 10 h), this often leads to a lower Co reduction degree for Co/Al_2O_3 catalysts relative to less interacting catalysts (e.g., Co/SiO_2). Except for Mo, B and K promoters [21,23,24], most other promoters were reported to significantly improve both Co reduction degree and, to a lesser extent, Co dispersion, thus increasing CO conversion rates on a per gram catalyst basis. The Co particle size effect is a recent area of interest. In some reports, constant Co site time yields have been reported for Co particle sizes within the range of 8-10 nm [26,27]. In one report, Co particle sizes of 7-8 nm were found to favor production of C_{5+} hydrocarbons [28].

However, an important issue regarding how the selectivities of hydrocarbons (HCs), including CH_4, C_{5+}, olefin/paraffin contents, and CO_2 selectivity depend on CO conversion (X_{CO}) over cobalt based catalysts has not been well established. Researchers have realized that selectivity varies with conversion and that comparison of selectivities over different catalysts should be performed at similar conversion levels [2,31]. Storsaeter et al. [32] conducted a study of the water effect at 210 °C and 20 atm over Co/Al_2O_3, Co/SiO_2 and Co/TiO_2 catalysts in a fixed-bed reactor. External water was added at about 40% CO conversion. Recently, Li et al. [33-34] studied the water impact on FTS over Co/Al_2O_3 and Co/SiO_2 catalysts at CO conversions in the range of 20-40%.

Likewise, most previous studies over supported Co catalysts [2,3,10,13,14,19,20,24,25,35] were often conducted at or below moderate conversion levels (i.e., moderate conversion being defined here as 35-60%), which does not provide reliable information about the relationship between catalyst conversion and selectivities. A few studies over Co based catalysts were conducted at higher CO conversion levels (e.g. 60-80%), but experiments were conducted either over different catalysts or by changing reaction temperature over the same catalysts [6,17,22,26], making it difficult to establish conversion and selectivity relationships. Recently, Borg et al. [2] conducted experiments over Re promoted Co/Al_2O_3 catalysts for CO conversions up to 72% and found that C_{5+} selectivities increased monotonically with increasing CO conversion levels. However, effects of CO conversion on methane and CO_2 selectivities, as well as catalyst stability, were not reported. Thus, there is uncertainty as to how selectivity changes with CO conversion, as well as how CO conversion influences the extent of secondary reactions of olefins, especially at high CO conversion levels. This will be a hindrance in choosing the proper reaction conditions for the cobalt based FTS reaction.

The present study was undertaken to study the effect of CO conversion on these parameters (i.e., CH_4 and C_{5+} selectivities, CO_2 selectivity, H_2/CO usage, and olefin/paraffin). CO conversions between 12 and 94% were achieved in four separate runs by using a wide range of space velocities (i.e. 0.3-15 Nl/h/gcat). The effect of CO conversion level on catalyst stability, and possible explanations on the observed effect of CO conversion on selectivities, are discussed, based on different water partial pressures generated at different CO conversion levels. In addition, The used catalysts were retrieved at different CO conversion levels from the slurry reactor, cooled under inert gas, and solidified for analysis by XANES. The data were utilized to determine if any structural changes occurred in the catalyst as a function of CO conversion (i.e., water partial pressure).

Experimental

Catalyst Preparation: The support used was Sasol Catalox-150 γ-Al_2O_3. The support was first calcined at 400°C in a muffle furnace for 4 h. The catalyst was prepared by a slurry impregnation method, whereby the ratio of the volume of loading solution used to the weight of alumina was 1:1, such that approximately 2.5 times the pore volume of solution was used to prepare the catalyst. Due to the solubility limit of cobalt, two impregnation steps were used. After each impregnation step, the catalyst was dried under vacuum in a rotary evaporator from 80 to 100°C. This large batch of dried cobalt nitrate on alumina was split into several equivalent batches in order to add different promoters. One of them was used to load 0.27% wt% Ru by incipient wetness impregnation. The Ru precursor was ruthenium (II) nitrosyl nitrate. After final drying, with gradual heating from 80 to 100°C, the catalyst was calcined at 350°C under flowing air for 4 h.

BET Measurement: BET measurements for the catalysts were conducted using a Micromeritics Tri-Star system to determine the loss of surface area with loading of the metal. Prior to the measurement, samples were slowly ramped to 160°C and evacuated for 24 h to approximately 50 mTorr.

Hydrogen Chemisorption with Pulse Reoxidation: Hydrogen chemisorption measurements were performed using a Zeton Altamira AMI-200 unit, which utilizes a thermal conductivity detector (TCD). The sample weight was always 0.220 g. The catalyst was activated at 350°C for 10 h using a flow of pure hydrogen and then cooled under flowing hydrogen to 100°C. The sample was then held at 100°C under flowing argon to prevent physisorption of weakly bound species prior to increasing the temperature slowly to the activation temperature. At that temperature, the catalyst was held under flowing argon to desorb the remaining chemisorbed hydrogen so that the TCD signal returned to the baseline. The TPD spectrum was integrated and the number of moles of desorbed hydrogen determined by comparing to the areas of calibrated hydrogen pulses. Prior to experiments, the sample loop was calibrated with pulses of nitrogen in helium flow and compared against a calibration line produced from gas tight syringe injections of nitrogen under helium flow.

After TPD of hydrogen, the sample was reoxidized at the activation temperature by injecting pulses of pure oxygen in helium referenced to helium gas. After oxidation of the cobalt metal clusters, the number of moles of oxygen consumed was determined, and the percentage reduction was calculated assuming that the Co^0 reoxidized to Co_3O_4. While the uncorrected dispersions (uc) are based on the assumption of complete reduction, the corrected dispersions (c), which are reported in this work, include the percentage of reduced cobalt as follows.

$\%D_{uc}$ = (# of Co^0 atoms on surface × 100%)/(total # Co^0 atoms)

$\%D_c$ = (# of Co^0 atoms on surface × 100%)/[(total # Co atoms)(fraction reduced)]

Temperature Programmed Reduction (TPR): TPR profiles of fresh catalyst samples were obtained using a Zeton Altamira AMI-200 unit. Calcined fresh samples were first heated and purged in flowing argon to remove traces of water. TPR was performed using 30 ml/min of a 10%H_2/Ar mixture referenced to argon. The ramp rate was 5°C/min from 50°C to 1100°C, and the sample was held at 1100°C for 30 min.

In-situ XANES spectroscopies: To study Co metal and Co oxide states during Fischer-Tropsch reaction, several used catalyst samples of the 0.273%Ru-25%Co/Al_2O_3 catalyst were withdrawn from the CSTR under inert gas during a catalyst testing run at different CO conversion levels (i.e., different water partial pressures), cooled and solidified in the waxy product, and analyzed by XANES at the Co K-edge.

Catalyst Testing: Typically, the catalyst (10-18 g) was ground and sieved to 80-325 mesh before loading into a fixed-bed reactor for 10 h of *ex-situ* reduction at 350°C and atmospheric pressure using a gas mixture of H_2/He (60 NL/h) with a molar ratio of 1:3. The reduced catalyst was then transferred to a 1-L continuously stirred tank reactor (CSTR) containing 315 g of melted Polywax 3000, under the protection of a N_2 inert gas.

The transferred catalyst was further reduced *in-situ* at 230°C at atmospheric pressure using pure hydrogen (30 NL/h) for another 10 h before starting the FT reaction. In this study, the FTS conditions that were used are 220°C, 1.5 MPa, $H_2/CO = 2.1$, and 0.3-15 NL/g-cat/h.

Results and discussion

Results of surface area measurements by adsorption of nitrogen at 77K and hydrogen chemisorption/pulse reoxidation are listed in Table 1. The BET surface area for the 0.27% Ru promoted 25% Co /γ-Al_2O_3 catalyst is 106 m^2/g. A wt.% loading of 25% metal is equivalent to 33% by weight for Co_3O_4. If the Al_2O_3 is the only contributor to the area, then the calculated area of the Co/Al_2O_3 catalysts should be $0.67\times140 = 94$ m^2/g, which is close to the experimental value. The average pore diameter of the catalyst is 9.0 nm, close to that of the support (10 nm). Therefore, the BET result suggests that significant pore blocking did not occur.

From the data in Table 1, a medium Co reduction degree (59%) and a high Co dispersion (21%) were obtained. This resulted in rather small average Co cluster size, 5 nm. It is therefore reasonable to assume that Co particles were located within the pores of the support.

Table 1 Results of BET surface area, porosity, and hydrogen chemisorption with pulse reoxidation

Description	N_2 physisorption			Chemisorption with pulse reoxidation		
	BET Surf. Area m^2/g	Pore Volume cm^3/g	Average Pore Diameter, nm	Reduction[(a)] Degree, %	Corrected Disp., %	Corrected Co Cluster size, nm
Catalox-150 γ-Al_2O_3	140.2	0.44	10.2			
0.27%Ru-25%/Co/γ-Al_2O_3	106.3	0.25	9.0	58.7	20.7	5.0

Because smaller cobalt oxide crystallites – those interacting with the support – tend to be more difficult to reduce [1, 36], reduction promoters such as Ru [37,38], Re [6, 39-41], and Pt [8,42,43] are often added at low loadings to the supported cobalt catalysts. There is a debate in the literature as to how these promoters operate – either by a chemical effect or by H_2 dissociation and spillover. In the latter case, the reduction promoter reduces first, allowing H_2 to dissociate on the promoter surface. Then, the dissociated hydrogen is suggested to facilitate the nucleation of reduced Co^0 sites in cobalt oxides.

Four different runs were conducted over the 0.27% Ru-25%Co/Al_2O_3 catalyst by using different space velocities of 0.3-15 NL/g-cat/h, but the same other reaction conditions, 220 °C, 1.5 MPa, H_2/CO =2.1, in order to achieve a wide range of CO conversions, 12-94%. The reaction results (activity and selectivity) of the four runs were obtained during the first 100 h at the space velocity of 10 or 6 NL/g-cat/h. Changes in HC selectivity (CH_4 and C_{5+}), CO_2 selectivity, H_2/CO usage ratio, and C_4 olefin to paraffin (OPT) ratios with CO conversion are shown in Figures 1a-e. Note that the data used for plotting the figures were obtained after 100 h of FTS. A measurable decrease in CH_4 selectivity (from 11.5 to 7.1%) with increasing CO conversion up to 70% was observed. After that, the extent of decrease in CH_4 selectivity became smaller (7.1-6.7%). For CO conversions higher than 80%, CH_4 selectivity started to increase (7.6-9.1% at X_{CO} of about 81 to 94%). Decreasing CH_4 selectivity is often accompanied by increasing C_{5+} selectivity. As shown in Figure 1b, C_{5+} selectivity increased from 79.5 to 85.5% as the CO conversion increased from 12 to 80%, but it remained nearly unchanged for CO conversions above 80%. Changes in CO_2 selectivity with CO conversion (Figure 1c) displayed a slightly different trend. First, it increased slowly monotonically (0.6 to 1.5%) with increasing CO conversion up to 70%. Beyond that point, the increase in CO_2 selectivity became more rapid. For CO conversion greater than 90%, the CO_2 selectivity was greater than 5.0 %. The H_2/CO usage ratio (Figure 1d) decreased monotonically with increasing CO conversion. At 12% CO conversion, the H_2/CO usage ratio was 2.26, and it decreased to 1.9 at a CO conversion of 94%. Finally, it was observed that increasing CO conversion up to 94% led to a linear decrease in the C_4 olefin-to-paraffin (OTP) ratio (Figure 1e). This is likely due to the competitive adsorption of CO and the alkenes.

The increase in C_{5+} selectivity and decrease in CH_4 selectivity with increasing CO conversion on different cobalt-based FTS catalysts were also reported in several studies [2,4,32,44-46]. The highest CO conversion in a previous study was 72% [2]. This phenomenon has been explained by the role of indigenous and externally added water in the reaction system. One explanation was based on a reaction transport model [27], which suggests that water enhances the rate of transport of CO and H_2 reactants and increases the accessibility of sites located within the least accessible porous regions. This model could explain enhanced CO conversions by the addition of water on some catalysts, but when the diffusion restriction in catalyst pores would be reduced by the water effect, CH_4 selectivity could increase, which is not what we observed. Another theory is based on the hypothesis that water inhibits secondary hydrogenation of olefins and increases the chain growth probability by inhibition of chain termination [27,32] or by water increasing active carbon coverage [47], which appears to be kinetic in nature. These assumptions were used to explain the increased C_{5+} selectivity and decreased CH_4 selectivity with increasing conversion [32]. Other explanations could also apply, such as the interaction of water and co-adsorbed CO lowering the barrier for CO dissociation [47]. In this study, increasing C_{5+} and decreasing CH_4 selectivities with increasing CO conversion were observed up to 80% CO conversion.

CH_4 selectivity on the $Ru\text{-}Co/Al_2O_3$ catalyst stops decreasing, and instead, tends to increase, at CO conversions above 80% (Figure 1a). Meanwhile, CO_2 selectivity was found to increase faster above CO conversion levels of 80%. These selectivity changes might reflect a change in the oxidation state of a fraction of Co at high CO conversions. Details about the oxidation state change of Co with variations in CO conversion are discussed later.

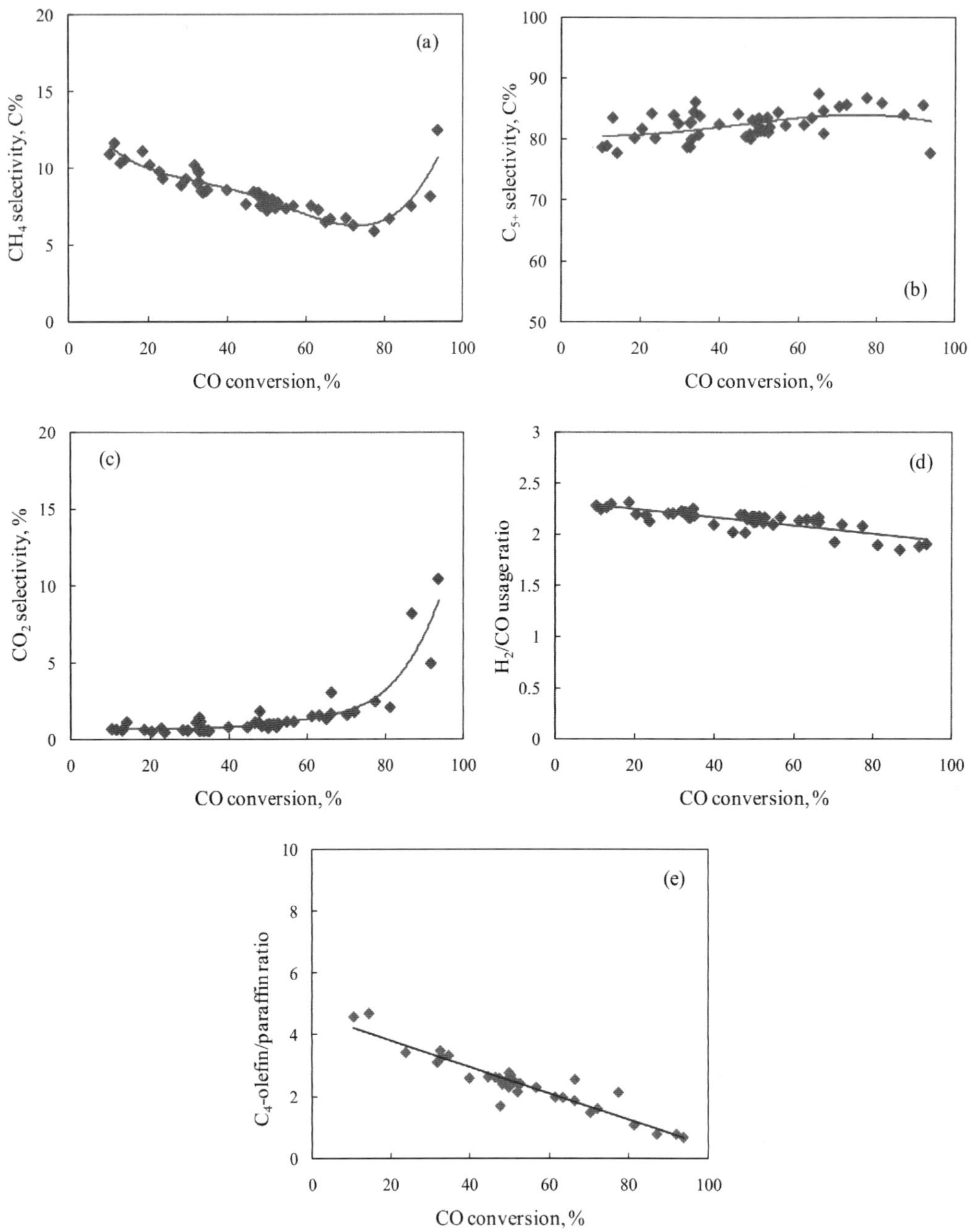

Figure 1: Change of (a) CH_4 selectivity, (b) C_{5+} selectivity, (c) CO_2 selectivity, (d) H_2/CO usage ratio, and (e) C_4 olefin/paraffin with CO conversion (0.27%Ru-25%Co/Al_2O_3, 220°C, 203.3 psig, H_2/CO =2.1 and 0.3-15.0 NL/g-cat/h).

Experimental and calculated P_{H_2O} / P_{H_2} values versus CO conversion are illustrated in Figure 2. The calculated values are based on the stoichiometry at a given CO conversion. The experimental and predicted values are in good agreement for CO conversion below 70%. However, deviations between the experimental and calculated values are higher for CO conversions greater than 70%.

CO conversions above 70% led to a rapid increase in the indigenous water partial pressure. When the CO conversion was about 70%, the experimental P_{H_2O}/P_{H_2} value was about 0.9, whereas at CO conversions of 80 and 90%, P_{H_2O}/P_{H_2} values in the reactor increased to 1.2 and 1.5, respectively. These P_{H_2O}/P_{H_2} values at 70% CO conversion and above 80% CO conversions are equivalent to the values present after adding 25% and more than 30% external water at 40% CO conversion. Adding >30% water at a 40% CO conversion level caused rapid and irreversible deactivation of 0.5%Pt15%Co/Al_2O_3 catalyst. EXAFS/XANES measurements on used Co catalysts solidified in wax suggest that the catastrophic deactivation was due to oxidation of a significant fraction of Co to an irreducible oxidized cobalt compound – e.g. a cobalt-support compound - due to water reoxidation of highly dispersed Co particles [15,48-51]. Using the same methodology for 25%Co/Al_2O_3 (with >10 nm average Co cluster size), high levels of H_2O (e.g., > 25%) resulted in the reoxidation of a fraction of cobalt to form CoO, which was re-reduced when the addition of H_2O was terminated. The oxidation-reduction cycle also appeared to lead, based on changes in Co-Co metal coordination in EXAFS, to growth in the average Co metal cluster size, leading to some irreversible deactivation from the oxidation-reduction cycle [52]. Krishnamoorthy et al. [30] reported that reoxidation of Co becomes a possibility when H_2/CO was about 0.8. Therefore, considering these earlier water co-feeding studies, water reoxidation of Co metals is possible as CO conversion reaches 70% and higher, and it becomes more severe for CO conversion above 80% due to very high P_{H_2O}/P_{H_2} (>1) values in the reactor. Furthermore, the increased oxidation of Co metal to oxidized Co species is expected to lead to increases in the WGS reaction rate.

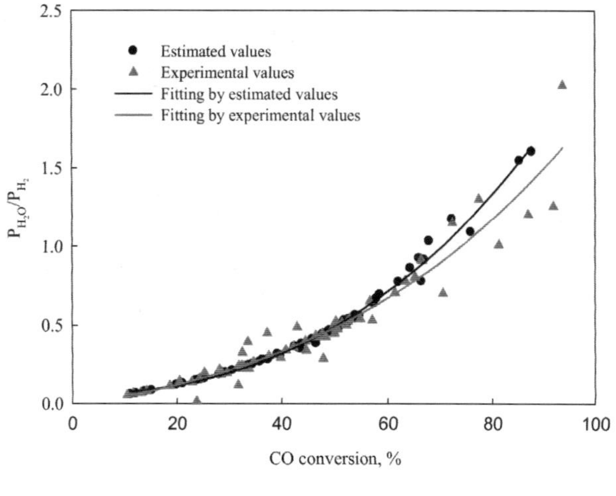

Figure 2: Change of experimental and calculated (based on CO conversion) P_{H_2O}/P_{H_2} with CO conversion (0.27%Ru-25%Co/Al_2O_3, 220°C, 203.3 psig, H_2/CO =2.1 and 0.3-15.0 NL/g-cat/h).

Increasing CO conversion led to a decrease in the H_2/CO usage ratio (Figure 1d). This might be caused by formation of more paraffins with increasing CO conversion (<70% CO conversion) and increased WGS. Decreasing olefin content with increasing CO conversion (Figure 1e) has been explained by CO conversion increasing olefin readsorption due to changes in competitive adsorption and water decreasing secondary hydrogenation reactions [2,27]. This was also reported by Storsaeter et al. [32] using different catalyst supports (Co/SiO_2, Co/TiO_2, and Co/Al_2O_3), with or without Re promoter.

The water effect of the FTS reaction has been found to be related to Co cluster size, which was demonstrated by EXAFS studies [25,27,50,53,54]. Thus, the value of CO conversion at which significant changes in CO_2 and CH_4 selectivities occur may vary from one Co catalyst to another. Generally, small Co particles (e.g., <6 nm), are more sensitive to water and are more easily reoxidized [2,53,54]. The data reported here is for a $0.27\%Ru25\%Co/Al_2O_3$ catalyst, which contains 5.0 nm average Co particles as determined by H_2 chemisorption measurements. This small Co particle size should thus be susceptible to reoxidation by water [2,15]. It is expected that cobalt catalysts with larger Co particles should be more resistant to water reoxidation of Co metal at higher CO conversion levels.

In a separate run, cobalt catalyst samples were retrieved from the reactor at different CO conversion levels in order to determine if structural changes occurred with the catalyst. Interestingly, a systematic trend was observed, whereby the higher conversion levels did lead to oxidation of a small fraction of the cobalt in the catalyst. XANES derivative spectra display changes in oxidation state in the clearest manner. With increasing conversion levels, there is a decrease in the XANES feature at lower energy corresponding to the metallic state (see down arrow, Figure 3 left), while at the same time there is an increase in the XANES feature at higher energy corresponding to an oxidized form of cobalt (see up arrow, Figure 3 left). References for Co^0 and CoO are also provided (Figure 3, right; note difference in scale). Though changes are readily detected, it should be noted that the extent of oxidation appears to be quite low, even at the highest conversion levels. However, if the oxidation is confined to the surface – and most likely impacting the smaller crystallites – then this could result in significant losses in FT activity due to losses in Co^0 surface sites, since the bulk Co does not participate in the catalysis. This result thus confirms that, under conditions where a high partial pressure of water is present in the reactor (e.g., high CO conversion or water co-feeding), oxidation of a fraction of cobalt does occur [32,48,50,55]. Note that this should not be confused with catalyst aging at moderate conversions, where other deactivation mechanisms may predominate [56]. Deactivation by oxidation is dependent on many factors, including the interaction of cobalt particles with the support [51] and cobalt particle size [54]. Thus, different cobalt research catalysts may display significantly different responses to the impact of water.

Co catalyst stability is related to catalyst structure and cobalt particle size. Larger cobalt particles, wider pore sizes, and moderate Co-support interactions have beneficial effects on the stability of Co catalysts [4]. However, higher water partial pressures favor the reoxidation of small supported Co particles and may decrease cobalt catalyst life. Our data indicate that the catalyst deactivation rate in terms of CO rate before and after attaining the highest CO conversions was 0.08 mmol CO/g-cat/h for the low CO conversion run (X_{CO} < 65%, Run #1); then higher deactivation rates were obtained for runs with higher CO conversions. The deactivation rates increased to 0.18, 0.23 and 0.45 mmol CO/g-cat/h for the runs which had peak conversions of 72% (Run #4), 86% (Run #2) and 94% (Run #3) CO conversion. Thus, attaining a higher CO conversion damages the catalyst and causes higher deactivation rates.

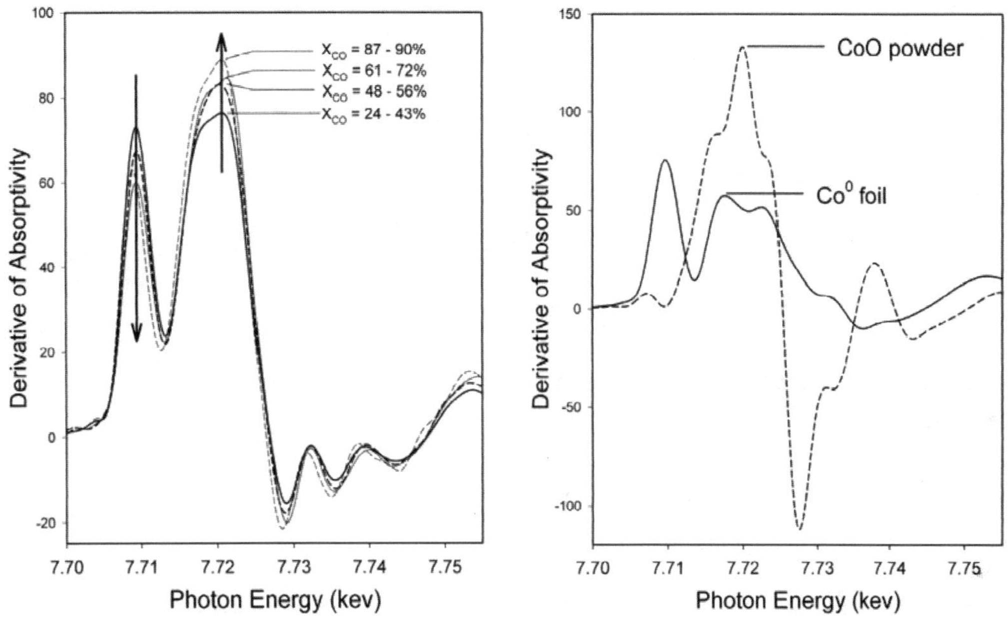

Figure 3: (left) XANES derivative spectra as a function of conversion indicate a greater extent of oxidation at higher conversion. Note that the downward arrow denotes a decrease in the XANES feature corresponding to Co^0, while the upward arrow shows an increase in the XANES feature of CoO oxide, most likely confined to the surface shell. (right) XANES derivative spectra of reference compounds.

Conclusions

In summary, the CH_4 selectivity decreased and the C_{5+} selectivity increased with increasing CO conversion. When the CO conversion was over 70%, the decrease in CH_4 selectivity and increase in C_{5+} selectivity decelerated. CO_2 selectivity started to increase rapidly at about 70% CO conversion for the Ru-Co/Al_2O_3 catalyst. It is presumed that Co metal was reoxidized at high CO conversions (>70%) due to the high water partial pressures (P_{H_2O} / P_{H_2} > 0.9), resulting in deactivation. This assumption is evident by the XANES results of catalyst samples withdrawn from the slurry reactor, solidifed in their in-situ-like state, at different CO conversion levels. Therefore, controlling CO conversion below 70% for long-term testing of Co/Al_2O_3 catalyst is suggested to minimize CH_4 and CO_2 selectivities and extend catalyst life. CO conversions higher than about 70% lead to smaller H_2/CO usage ratios and lower olefin content. In this study, the H_2/CO usage ratio was found to decrease from 2.26 to 1.9 when CO conversion increased from 12 to 94% at a H_2/CO feed ratio of 2.1. The decreasing H_2/CO usage rate with CO conversion likely resulted from an increase in the WGS rate.

Acknowledgments

We acknowledge the financial supports of Qatar National Research and the Commonwealth of Kentucky.

References

1. G. Jacobs, T.K. Das, Y.Q. Zhang, J.L. Li, G. Racoillet, B.H. Davis, Appl. Catal. 233 (2002) 263.
2. Ø. Borg, S. Eri, E.A. Blekkan, S. Storsæter, H. Wigum, E. Rytter, A. Holmen, J. Catal. 248 (2007) 89-100.
3. S. Storsæter, B. Tøtdal, J.C. Walmsley, B.S. Tanem, A. Holmen, J. Catal. 236 (2005) 139-152.
4. A.Y Khodakov, W. Chu, P. Fongarland, Chem. Rev. 107 (2007) 1692-1744.
5. J.H. Oh, J.W. Bae, S.J. Park, P.K. Khanna, K.W. Jun, Catal. Lett. 130 (2009) 403-409.
6. G. Jacobs, J.A. Chaney, P.M. Patterson, T.K. Das, B.H. Davis, Appl. Catal. 264 (2004) 203-212.
7. S. Vada, A. Hoff, E. Adnanes, D. Schanke, A. Holmen , Topic catal. 2 (1995) 155-162.
8. D. Schanke, S. Vada, E.A. Blekkan, A.M. Hilmen, A. Hoff, A. Holmen J. Catal. 156 (1995) 85-95.
9. L.H. Yu, S.M. Zhang, X.Z. Guo, D. Wang, S.R. Wang, S.H. Wu, Central European J. Chem. 5 (2007) 144-155.
10. D.Y. Xu, W.Z Li, H.M. Duan, Q.J. Ge, H.Y. Xu, Catal. Lett. 102 (2005) 229-235.
11. S.A. Hosseini, A. Taeb, F. Feyzi, Catal. Commun.6 (2005) 233-240.
12. Y. Li, T. Wang, C. Wu, X. Qin, N. Tsubaki, Catal. Commun. 10 (2009) 1868-1874.
13. J. Panpranot, Jr., J.G. Goodwin, A.J. Sayari, A. J. Catal. 211 (2002) 530-539.
14. E. Iglesia, S. L. Soled, R.A. Fiato. J. Catal. 137 (1992) 212-224.
15. B. Jongsomjit, J. Panpranot, J.G. Goodwin, Jr., J. Catal. 204 (2001) 98-109.
16. J.H. A. Martens, V.H.F.J. Blik, R. Prins, J. Catal. 97(1986) 200-209.
17. J.W. Bae, S.M. Kim, S.J. Park, Y.J. Lee, K.S. Ha, K.W. Jun, Catal. Commun. 11 (2010) 834-838.
18. G. Jacobs, M.C. Ribeiro, W. Ma, Y. Ji, S. Khalid, P.T.A. Sumodjo, B.H. Davis, Appl. Catal. 361 (2009) 137-151.
19. J. Gaube, H.F. Klein, Appl. Catal. 350 (2008) 126-132.
20. J.K. Jeon, C.J. Kim, Y.K. Park, S.K. Ihm, Korean J. Chem. Eng. 21 (2004) 365-369.
21. Y.Z. Chen, T.H. Wang, Catal. Lett. 22 (1993) 165-177.

22. Y.Y. Liu, T. Hanaoka, T. Miyazawa, K. Murata, K. Okabe, K. Sakanishi, Fuel Proc. Tech. 90 (2009) 901-908.
23. P. Tupabut, B. Jongsomjit, P. Praserthdam, Catal. Lett. 118 (2007) 195-202.
24. E. van Steen, E.L. Viljoen, J. van de Loosdrecht, M. Claeys Appl. Catal. 335 (2008) 56-63.
25. A. Feller, M. Claeys, E. van Steen, J. Catal. 185 (1999) 120-130.
26. G.L. Bezemer, J.H. Bitter, H.P.C. E. Kuipers, H. Oosterbeek, J.E. Holewijn, X.D. Xu, F. Kapteijn, J. van Dillen, K. P. de Jong, J. Am. Chem. Soc. 128 (2006) 3956-3964.
27. E. Iglesia, Appl. Catal. 161 (1997) 59-78.
28. Ø. Borg, D.C. Pascal, A.I.Dietzel, E.Z. Spjelkavik, J.C.Tveten, S.D. Walmsley, S. Eri, A. Holmen, R.J. Erling, Catal. 259 (2008), 161-164.
29. J.P. den Breejen, J.R.A. Sietsma, H. Friedrich, J.H. Bitter, K.P. de Jong, J. Catal. 270 (2010) 146-152.
30. S. Krishnamoorthy, M. Tu, M.P. Ojeda, D. Pinna, E. Iglesia, J. Catal. 211 (2002) 422-433.
31. R.J. O'Brien, L. Xu, R.L. Spicer, S. B, D.R. Milburn, B.H. Davis, Catal. Today 36 (1997) 325.
32. S. Storsæter, Ø. Borg, E.A. Blekkan, A. Holmen, J. Catal. 231 (2005) 405-419.
33. J.L. Li, X. Do. Zhan, Y.Q. Zhang, G. Jacobs, T. Das, B.H. Davis, Appl. Catal. 228 (2002) 203–212.
34. J.L. Li, G. Jacobs, T.K. Das, Y.Q. Zhang, B.H. Davis, Appl. Catal. 236 (2002) 67-76.
35. J. Hong, W. Chu, P.A. Chernavskii, A.Y. Khodakov, J. Catal. 273 (2010) 9-17.
36. A. Yu Khodakov, J. Lynch, D. Bazin, B. Rebours, N. Zanier, B. Moisson, P. Chaumette, J. Catal. 168 (1997) 16.
37. A. Kogelbauer, J.G. Goodwin, Jr., R. Oukaci, J. Catal. 160 (1996) 125.
38. E. Iglesia, S.L. Soled, R.A. Fiato, G.H. Via, J. Catal. 143 (1993) 345.
39. A.M. Hilmen, D. Schanke, A. Holmen, Catal. Lett. 38 (1996) 143.
40. D. Bazin, L. Borko, Zs Koppany, I. Kovacs, G. Stefler, L.I. Sajo, Z. Schay, L. Guczi, Catal. Lett. 84 (2002) 169.
41. M. Ronning, D.G. Nicholson, A. Holmen, Catal. Lett. 72 (2001) 141.
42. G. Jacobs, J.A. Chaney, P.M. Patterson, T.K. Das, J.C. Maillot, B.H. Davis, J. Synch. Rad. 11 (2004) 414.
43. L. Guczi, D. Bazin, I. Kovacs, L. Borko, Z. Schay, J. Lynch, P. Parent, C. Lafon, G. Stefler, Z. Koppany, I. Sajo, Top. Catal. 20 (2002) 129.
44. H. Schulz, E. van Steen, M. Claeys, in: H.E. Curry-Hyde, R.E Howe (Eds.), Natural Gas Conversion II, Elsevier, Amesterdam, 1994, p. 455.
45. Eur. Pat. Appls. 201 557 (1983).
46. C.J. Kim, Eur, Pat. Appls. 339923 (1988).
47. C.J. Bertole, Charles A. Mims, Gabor Kiss, J. Catal. 210 (2002) 84-96.
48. P. J. Van Berge, J. van de Loosdrecht, S. Barradas, A.M. van der Kraan, Catal. Today 58 (2000) 321-334.
49. G. Jacobs, T.K. Das, P.M. Patterson, J.L. Li, L. Sanchez, B.H. Davis, Appl. Catal. 247 (2003) 335-343.
50. G. Jacobs, P.M. Patterson, T.K. Das, B.H. Davis, Appl. Catal. 270 (2004) 65-76.
51. G. Jacobs, T.K. Das, J.L. Li, M.S. Luo, P.M. Patterson, B.H. Davis, Stud. Surf. Sci. Catal. 163 (2007) 217-253.

52. G. Jacobs, A. Sarkar, M.S. Luo, A. Dozier, B.H. Davis, Ind. Eng. Chem. Res. 47 (2008) 672-680.
53. W. Ma, G. Jacobs, D. E. Sparks, M. K. Gnanamani, V. R. R. Pendyala, C.H. Yen, J. L. S. Klettlinger, T.M. Tomsik and B. H. Davis, Fuel. 90 (2011)756-765.
54. E. van Steen, M. Claeys, M.E. Dry, J. van de Loosdrecht, E.L. Viljoen, J.L. Visagie, J. Phys. Chem. B.109 (2005) 3575-3577.
55. D. Schanke, A.M. Hilmen, E. Gergene, K. Kinnari, E. Rytter, E. Adnanes, A. Holmen, Catal. Lett. 34 (1995) 269.
56. J. van de Loosdrecht, B. Balzhinimaev, J.-A. Dalmon, J.W. Niemantsverdriet, S.V. Tsybulya, A.M. Saib, P.J. van Berge, J.L. Visagie, Catal. Today 123 (2007) 293.

(68e) Modeling the Start-up Phase of Fischer Tropsch Synthesis In a Fixed Bed Reactor: Effect of Pore Filling and Heat Transfer Through the Catalyst Bed

Syed Ali Zeeshan Gardezi[1], Babu Joseph[1], D. Yogi Goswami[2] and John Wolan[1], (1)Chemical & Biomedical Engineering, University of South Florida, Tampa, FL, (2)Clean Energy Research Center, University of South Florida, Tampa, FL

Fischer Tropsch Synthesis is a highly exothermic process. During the startup, when the catalyst support pores are empty; there is a strong possibility of catalyst sintering due to rapid heat generation and poor heat transfer. As the reaction proceeds, these pores begin to fill with product hydrocarbons. As a result, the overall heat transfer improves, first, due to diffusion limitations imposed by the liquid in the pores and second, due to the change in the heat capacity of the bed as the pores are filled.

Finite difference approximation is used to model the intra-pellet reaction and heat transport. However, finite element analysis was used to model the inter-pellet heat and mass transport. Vapor-liquid equilibrium was assumed to be attained at each pore, thus virtual pressure and fugacity coefficients were used to develop a quantitative correlation. The goal was to estimate the time required to fill the catalyst pores. This time represents the unsteady transient phase which requires fine temperature adjustments to avoid reaction run away.

The resulting kinetic model, coupled material and energy balance are used to assess the optimum reactor startup procedures. Another goal is to identify the impact of different types of inert packing used to control heat transfer and conversion in the fixed bed reactor with a view towards avoiding the hot spots and steep temperature gradients. The combined effect of heat transfer and kinetics present a challenging problem for both the bed design and for safe startup sequence of the highly exothermic FTS reactions in jacket cooled fixed bed reactors.

See more of this Session: Coal, Biomass and Natural Gas to Liquids II

See more of this Group/Topical: Topical 6: 11th Topical Conference on Gas Utilization

(68f) Modeling Diffusional Limitations In Fischer Tropsch Synthesis Catalyst at near Critical and Supercritical Solvent Conditions

Aswani K. Mogalicherla, Elfatih E. Elmalik and Nimir O. Elbashir, Chemical Engineering, Texas A&M at Qatar, Doha, Qatar

Modeling Diffusional Limitations in Fischer Tropsch Synthesis Catalyst at Near Critical and Supercritical Solvent Conditions

Aswani K Mogalicherla, Elfatih E.Elmalik, Nimir.O.Elbashir*

Chemical Engineering Program, Texas A&M University at Qatar, PO Box 23874, Doha Qatar

*Corresponding author: nelbashir@tamu.edu

Abstract: Conducting Fischer-Tropsch synthesis (FTS) reaction in near critical and supercritical fluid (SCF) media has been demonstrated to have certain advantages over the traditional routes because of the unique characteristics of the supercritical phase. It combines the desirable properties of gas-like diffusion along with liquid-like heat transfer and solubility to overcome many of the limitations of the current industrial FTS reactors.[1] In the process of tuning transport properties, there is a possibility that the reaction moves from a kinetic control regime to diffusion control regime. Therefore, analyzing only the variations in transport properties (diffusivity, density and viscosity) with operating conditions is not a sufficient task for optimal reactor design. The relative change in the diffusion rate with respect to the rate of reaction has to be determined for identifying the controlling regime.[2] Very few studies have been reported on diffusional limitations in FTS catalysts at supercritical conditions.[3] Those studies were not accounting the details of reaction kinetics, product distribution and density dependency of diffusivities. In most of the cases first order reaction kinetics were assumed, which is not a valid assumption. Even though a large number of kinetic models for FTS have been developed, these models were not used in estimating diffusional effects in FTS catalysts.

The objective of the present work is the assessment of diffusional limitations in fixed bed tubular reactors which typically contain catalyst particles of a few millimeter diameters. For this a conventional theory of diffusion-reaction in porous pellets is employed. To account the details of reaction kinetics, LHHW model developed by Mogalicherla et al. (2010) from cobalt catalyst at near critical and supercritical fluid (n-hexane) conditions was used.[4] The binary diffusional coefficients of all reactants and products in supercritical solvent were estimated from correlation proposed by Hong He (1998) which accounts the molecular weights and critical properties of all compounds, reactor operation conditions and reaction mixture density.[5] The following conditions have been considered in developing this model: temperature range from 220 - 260°C, pressure range from 35-80 bar, and H_2/CO feed ratios of 1-2, with supercritical solvent hexanes/syngas molar ratio of 3:1. The catalyst effectiveness factor and concentration profiles of reactants and products along the pellet radius were estimated in three different size catalyst particles (0.5mm,1 mm and 2 mm) and at three CO conversion levels (30, 60, 80%). A single chain growth probability (α-value) model was assumed in all the simulations. The variation of α and product distribution with conversion and operating conditions was taken into account from the experimental data of Elbashir and Roberts (2005).[6] The other catalyst properties such as porosity, pore volume, surface area, and tortuosity factor were adopted from the properties of commercial Co/Al_2O_3 pellets. The effectiveness factor for LHHW type kinetics was estimated using an approximation method proposed by Hayes et al.(2007).[7] The optimal operating conditions (temperature pressure, density, one pass conversion and catalyst particle size) for SCF-FTS operated in a fixed bed reactor were identified.

References

1. Elbashir, N.O; Bukur, D.B.; Durham, E.; Roberts, C.B. Advancement of Fischer-Tropsch synthesis via utilization of supercritical fluid reaction media. *AIChE Journal* **2010** *56*(4), 997-1015.
2. Zhan, X.; Davis, B.H. Assessment of internal diffusion limitation on Fischer-Tropsch product distribution. *Applied Catalysis A: General* **2002**, *236*, 149-161
3. Fan, L.; Yoshii.K.; Yan.S.; Zhou.J.; Fujimoto, K. Supercritical phase Fischer-Tropsch Synthesis: Catalyst pore-size effects. *AIChE Journal* **1992** *38*(10), 997-1015.
4. Mogalicherla, A.K.; Elfatih, E. Elmalik; Elbashir, N.O. A fugacity based kinetic model for supercritical Fischer-Tropsch synthesis over cobalt based catalytic systems. *AIChE Annual Meet* **2010**, November 7-12th, Slat lake City, Utha.
5. Hong He, C. Infinite-dilution diffusion coefficients in supercritical and high-temperature liquid solvents. *Fluid Phase Equilibria* **1998**, *147*, 309-317
6. Elbashir, N.O; Roberts, C.B. Enhanced incorporation of α-olefins in the Fischer-Tropsch synthesis chain growth process over an alumina –supported cobalt catalyst in near-critical and super critical hexane media. *Industrial and Engineering Chemistry Research* **2005**, *44*(3), 505-521.
7. Hayes, R.E.; Mok, P.K.; Mmbage, J.; Votsmeier ,M. A fast approximation method for computing effectiveness factor with non-linear kinetics. *Chemical Engineering Science* **2007**, *62*, 2209-2215

See more of this Session: Coal, Biomass and Natural Gas to Liquids II

See more of this Group/Topical: Topical 6: 11th Topical Conference on Gas Utilization

(68g) Fischer-Tropsch Cobalt Catalyst Activation and Handling through Wax Enclosure Methods

Jennifer L.S. Klettlinger, NASA Glenn Research Center, Cleveland, Ohio

(Judy) Chia H. Yen, University of Toledo, Cleveland, Ohio

Leah Nakley, Sierra Lobo Inc., Cleveland, Ohio

Angela D. Surgenor, NASA Glenn Research Center, Cleveland, Ohio

Acknowledgments

We would like to express our special appreciation to the NASA Fundamental Aeronautics Subsonics Fixed Wing Project for funding this research. We would also like to express our appreciation to the Center of Applied Energy Research at the University of Kentucky for their help and guidance in this effort and to Terry Mccue for his expertise with the scanning electron microscope.

Abstract

Fischer-Tropsch (F-T) synthesis is considered a gas to liquid process which converts syn-gas, a gaseous mixture of hydrogen and carbon monoxide, into liquids of various hydrocarbon chain length and product distributions. Cobalt based catalysts are used in F-T synthesis and are the focus of this paper. One key concern with handling cobalt based catalysts is that the active form of catalyst is in a reduced state, metallic cobalt, which oxidizes readily in air. In laboratory experiments, the precursor cobalt oxide catalyst is activated in a fixed bed at 350°C then transferred into a continuous stirred tank reactor (CSTR) with inert gas. NASA has developed a process which involves the enclosure of active cobalt catalyst in a wax mold to prevent oxidation during storage and handling. This improved method allows for precise catalyst loading and delivery into a CSTR. Preliminary results indicate similar activity levels in the F-T reaction in comparison to the direct injection method. The work in this paper was supported by the NASA Fundamental Aeronautics Subsonics Fixed Wing Project.

Introduction

Alternative fuels research plays an important role in emissions[1,4] reduction and in the effort to decrease the country's energy dependencies. The Fischer-Tropsch process provides the ability to convert the nation's natural energy resource such as coal, natural gas, and biomass into a cleaner and more economical alternative to traditional commercial jet fuel. Fischer-Tropsch jet fuel is considered a synthetic jet fuel, which is synthesized from hydrogen and carbon monoxide in the presence of a catalyst. F-T jet fuel composition differs from petroleum-based, conventional commercial jet fuel because of differences in feedstock and production methodology. Fischer-Tropsch fuel typically has a lower aromatic and sulfur content and consists primarily of iso and normal paraffins. This alternative fuel is an active area of research across the world because of its reduction in emission particulates[1,4].

The Fischer-Tropsch process has three distinct steps: gasification, synthesis, and product upgrade. The gasification step produces syngas (hydrogen and carbon monoxide) from many hydrocarbon resources. The synthesis involves the conversion of syngas to syn-crude. Product upgrade processes the syn-crude and separates it into useable liquid fuels. The synthesis step can be optimized to increase yields and reduce energy inputs into the overall process.

Cobalt-based Fischer-Tropsch catalyst is typically cobalt oxides on various ceramic supports (e.g., alumina, silica, titanium oxide, etc). This cobalt catalyst requires a reduction treatment to convert cobalt oxides to metallic cobalt which catalyzes F-T synthesis reactions. This reduction step generally requires an operating temperature around 350°C, which is beyond the design temperature of conventional F-T slurry reactors; consequently it is carried out in a separate activation reactor. After reduction, the cobalt catalyst is preserved in an inert atmosphere in the reactor. Conventional test procedure is to pneumatically load the reduced catalyst from the activation reactor directly into the F-T synthesis reactor by using an inert gas. The activation reactor is weighed before and after the catalyst is transferred to the F-T reactor to deduce the catalyst charge. The pneumatic conveyance method has the disadvantage of insufficient catalyst quantification and the potential for losing catalyst in the inactive reactor volume. This paper proposes an alternate method to encase the reduced catalyst in a wax mold while in an inert glove box. The alternate method assures that the prescribed amount of catalyst is loaded into the reactor without re-oxidizing the active cobalt catalysts before testing.

Experimental

Activation Procedure

The activation reactor is a fixed-bed reactor designed by GRC[2] based on requirements of a similar unit used by the Center for Applied Energy Research (CAER). Figure 1 shows this reactor with and without insulation which is used for the activation of cobalt catalysts. The reactor is fabricated from a 1" OD, seamless 316 stainless steel tube with a wall thickness of 0.083", is 12" long, and is designed as a simple gas heat exchanger. Two tube socket weld unions are TIG welded to the ends of the tube to produce end connections that allow the reactor to be removed from the overall system.

There are no internal baffles or supports for the catalyst materials. The reactor includes instrumentation both internally and externally for monitoring temperatures and pressures.

Figure 1: Activation Reactor (with and without insulation)

The reactor is designed to activate catalyst materials up to 225 psig and 430 °C with controlled flows of gaseous hydrogen, carbon monoxide, argon or nitrogen. Flow rates to the reactor can range from 0-120 SPLH. The reactor temperature is controlled by Labview software, which provides automatic alarms and shutdowns for high temperatures and pressures. The Labview software is also used as a data acquisition system. A mighty-band coil heater surrounds the reactor with a heater power capacity of 940 W. The heater also includes an integrated type K thermocouple and can either be in a manual or automatic ramp mode with the use of the Labview control software. Two type K thermocouples are inserted from the bottom of the reactor and are at fixed lengths. One thermocouple is near the bottom of the catalyst bed while the other is centralized. Multiple thermocouple locations verify that the catalyst bed is uniformly heated to enable full catalyst activation.

Before catalyst can be loaded, the reactor must be correctly assembled with the thermocouples in their specific positions and glass wool covering the bottom of the reactor. A piece of 1/4" steel tubing is temporarily inserted over the thermocouples and inserted from the bottom of the reactor while securing the bottom reactor nut. This steel tube keeps the internal thermocouples centered while inserting the glass wool. Approximately three inches of woven glass wool is inserted from the top of the reactor with a ¼" metal rod, around the steel tube and pushed down to the bottom of the reactor. Once the glass wool is evenly inserted around the steel tube, the steel tube is removed from the top of the reactor. The metal rod is then used to even out the glass wool around the thermocouples to fill in any voids. A visual inspection of thermocouple location is done from the top of the reactor to verify they are centered. Keeping the reactor in the vertical position, a weighed amount of catalyst is slowly poured into the top of the reactor and the top nut is secured. The amount of catalyst that is loaded into the reactor varies according to the test parameters. Catalyst weight loss during activation is usually 10% or less.

The loaded reactor is then vertically mounted by a wall support bracket (see Figure 1). All inlet and outlet gas lines are connected and electrical wiring is attached. Once the reactor is installed onto the wall, a low pressure leak check is performed with nitrogen to verify the system is sealed. The activation process is started after a successful leak check. Inert gas flow to the activation reactor is initiated, followed by activation gas feed (typical 33% H2, 67% inert). The activation gas flow rates are calculated based on a total flow rate of 30 SLPH/10 g-catalysts. This activation is carried out near atmospheric pressure. Figure 2 shows the reactor setup, where flow of gasses pass from top to bottom of this reactor.

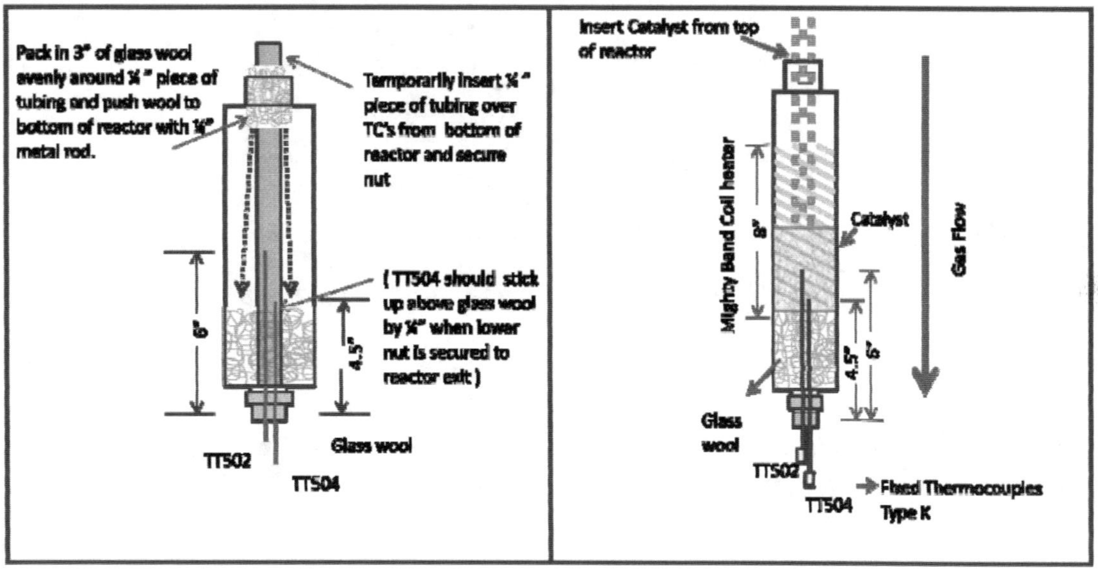

Figure 2: Fixed Bed Activation Reactor

Using the ramp-soak capability of Labview software, the catalyst is heated at rate of 1°C per minute to a pre-set target activation temperature (typically 350°C). The catalyst is held at the designated temperature and gas flow rates for 10 hours. The reactor heater automatically turns off after 10 hours and the catalyst is cooled to room temperature with continuous gas flow. The remaining hydrogen in the reactor is purged out of the system with nitrogen and pressurized to approximately 2psig. This slightly positive pressure is maintained by closing isolation ball valves on the reactor. Keeping the reactor under pressured nitrogen prevents the catalyst from being exposed to oxygen, thus inhibiting catalyst oxidation during handling.

Mold Making

Once activation is complete, the pressurized reactor is taken to a VAC NEXUS glove box where the catalyst is removed and weighed under inert conditions. **Error! Reference source not found.** shows the VAC NEXUS glove box used in this study. The reactor is loaded into a chamber where the atmospheric gases are evacuated, then transferred into the glove box's inert environment. Once inside the glove box, the activation reactor is opened and the catalyst is removed. A known amount of catalyst is weighed out and recorded.

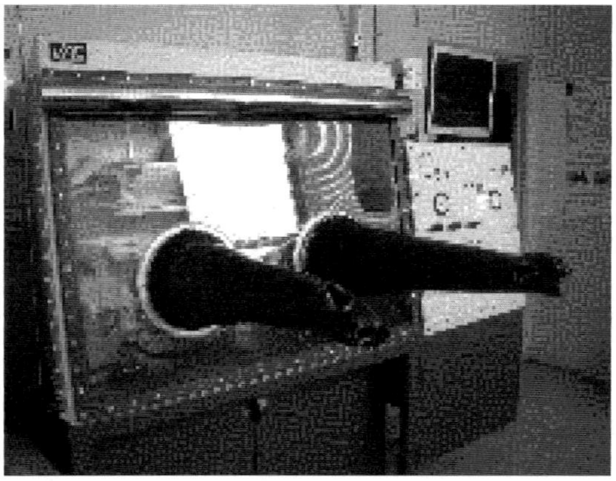

Figure 3: VAC NEXUS Glovebox used for wax enclosure preparation

This catalyst sample is then submerged in Polywax500[3] to create a solid wax enclosure. The goal is to encapsulate catalyst in a non-permeable wax mold (see Figure 4). This is accomplished by melting Polywax500[3] in a beaker and proceeding with a liquid wax/catalyst layering technique (see

Figure 5). A layer of melted Polywax500[3] is first poured into a silicone mold (see **Error! Reference source not found.**).

Figure 4: Empty Wax Mold

Activated catalyst is then poured on top of the molten wax followed by additional molten wax. During the layering process, the liquid polywax[3] infiltrates the catalyst pushing out any gas trapped between catalyst particles and creating a solid wax piece. Both catalyst weight and wax weight are recorded and accounted for in mass balance calculations. Once the catalyst enclosure cools and solidifies, it is removed from the mold and is taken out of the glove box for placement in the CSTR (see

Figure 6)

Figure 5: First layer of Molten Wax in Mold (top left), Catalyst inserted on top of Molten Wax in Mold (top right), Molten

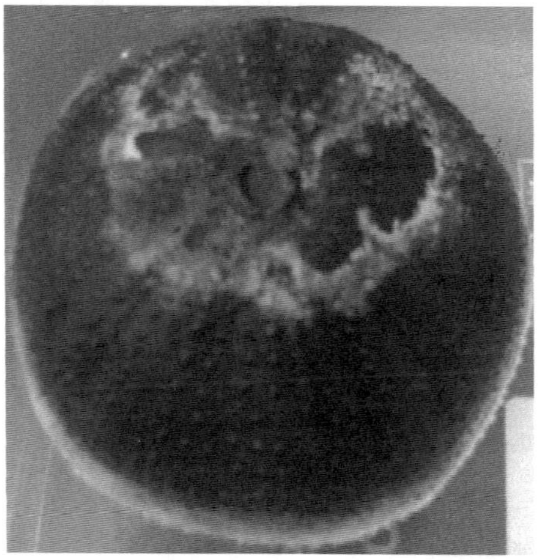

Figure 6: Final Wax Enclosure Solidified

Wax infiltrating Cobalt Catalyst (bottom left), Final Molten Wax on top of Cobalt Catalyst (bottom right)

The wax mold is then transported to the CSTR, where it is loaded into the top of the reactor vessel just before closing and sealing the reactor. After the autoclave reactor is sealed, a timed inert purge is carried out to remove all oxygen in the system. Inert purge must be completed before the wax enclosure is melted to avoid potential oxidation. Normal reactor start-up procedure begins after inert purge.

Results and Discussion

Slurry Reactor Comparison

Table 1 summarizes the comparison of three reactor runs using the same base catalyst. A baseline cobalt catalyst was prepared by CAER in 2007 for NASA testing in the newly established test facility. The purpose of NASA testing this catalyst was to verify facility operability. CAER's test results serve as the baseline for gauging the alternate catalyst handling method.

Table 1: Raw Data Comparison

Run	GRC (COC-001)	GRC (INI-001)	CAER (TKD-007)
Catalyst ID	ZYQ036	Same	Same
Catalyst Composition	0.5%Pt 15%Co on Alumina support	Same	Same
Catalyst Amount (g)	21.0	21.0	20.0
GHSV (slph/g)	5.1	5.2 (0 to18 hrs) 2.6 (18hrs to end)	5.5 (0 to 676 hrs) 2.0 (676 hrs to end)
H2 : CO Molar Ratio	2.0	1.8	2.0
Reactor Pressure (psig)	268	275	270
Reactor Temperature (^{o}C)	220	220 to 223	220
Total Run Time (Hours)	381	138	628 @ 5.5 960 total
Mass Balance (Out/In)	90.6%	98.3%	

The reactor performance in NASA Run INI-001 was inferior to those achieved in the CAER's run TKD-007. The reactor conversions in the NASA GRC run (COC-001) are comparable to the reactor performance achieved by CAER considering the variation in space velocity and the state of operation.

Run INI-001, the initial test performed in the Alternative Fuels Research Lab, experienced many operational problems which led to poor catalyst performance. Process variables in Run INI-001 were adjusted to compensate for poor catalyst activity: higher reactor operating temperature (223°C vs. 220 °C in CAER run) and lower space velocity (2.6 SLPH/g-catalyst vs. 5.5 SLPH/g-catalyst in CAER run). The reasons for poor catalyst activity include the following:

- Catalyst was likely exposed to air and partially oxidized after activation. The catalyst was encased in a Polywax 3000[3] mold in an inert atmosphere in the glove box. The wax mold was then left in a sealed glass in ambient air for 12 days. Catalyst wax mold cracked during storage and was still loaded for this run.
- Catalyst activation may have been incomplete. Target activation temperature was 350 °C. The observed catalyst bed inlet/outlet temperatures were 352 °C and 247 °C respectively. Non-uniform temperatures may have led to partial activation.
- Gas feeds (i.e., H2 and CO) flows were interrupted due to flow meter malfunctioning.

Run COC-001 operated for 381 hours with consistent feed flows and reactor temperature/pressure control. The main objective for this run was to confirm hardware/software reliability after an extended maintenance shutdown. This run was initiated with known issues including: several instruments being out of calibration; expired GC calibration gases were used for first 320 hours of test; product gas compositions were sometimes extrapolated outside the calibration gas range; feed gas purity may vary slightly from bottle to bottle. The data scatter of run COC-001 in

Figure 7 and poor mass balance in summary Table 1 are consistent with the state of operation. Never the less, the data collected during Run COC-001 confirms that the cobalt catalyst used was active and it had substantially better performance than that was achieved with an inactive catalyst during Run INI-001. Run COC-001 operated at lower operating temperature, higher space velocity while achieving similar per pass conversions as in Run INI-001. The extent of reactor conversions during Run COC-001 is also confirmed by comparing feed flow rate to the reactor effluent gas flow rate, see Figure 9A. Run COC-001 result is supportive of the effectiveness of wax mold prevent active cobalt catalyst from being oxidized.

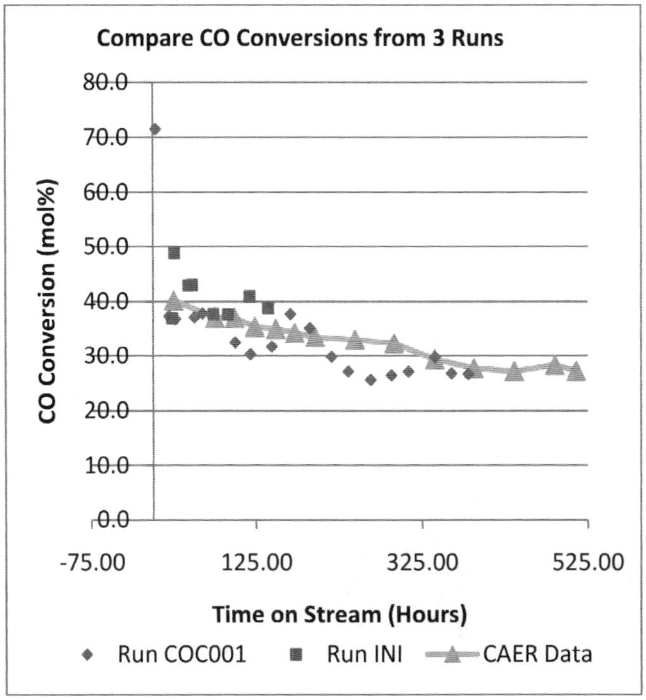

Figure 7: CO Conversion Comparison of Three Separate Runs

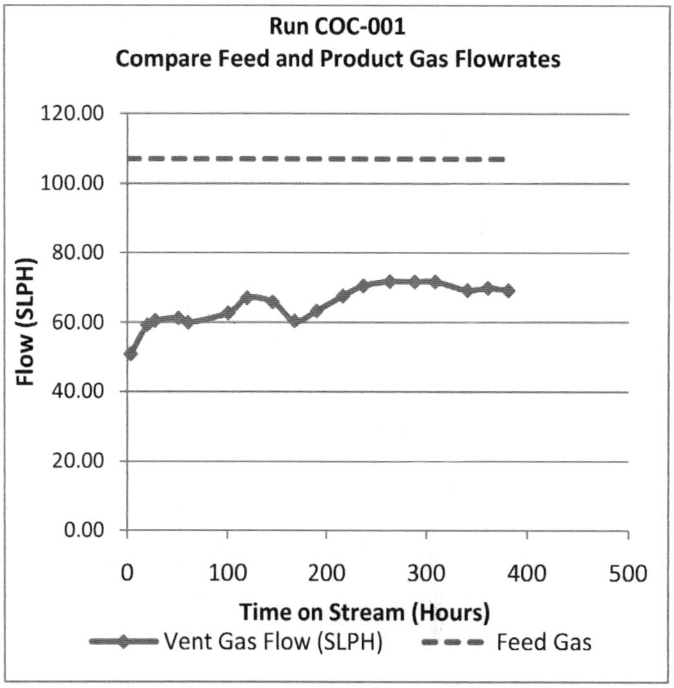

Figure 8: Comparison of Feed Gases and Product Gas Flow rates

Scanning Electron Microscopy (SEM)

The mold making procedures previously described were followed in order to create a sample for SEM verification. These samples were made of an inactive alumina supported cobalt catalyst (25% cobalt, by weight). The wax enclosure mold (see

Figure 6) was too large to fit into the SEM instrument, so a portion of the sample, was cut and used for analysis. Figure 9 shows the setup of the SEM sample used and mounting of these pieces to SEM sample plate.

Figure 9: Wax Enclosure Sample Preparation for SEM

The sample was analyzed in multiple locations, which included the interior portion of the wax enclosure and the outside surface of the wax enclosure. Figure 10 shows each sample individually, the left image shows the external surface portion and the right image displays the internal surface portion. The white markings indicate the areas examined in the SEM. A Hitachi S-4700 Scanning Electron Microscope (SEM) was used to analyze the cobalt catalyst wax enclosure's integrity.

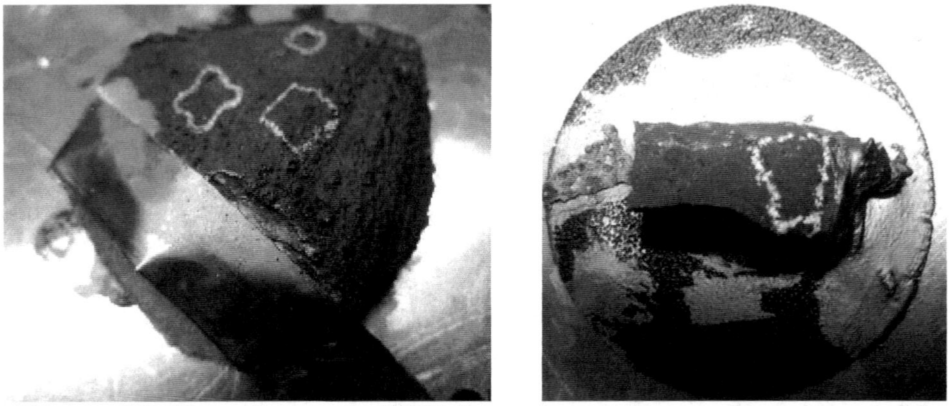

Figure 10: Two Separate SEM Samples: wax enclosure outside surface (left) and internal wax enclosure surface (right)

Figure 11A and Figure 11B show the SEM images of inactive plain alumina supported cobalt catalyst particles (not wax enclosed). Figure 11C shows the SEM image of the wax enclosed alumina supported cobalt catalyst, this image is for an interior location. Figure 11A and Figure 11C have the most similar SEM settings; as indicated on the images, Figure 11A was taken at 12.2mm and Figure 11C was taken at 10.5mm. The distance differences between these two images was negligible, therefore it was disregarded in these interpretations. It is clear that the general morphology of the particles is different in these two images. Figure 11A shows more defined edges to the particles. It is reasonable to suggest that the morphology differences are due to the presence of wax coating the catalyst surface. There is also no indication of void space in Figure 11C, this may prove that there are no gas bubbles located throughout the middle of the wax mold. The lack of gas bubbles indicates that the wax has properly infiltrated the catalyst sample and likely that each catalyst particle has been coated by wax.

A: SEM Image of Cobalt Catalyst Particles

B: SEM Image of Cobalt Catalyst Particles

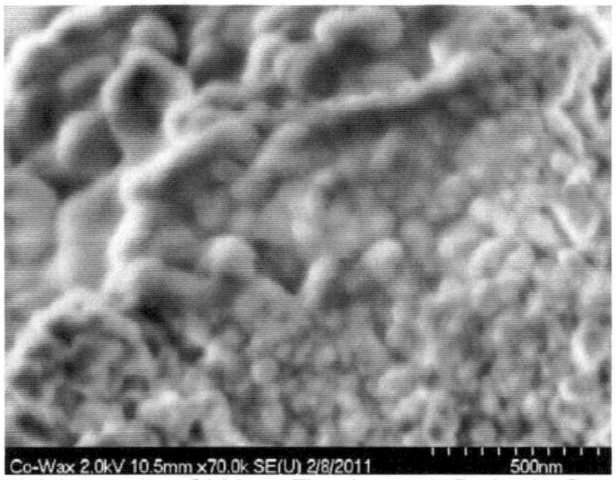

C: SEM Image of Wax Enclosed Cobalt Catalyst

Figure 11: SEM Results

Figure 12 shows the SEM image of the outside layer of wax enclosure at a higher magnification. It is interesting to note the texture of the surface and point out that there is little indication of separate particles. There is no clear indication of separate cobalt particles in comparison to the wax surface.

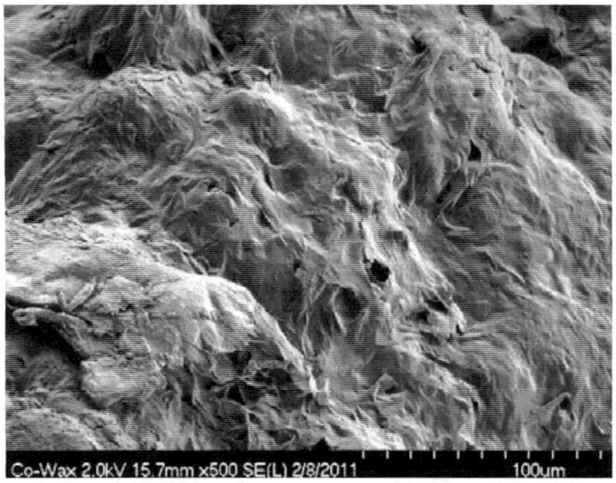

Figure 12: SEM Image of the outside layer of wax enclosure

Conclusions

The methods described in this paper for mold making were intended to preserve activated cobalt catalyst, and facilitate precise catalyst loading in CSTR experiments. The wax mold technique prevents the reduced cobalt catalyst from being oxidized during the loading process. The F-T reactor conversions observed at the NASA Glenn Research Center are comparable to the data collected at the Center for Applied Energy Research (University of Kentucky) using the same catalyst under similar process conditions. As such, we believe that the reduced cobalt catalyst remains active in the wax mold. NASA GRC experiments have shown some variability and lacked the stable results that CAER reported. Another CSTR experiment with the same catalyst will be conducted in the near future as operating issues at the NASA facility have been resolved. Ultra high purity gasses have been purchased and mechanical components have been calibrated and upgraded in order to reduce this oscillation in data.

The SEM images of catalyst-wax molds were collected as preliminary data. The focus of this SEM study was on the interior surface of the mold in secondary electron mode. This provided us with useful information regarding the morphology of the Co/Al_2O_3 catalyst particles coated with wax in comparison to uncoated catalyst particles. The clear differences in morphology lead us to believe that the catalyst particles are fully coated in wax. We will further investigate and verify our wax mold enclosure technique by collecting additional SEM images. We will examine the sample's cross section cut morphology to see how the catalyst particles relate to the surface and verify that there is no evidence of gas bubbles throughout the depth of the mold. We will look at the x-ray (EDS) chemistry at different locations throughout the mold. These locations will include the sample surface, as well as the internal cross section of this sample. We will compare x-ray (EDS) chemistry of a wax mold loaded with active cobalt catalyst versus inactive cobalt catalyst. This data will allow us to quantify and compare the oxygen content of the samples. As data is collected, proof of wax enclosure will be reported.

References

1. Edwards, T., Minus, D, Harrison, W., Corporan, E., "Fischer-Tropsch Jet Fuels-Characterization for Advanced Aerospace Applications

2. Surgenor, A, Klettlinger, J, Yen, C., Nakley, L., "Alternative Fuels Research Facility," AIChE Spring Meeting & 7th Global Congress on Process Safety paper ID#211849 , March 2011.

3. Baker Hughes, (February 2011), POLYWAS Polyethylenes. Retrieved from http://www.bakerhughes.com/assets/media/technicaldatasheets/4c97b7f7fa7e1c37f1000015/file/28729_polywax-sheet-2-5-10.pdf.pdf&fs=302313

4. Bulzan, Dan, Bruce Anderson, et. al. "Gaseous and Particulate Emissions Results of the NASA Alternative Fuel Experiment (AAFEX)", Preceedings of ASME Turbine Expo 2010: Power for Land, Sea and Air, GT2010, June 14-18, Glasgow, Scotland.

(72a) Using OpenFOAM for Modeling Vapor Cloud Explosions In the Chemical and Petrochemical Industry

Timothy Melton, John Cornwell and Benjamin Ishii, Quest Consultants Inc., Norman, OK

As computing power and storage have become more affordable over the years, it has been possible to use more sophisticated models to calculate the effects of hazardous releases in the chemical and petrochemical industries. Until recently, modelers generally had to use either simple, analytic models that were written by others and were virtual "black boxes" or write their own software to implement the numerical methods necessary to solve the equations that defined the model. With the release of the free, open source CFD Toolkit, OpenFOAM, modelers now have a powerful tool, that enables them to model a wide range of situations and complex interactions that were too expensive and time-consuming to model in the past. In this paper, we discuss how Quest is using the OpenFOAM toolkit to model vapor cloud explosions in industrial settings and compare the results of our validation studies with experiments published in the open literature.

See more of this Session: Advances In Aromatics and Petrochemicals Processing III
See more of this Group/Topical: Fuels and Petrochemicals Division - See Also Topicals 4, 6, and 7

The Use of Overpressure Exceedance Curves in Building Siting

Jeffrey D. Marx and Kendall M. Werts
Quest Consultants Inc.
Norman, OK
www.questconsult.com
jdm@questconsult.com, kmw@questconsult.com

ABSTRACT

Compliance with the American Petroleum Institute (API) Recommended Practice (RP) 752 has become an industry standard. Design and layout of new facilities can incorporate most of the principles and spacing requirements presented in API RP 752 with a basic consequence modeling study. Yet many existing facilities are finding that a basic consequence modeling study indicates major problems for onsite buildings when evaluating vapor cloud explosion impacts. The worst-case explosion scenarios, and often the "more likely" scenarios, typically show unacceptable overpressure impacts at buildings within or near hydrocarbon processing units. API RP 752 does allow the analyst to perform a risk-based analysis to evaluate the acceptability of building impacts. One of the most useful forms of a risk-based study is the generation of overpressure exceedance curves. These measures of potential impact to a building incorporate a wide range of potential explosion scenarios, and the probability of each of those events. This paper will explore the construction of overpressure exceedance curves for a typical hydrocarbon processing facility, and will highlight some of the benefits, as well as some of the potential problems, of their application.

INTRODUCTION

When performing the design and layout tasks for a new or proposed onshore hazardous materials facility, application of American Petroleum Institute (API) Recommended Practice (RP) 752 provides a means to locate occupied buildings at safe distances from potential fire, toxic, and explosion overpressure events. When coupled with the layout process, the use of RP 752 can provide valuable insight into providing either safe locations for personnel or protective structures for vulnerable work areas. This advantage available to facilities in the design stage is obviously not available to older, existing ones. When RP 752 is applied to older facilities, the results for many occupied buildings indicate unacceptable consequences. Oftentimes, the potential fire impacts on a building and the potential toxic vapor cloud impacts on building occupants are understood by the facility's operator, and can be addressed through emergency response plans and personnel protective equipment. It is the potential vapor cloud explosion impacts that are of concern in petrochemical facilities, and those hazards may not be fully defined or understood. It is these potential explosion impacts that are the focus of this paper.

The standard method for evaluating vapor cloud explosions is to use a blast-curve correlation model. These are the "simple" models such as Baker-Strehlow-Tang (BST), congestion assessment method (CAM), and TNO multi-energy, among others. While these models are relatively easy to use, they are not easy to use correctly. Modeling explosion scenarios with these models requires a proper description of the explosion source. Parameters such as the fuel reactivity, degrees of confinement or congestion, and volume occupied by the flammable vapor cloud are all important in evaluating each unique explosion scenario.

Results from blast curve models provide overpressure versus distance from the center of the explosion. This data is used to define the peak side-on overpressure that a building may experience. While impulse is also calculated (as a function of distance from the center of the explosion) by these models, overpressure is the primary measure of impact used to evaluate building damage. Results from these models are typically considered conservative, assuming that the correct modeling parameters were applied.

More advanced models such as computational fluid dynamics (CFD), can also be used to predict overpressure impacts. But because these models require considerably more resources to set up and run even one calculation, they are typically only used for evaluating specific scenarios. CFD approaches may be useful for offshore facilities where local geometries can cause significant variations in the near-field results. These models are also applied to locations where blast wave directionality, reflections, or channeling are important, or for very odd geometries in the explosion source area.

The explosion modeling methodology employed in the example in this paper is the Quest model for estimating flame speeds (QMEFS) [1]. This model is based on the BST model, but allows the user to specify the defining parameters in more detail. This avoids the step changes inherent in the typical low/medium/high categories defined by the BST model and gives the analyst more flexibility in describing an explosion scenario.

RISK-BASED EXPLOSION ANALYSIS

In the case where a consequence analysis shows the potential overpressure impacts to a particular building to be unacceptable, an analyst can turn to several different risk-based methods for a more complete assessment. This assumes that care has been taken to verify the applicability of the explosion modeling scenarios that are the root of the concern. The most common risk-based options include:

- Risk ranking (risk matrix approach)
- Qualitative risk assessment
- Quantitative analysis
    - Overpressure probability contours
    - Overpressure exceedance curves

In a risk ranking system, explosion events are evaluated based on their perceived likelihood. Assuming that this method is based upon previous explosion modeling, it represents a semi-quantitative approach. This method provides the analyst with a risk-based scale on which multiple accident (and impact) scenarios can be judged. Scenarios with moderate to high risk can be assigned risk reduction or mitigation measures. Low risk scenarios (those that have minimal consequences or those that have very low likelihoods) can often be discounted and not considered in further evaluation.

A qualitative risk assessment is similar to a risk ranking system in that it involves a semi-quantitative assessment. The most common way to implement this approach is to select the maximum flammable material release magnitude on which to base the explosion calculations (e.g., a release from a 2-inch diameter hole). This serves to (potentially) limit the extent of the explosion impact, and creates a qualitative cut-off for the maximum "credible" event.

If semi-quantitative methods are not helpful or produce results deemed unacceptable, the analysis can proceed to a fully quantitative approach. This process uses quantitative methods to describe the hazards (explosion impacts) and quantitative methods to describe the probability of each event. Probabilities are developed using tools such as failure rate databases, ignition models, and probabilistic weather data. Quantitative approaches for determining occupied building explosion impacts follow the methodology of a quantitative risk analysis (QRA), but cannot be referred to as a measure of risk, since they do not fully describe the total risk to a building. The output of a quantitative explosion overpressure analysis is typically in the form of one of two approaches: overpressure probability contours (similar to risk contours) or overpressure exceedance curves. Probability contours typically can only show the spatial location of the probability of being exposed to a specified overpressure value. While this has some value, especially if there is a target threshold value for one particular building, the analyst often needs more information.

The generation of overpressure exceedance curves (OECs) provides a substantial amount of information for specific locations that may be affected by explosion events. OECs represent a cumulative summary of the overpressure and probability of a full range of events that could affect a building. The OEC technique has been applied in facility or building siting studies worldwide [2], and is accepted as a viable means of making decisions concerning the suitability of buildings in and near flammable materials facilities, for compliance with API RP 752..

CONSTRUCTION OF OVERPRESSURE EXCEEDANCE CURVES

To generate OECs, a quantitative analysis is required. Adhering to the methodology used in a QRA, the following steps are completed:

- Accident scenario selection
  - Multiple release locations
  - Multiple release sizes
  - Defined set of potential explosion sites (PESs)
- Dispersion analysis
  - Multiple atmospheric stabilities
  - Multiple wind speeds
  - Multiple wind directions
- Explosion modeling
  - Comparison of flammable cloud volume to PES volume
  - Calculation of source strength (energy)
  - Overpressure calculation
- Probability analysis
  - Release event frequencies based on equipment failure rates
  - Ignition probabilities
  - Weather condition probabilities
- Combination of consequences and probabilities to generate OECs

This methodology is consistent with that published by CCPS [3] for chemical process QRAs. It begins with accident scenario selection, and in the context of vapor cloud explosions, must consider all potential release sources and each potential explosion site (PES). Because most flammable materials do not produce damaging overpressures when ignited in open areas [4], PESs must be defined in regions of confinement or congestion. Each PES is assigned several parameters that are used in the explosion model when calculating the strength of a potential explosion event.

Release events are selected to include a wide range of potential flammable vapor clouds, from all possible release locations. It is important to remember that a PES can be filled with flammable gas from many different sources, including those evolving in adjacent process units. Multiple releases (i.e., hole sizes) are evaluated for each release location. This provides a probabilistic distribution for the release magnitudes, where large events (e.g., pipe ruptures) have low probabilities, and small events (leaks) have higher ones.

Release and dispersion modeling is conducted for each potential flammable fluid release selected. The dispersion behavior of flammable vapors is calculated using the lower flammable limit (LFL) as the endpoint. For each release scenario, modeling is carried out for multiple weather conditions (atmospheric stability, wind speed, and wind direction). This allows a determination of which PESs a particular flammable cloud may reach. This can include the PES in which a release originates, and those that are some distance away. A flammable cloud may have sufficient mass to be involved in multiple PESs simultaneously. Provided that the PESs are separated by approximately 3-5 meters of open space, ignition of the flammable cloud will result in independent explosions from each PES [5].

The end product of these first two tasks is the creation of a list of all potential explosion events. Each event will be characterized by the fuel reactivity (specific to the released material), the volume involved (all or part of a PES), and the confinement/congestion parameters that describe the PES. In a typical petrochemical process plant, the list of explosion events will number in the tens (or sometimes hundreds) of thousands due to the many combinations of release events, hole sizes, weather conditions, and available PESs.

To construct an OEC, the list of potential explosion events is evaluated in reference to one specific location. The explosion model is run for each event, and the resulting overpressure at that point is calculated. For any one building, multiple points may need to be evaluated in order to account for shorter distances between the different sides of the building and the various PESs that exist around it.

The construction of an OEC involves creating a data set of predicted overpressure (for the specified location of concern) and probability pairs for the entire range of explosion events considered in the analysis. Each data point represents a unique explosion event and its specific probability of occurrence. The data set is then sorted by peak overpressure from high to low. An exceedance curve is generated by creating a cumulative sum of the probability values, beginning with the highest overpressure value. Thus, for any overpressure value represented by the exceedance curve the probability is the summation of the probability of that particular overpressure plus the probabilities of all events resulting in higher overpressures at the specified location. The exceedance curve then represents the probability of experiencing or exceeding a particular overpressure value, for the location specified.

EXAMPLE ANALYSIS

As an example, consider a small hydrocarbon processing facility with eight small equipment groups. The layout of this fictional facility is shown in Figure 1. In the figure, the facility's four occupied buildings are labeled, and the PESs (regions of congestion due to process equipment) are shown as bold-outlined polygons. The PESs have various properties of confinement and congestion due to the obstacle density and restriction of flame expansion that is present in each particular area. All the flammable materials evaluated in this example can be classified as medium reactivity.

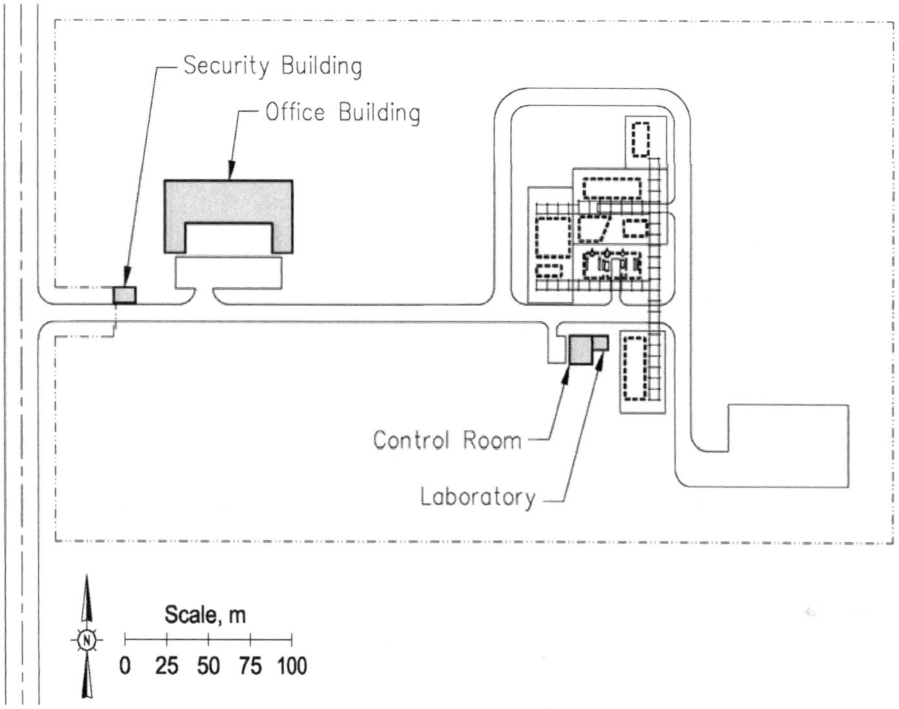

Figure 1
Example Facility Layout

When the potential release scenarios are identified and evaluated with the PESs chosen for this facility, a set of potential explosion scenarios is developed. Using the methodology described above, overpressure exceedance curves can be generated for each of the four occupied buildings identified in Figure 1. The locations chosen for evaluation on the buildings were the closest faces to the process area. For the office building and security building, the eastern faces were evaluated; the northern faces of the control room and laboratory building were evaluated. In addition, the eastern face of the laboratory building was evaluated due to its proximity to one PES.

Figure 2 shows the exceedance curve results for the example facility. To use these curves, either the overpressure or probability axis may be used as a basis. For example, using the laboratory east face curve in Figure 2, an overpressure of 1.5 psi or greater is expected to occur with a probability of about 7 x 10^{-4} per year. Likewise, the overpressure associated with a 1 x 10^{-4} per year exposure is about 1.6 psi. As can be seen in Figure 2, for lower probabilities (less than about 4 x 10^{-5} per year) the east face of the laboratory building is vulnerable to overpressures about 0.7 psi higher than the north face.

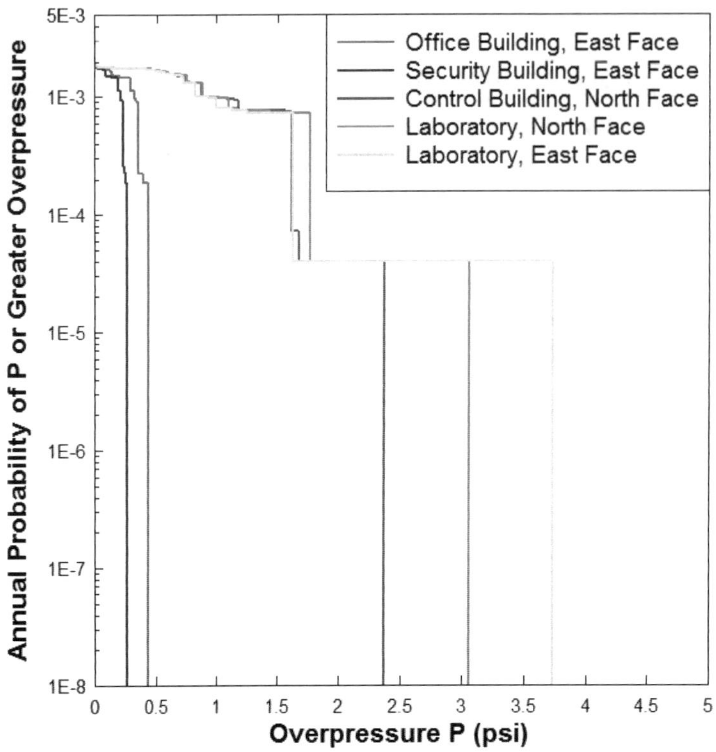

Figure 2
Overpressure Exceedance Curves for the Buildings in the Example Facility

APPLICATION OF OVERPRESSURE EXCEEDANCE CURVES

Once one or more overpressure exceedance curves have been developed for a specific location (typically an occupied building), the building's risk must be assessed. This usually involves application of risk acceptability or risk tolerance criterion for the occupied building, and an evaluation of the building's ability to withstand the calculated overpressure. A common approach to this portion of the analysis is to evaluate the overpressure at a building corresponding for a specific probability of exposure. For example, if a company sets 1.0 x 10^{-4} probability of building occupants being injured as their tolerable risk criterion, then the overpressure corresponding to that probability is compared against the structural response of the building and its ability to withstand that loading. If that level of overpressure is sufficient to cause damage to the building which results in injuries to its occupants, then the probability of explosion impacts becomes intolerable and some form of risk mitigation is required. If the overpressure corresponding to the 1.0 x 10^{-4} probability is lower than the damage threshold (i.e., a level of damage that is incapable of injuring building occupants), the risk is viewed as tolerable.

If a 1.0 x 10^{-4} criterion is applied to the example presented above, then an overpressure less than 0.5 psi must meet the damage threshold for the office and security buildings in order for an intolerable condition to exist. For most buildings of "ordinary" construction, this magnitude of overpressure is incapable of causing significant building damage or injuring building occupants. Some windows may be broken and other minor damage may result, but it is unlikely that the building will suffer structural failure.

If a 1.0 x 10^{-4} criterion is applied to the control and laboratory buildings, overpressures of about 1.6 to 1.7 psi are used for an tolerability analysis. If we assume that the buildings are steel-framed with metal siding, 1.7 psi is capable of causing significant damage, to the extent that building occupants may be injured. The buildings are not expected to collapse, but could experience significant deformation, and will likely experience loss of much of the exterior metal sheathing. This type of result indicates that the control and laboratory buildings either need to be moved to a more remote location or upgraded to withstand overpressures up to 1.7 psi in order to protect occupants from injury.

Because an OEC analysis provides a quantitative approach to determining potential explosion impacts, it is a significant improvement over a consequence-only approach. The coupling of probability to consequence (thus a risk-based approach) allows the large consequence events to be viewed in light of their low probability, and shows that the most probable events are the ones with small overpressure impacts. In many cases, it shows that the "worst-case" events (those producing the largest overpressures) that are traditionally used for siting purposes have a frequency low enough to fall below typical acceptability guidelines even when considered in a cumulative sense in the OEC.

One potential problem with the OEC approach is that it does not take into account the impulse on the building. Explosion modeling (e.g., the QMEFS model) does predict both overpressure and impulse for any given explosion. Yet the OEC methodology effectively de-couples the overpressure and impulse when it creates the cumulative probability curve in terms of overpressure only. A similar curve can be generated for impulse (an impulse exceedance curve), but it cannot be used with or compared to the OEC in any meaningful way.

CONCLUDING REMARKS

The application of overpressure exceedance curves represents a significant improvement to basing building siting solely on consequence analysis results. An OEC analysis brings probability into the analysis in a quantitative fashion, accounting for weather conditions, differing release sizes, and the specific layout of the facility. Once a probability threshold is selected, use of the OEC information can be applied much like in a consequence-only evaluation. This type of analysis satisfies the requirements of API RP 752 for the explosion overpressure impact portion of an assessment. If there are no significant toxic or fire hazards for the building, this type of analysis can form the basis for building siting decisions. In addition, the construction of OECs can be used to site temporary or portable buildings. In this case, the generation of OECs constitutes a "detailed analysis" – one of the options for the evaluation required by API RP 753. Care must be taken to evaluate temporary or portable buildings with regard to their lighter construction, applying the guidance presented in RP 753.

REFERENCES

[1] Melton, T. A. and J. D. Marx (2009), "Estimating Flame Speeds for Use with the BST Blast Curves." *Process Safety Progress*, Volume 28, No. 1, p. 5, March 2009.

[2] Chamberlin, G. A., and J. S. Puttock, *Vapor Cloud Explosion Risk Management in Onshore Plant Using Explosion Exceedance Techniques*, Hazards XIX: Process Safety and Environmental Protection, IChemE, 2006.

[3] CCPS (2000), *Guidelines for Chemical Process Quantitative Risk Analysis*. 2nd Edition, Center for Chemical Process Safety of the American Institute of Chemical Engineers, 2 Park Avenue, New York, New York, 2000.

[4] Harris, R. J., and M. J. Wickens (1989), *Understanding Vapour Cloud Explosions—An Experimental Study*. The Institution of Gas Engineers, Communication No. 1408, 1989.

[5] van den Berg, A. C., and A. L. Mos, *Research to Improve Guidance on Separation Distance for the Multi-Energy Method (RIGOS)*, HSE Research Report 369, 2005.

(72c) Maximize Ethylene Gain In Acetylene Removal Units - Optimization Strategies and New Generation Catalysts to Increase Production, Profit and Flexibility

Mingyong Sun, Sud-Chemie, Inc, Louisville, KY, **Denise B. Cooper**, Sud-Chemie, Inc., Houston, TX, Michael Urbancic, R&D Manager, Sud-Chemie, Inc, Louisville, KY and Steven Blankenship, R&D Chemist, Sud-Chemie, Inc, Louisville, KY

Acetylene is a troublesome byproduct in the manufacture of ethylene. It must be removed from the ethylene product before it can be sold, as it is considered a poison to downstream users even in very low concentrations. Selective hydrogenation of acetylene to ethylene is the industry preferred method of removing acetylene. Maximizing the conversion of acetylene to ethylene and minimizing side reactions (specifically the hydrogenation of ethylene to ethane) maximizes production and increases plant profitability.

This paper will address details on improving operation of tail-end acetylene converters to maximize profit. There are several parameters in operation of a tail-end converter that can be optimized to improve ethylene make, as well as plant reliability and flexibility; these operating parameters will be reviewed.

Additionally, the paper will discuss several common operational problems with tail-end acetylene converters. The most common of these operational problems is rapid deactivation, which leads to premature acetylene breakthrough in the reactor effluent and short cycles. Another common problem with many reactors is non selective operation, which can actually result in off-spec product if not recognized early in the operating cycle. Lack of flexibility is also cited by ethylene producers as problematic; operators need to be free to make changes to feed slate and production rate without concern that the changes will negatively affect the operation of the acetylene converter.

One key area to enable achievement of optimum performance and avoid operational problems is through proper catalyst selection to match the specific site operational needs (e.g., high activity, flexibility, CO-free operation, etc.). Recent improvements in tail-end acetylene converter catalyst now allow ethylene manufacturers to operate with a highly stable, selective catalyst that also provides enough activity to fully hydrogenate the acetylene despite feed slate or production rate changes. There is no longer a need to sacrifice selectivity for activity with these new catalysts. Included in the discussion will be a review of latest generation catalyst offerings from Süd-Chemie detailing laboratory test results, commercial operational experience and the benefits users have realized due to operation of these catalysts.

Explosion Hazards in Membrane Reactors

Niko Holst
Technische Universität Berlin
Straße des 17. Juni 135, 10623 Berlin, Germany
niko.holst@tu-berlin.de

Jörg Steinbach
Technische Universität Berlin
Straße des 17. Juni 135, 10623 Berlin, Germany
joerg.steinbach@tu-berlin.de

Prepared for Presentation at
American Institute of Chemical Engineers
2011 Spring Meeting
7th Global Congress on Process Safety
Chicago, Illinois
March 13-16, 2011

UNPUBLISHED

Explosion Hazards in Membrane Reactors

Niko Holst
Technische Universität Berlin
Straße des 17. Juni 135, 10623 Berlin, Germany
niko.holst@tu-berlin.de

Jörg Steinbach
Technische Universität Berlin

Keywords: Membrane reactor, Oxidative coupling of methane (OCM), explosion

Abstract

Membrane reactors are a technology to increase reaction selectivity and yield of methane oxidative coupling (OCM). The use of porous membranes can lead to back permeation of methane and hydrocarbon products into the reactor shell, where flammable gas mixtures can form. A model-based analysis is performed to investigate the influence of the key operating parameters on the shell side concentration profiles. An outlook on a procedure to determine a suitable reactor geometry is presented.

1. Introduction

Catalytic oxidation processes have attracted significant attention of researchers in academia and industry [1]. One trend concerning these reactions aims towards increasingly severe operating conditions, which can cause major safety problems. The oxidative coupling of methane (OCM) to form ethylene is bound to be operated at process temperatures from 700-950°C [2]. Concepts to overcome the selectivity and safety problems of this reaction are membrane reactors consisting of two concentric tubes separated by a ceramic membrane [3].

This configuration allows separate feeds of the reactants and therefore higher oxygen concentrations than tolerable in conventional packed-bed reactors. The explosion hazard is therefore drastically reduced in terms of normal operation. However, a major drawback of these reactors is that unreacted hydrocarbons can permeate to the shell side of the reactor, if non-selective porous membranes are used.

Little attention has been given to the fact that a flammable gas mixture can form in the hot oxygen-rich atmosphere. An ignition on the shell side of the reactor is particularly critical, as no solid catalyst is present to extinguish flames or to hinder propagation in the reactor and into down streaming units [4, 5].

A Hazop analysis was performed for a membrane reactor system to identify the main operational hazards, followed by a model-based analysis to investigate the dependency of the shell side concentration profiles on key reactor design parameters. The influence of reactor geometry, membrane permeability and reactor operating conditions is illustrated. An outlook on a method for prediction of the formation of flammable gas mixtures in a membrane reactor system is given.

2. Reactor model & methodology

2.1 Reactor Model

A functional scheme of the membrane reactor is shown in figure 1. A packed-bed of catalyst in the inner tube is surrounded by an empty annulus between membrane and outer wall.

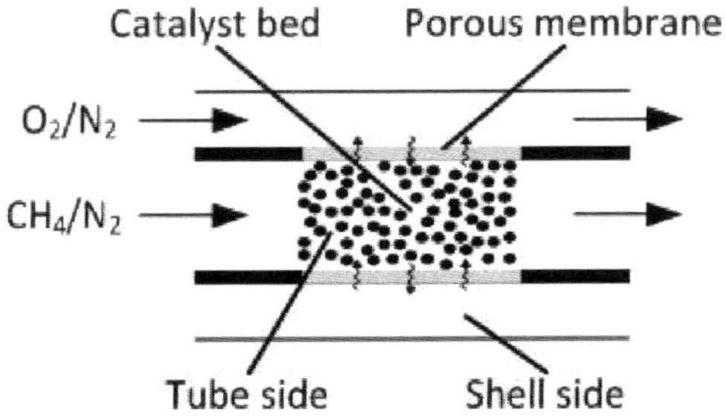

Figure 1: Membrane reactor

A kinetic scheme by Liu et al. for a widely used Li/MgO catalyst is incorporated in this model. The simplified network covers only the main species and does not include side products. Ethane and Ethylene are lumped as C_2H_6, the carbon oxides as CO_2. A scheme is shown in figure 2.

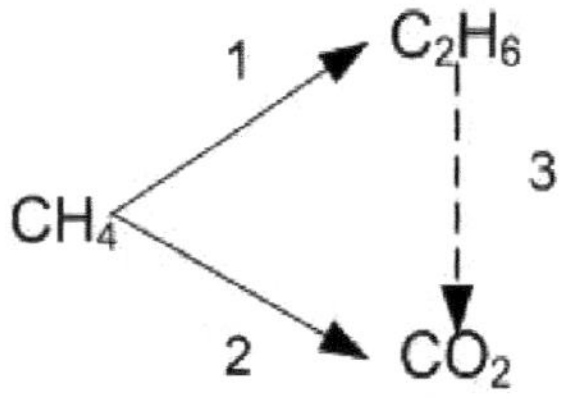

Figure 2: OCM kinetic scheme [6]

Homogeneous reactions in the reactor shell side, including gas phase coupling and combustion, are considered with a model by Lane et al. [7].

Individual mass balances are solved for shell and tube side for each species [Eq.1, 2].

$$\frac{\partial c_{i,shell}}{\partial z} = \frac{1}{v_z}\left(\sum_i r_{i,gas} + J_i 2\pi r_t / A_t\right) \text{ [Eq. 1]}$$

$$\frac{\partial c_{i,tube}}{\partial z} = \frac{1}{v_z}\left(\sum_i r_{i,catalytic} - J_i 2\pi r_t / A_t\right) \text{ [Eq. 2]}$$

The balances are coupled by mass transport through the mesoporous membrane, which is governed by Knudsen diffusion. The effective permeability J [mol/m^{2}s] combines membrane properties and the driving force for the mass transport, the partial pressure difference between reactor tube and shell [Eq.4].

$$J_i = \frac{D_{i,eff}(p_{i,shell} - p_{i,tube})}{RT d_M} \text{ [Eq.4]}$$

The membrane properties are summarized in table 2.

Table 2: Membrane properties

Material	α-alumina	**Porosity [-]**	0.5
Average pore size [m]	$2.2*10^{-9}$	**Tortuosity [-]**	2.95

The pressure drop in the packed bed of catalyst is considered according to Ergun equation [Eq.3].

$$\frac{dp}{dz} = -\frac{\rho_G u_0^2}{d_K}\frac{(1-\varepsilon_G)}{\varepsilon_G^3}\left[\frac{150(1-\varepsilon_G)}{Re} + 1{,}75\right] \text{[Eq. 3]}$$

The resulting set of differential equations is solved numerically using an adaptive Runge-Kutta scheme.

2.2 *Procedure and assessment*

2.2.1 Flammability limit of gas mixture

Knowledge about ignition behavior at OCM operating conditions is sparse [4]. No data is available that reflects the entire set of conditions in the reactor e.g. flow conditions, reactor internals and local mixing of the gases. Upper (UFL) and lower flammability limits (LFL) for mixtures of methane and oxygen over a wide temperature range are shown in figure 3 [8].

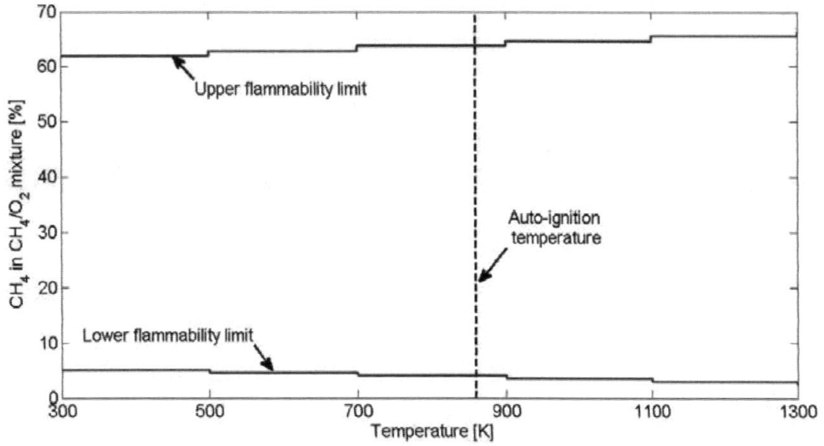

Figure 3: Flammability limits of undiluted methane-oxygen mixtures at 101.3kPa [8]

Data for C_2-hydrocarbons was not found for these conditions. Permeation of these species is small in comparison to methane transport, while the C_2-hydrocarbons are more likely to undergo spontaneous combustion in the gas phase. C_2-components and methane in the shell are thus lumped as hydrocarbons.

OCM reactors are commonly operated at atmospheric pressure with temperatures up to 950°C. In the membrane reactor setup, the LFL is of special interest, as the permeating hydrocarbons mix into the oxygen-rich atmosphere within the reactor shell. It is almost independent of the ratio of oxygen to nitrogen [9]. The main reason is that an excess of oxygen is present under these conditions. Moreover, similar heat capacities of nitrogen and oxygen lead to equal combustion

temperatures for varying compositions. The value at 1300K of xCH_4=2mol-% is assumed as a reference value for the LFL over the entire temperature range under consideration.
Generally, higher hydrocarbon concentrations in the shell side do not necessarily lead to an ignition of the gas mixture. They do however reflect a critical situation where an ignition is generally possible.

2.2.3 Procedure

Reactor operating parameters were varied to determine the key factors for the occurrence of hydrocarbon accumulation inside the reactor shell. The model variables are shown in figure 4.

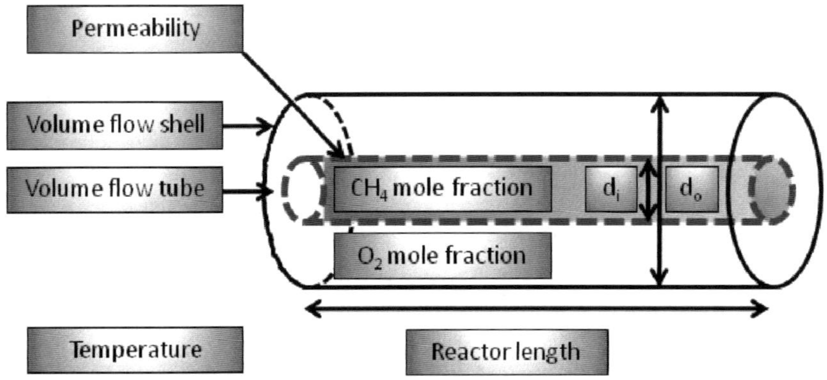

Figure 4: Model variables

The parameter range was chosen in a way to cover a broad range of operating conditions. The data range is listed in table 3.

Table 3: Parameter range

Volume flow tube (STP) v_{tube} [cm³/s]	5-45	**CH_4 / O_2 molar inlet fraction [mol-%]**	0-1
Volume flow shell (STP) v_{shell} [cm³/s]	5-45	**Reactor length [m]**	0-1
Inlet temperature [K]	973-1173	**Tube diameter d_i [cm]**	0.5-1.4
Membrane thickness [m]	10^{-4}-10^{-6}	**Shell diameter d_o [cm]**	1.8-3.2

Typical OCM membrane reactors range from 10 to 20cm in length to ensure low residence time at the extreme operating conditions [10, 11]. A value of 1m in length is chosen as the maximum to cover all possible setups. All other values reflect typical OCM membrane reactor conditions and geometry.
Reactor simulations were performed for the entire parameter range given in table 3. Simulations were aborted when methane was completely converted or when the maximum reactor length was reached.

3. Results and discussion

3.1 Hazard analysis

A hazard and operability study (Hazop) was performed for a membrane reactor system to identify critical operating parameters. The separate reactant feed and treatment reduces the explosion hazard in a way that both, premixing and joint preheating of the gases are avoided.
Direct contact of the gases is nevertheless possible if a leak forms or a total rupture of the membrane occurs. Even though the ceramic materials employed are very resistant towards the high temperatures, they are fragile to exertion of mechanic stress. The risk of membrane damage can be reduced by appropriate reactor design, including membrane bearing and sealing, but damages cannot be eliminated entirely. An explosion-pressure resistant design with flame arresters and non-return valves for supply and product lines is therefore indispensable.
This does however not rule out the necessity to avoid explosions inside the reactor during periods of normal operation. A minor pressure peak, caused by an ignition, can already damage the membrane. This can lead to feed gas mixing, which necessitates a complete shutdown and renewal of the system.

3.2 Shell gas phase composition

To avoid hazardous accumulations, an understanding of the position of the maximum accumulation, as well as on its magnitude is needed. Hence, the following analysis is split into two parts. The axial hydrocarbon profiles in the reactor shell are compared first. Subsequently, an analysis of the global hydrocarbon maximum allows to constrain the operability range.

<u>3.2.1 Axial concentration profiles</u>

Hydrocarbon concentration profiles in the reactor shell are shown in figure 5. All profiles show a steep increase near the reactor inlet, where the partial pressure difference, as the driving force for mass transport, is highest. In general, two main types can be distinguished: One the one hand, there is a profile with a distinct global maximum, where the hydrocarbon content declines towards the reactor outlet (5b). On the other hand, one can identify a profile, where the steep increase is followed by an almost constant, in some cases even strictly increasing, level (5a)

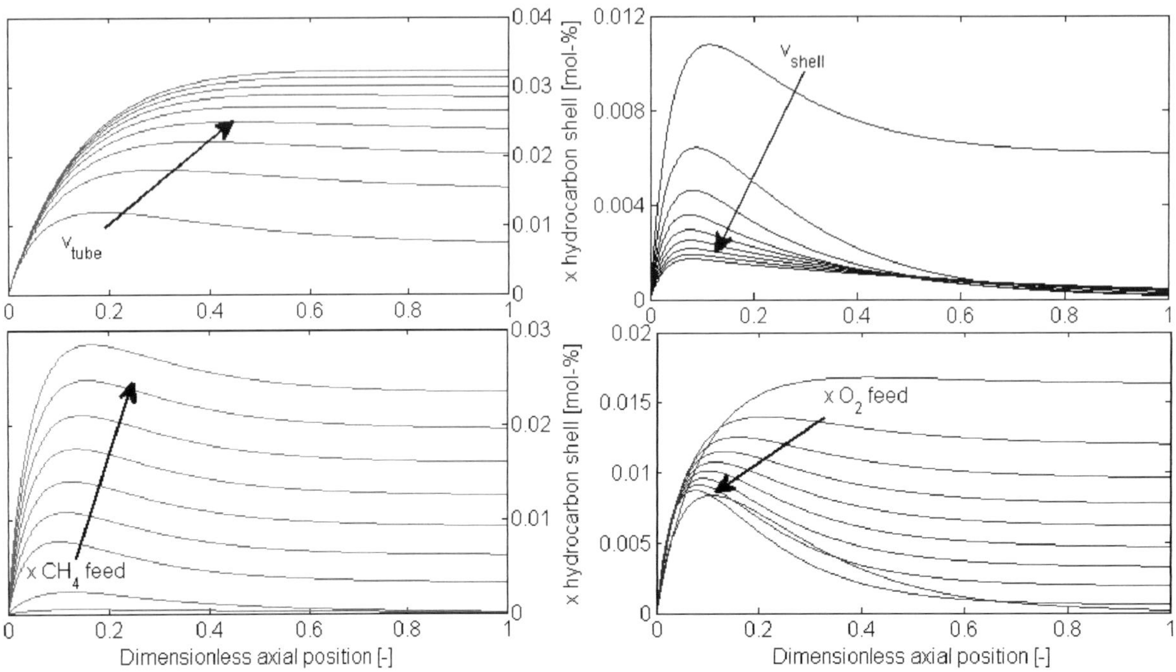

Figure 5: Influence of v_{tube} (a), v_{shell} (b), xCH_4 (c) and xO_2 (d) on shell concentration profiles

An increase in the tube flow rate leads to a higher hydrocarbon content inside the reactor shell (5a). A transition occurs between a profile with a distinct maximum and a monotonically increasing one. As v_{tube} is increased, the methane partial pressure gradient across the membrane remains high over a longer portion of the bed.

The flow rate in the reactor shell has an inverse effect on the accumulation (5b). An increase leads to a reduction of the maximal value. Then, two effects occur: On the one hand, the permeating hydrocarbons are diluted in the sweep gas. On the other hand, more fresh gas is available at any axial position to convert the hydrocarbons. The location of the maximum is only slightly shifted towards the reactor inlet, while the basic shape remains unchanged.

An increase in methane feed concentration yields in higher accumulation inside the reactor shell, as the driving force for mass transport through the membrane is increased. A global maximum is reached in the front part of the reactor, which is moved more towards the middle at higher values. The outlet concentration changes linearly with feed content, since the amount of hydrocarbons that is depleted by gas phase reactions remains almost constant.

A similar effect to a decrease in tube flow rate occurs, when the oxygen content in the shell feed is raised (5d). At low values, the accumulation reaches a maximum and is not considerably reduced any more. At higher values, a distinct global maximum forms as some hydrocarbons are consumed in the rear part of the reactor. In all of the cases shown, only one maximal value exists.

3.2.2 Maximal accumulation in reactor shell

In order to operate the membrane reactor without the risk of an ignition during normal operation, the maximum of the shell side hydrocarbon concentration must remain below the lower explosion limit. Thus, only the maximum value is analyzed in the following section. Contour plots of the maximum hydrocarbon concentration in the shell are shown as function of two variables which are altered over their entire operation range. All other parameters are kept constant for these considerations so that the results lead to a statement that combines the individual parameter values with the reactor behavior over a wider range of conditions.

Figure 6 shows the influence of methane and oxygen feed inlet molar fractions (left) and tube and shell flow rates (right). A strong influence of the methane feed content can be derived. The mass transport is driven by the gradient across the membrane and increases with methane concentration. The influence of the oxygen feed fraction is comparatively small. Lower hydrocarbon concentrations are attained, when higher values are employed. This effect can be attributed solely to an increased portion of gas phase reactions that occur inside the reactor shell.

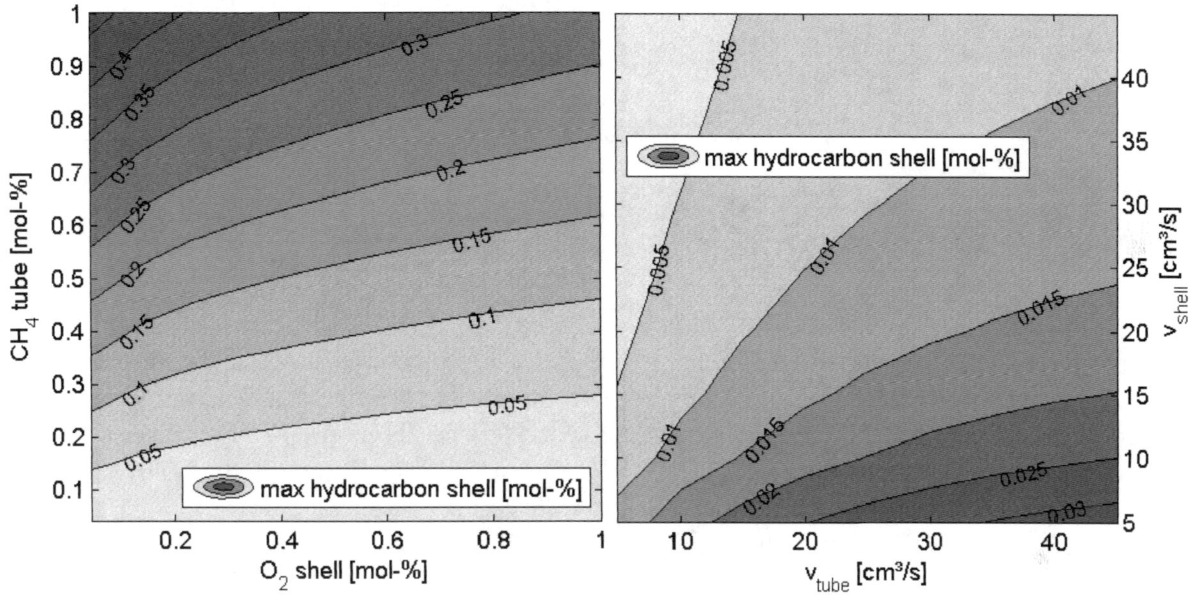

Figure 6: Influence of CH_4 and O_2 inlet molar fractions (a) and tube and shell flow rates (b)

Figure 6b reveals a strong dependency of the accumulation on the shell side volume flow. Increased tube flow rates lead to an increase of the accumulation in the shell. Thus, high flow rates in the shell and low flow rates in the tube are beneficial for the reactor operation.

Figure 7a shows the effect of the reactor geometry. A change in the diameters results in varying contact times. Tube and shell side diameters are therefore altered in a way that the ratios of the volumetric flow rate to the cross-sectional area, v_t/A_t and v_{sh}/A_{sh}, are kept constant in order to stress only on the effect of the manipulated variable. The tube diameter directly affects the membrane surface that is available for gas permeation. Thus, an increase results in higher permeation and more hydrocarbons in the reactor shell. This limits the maximal inner diameter of a membrane reactor.

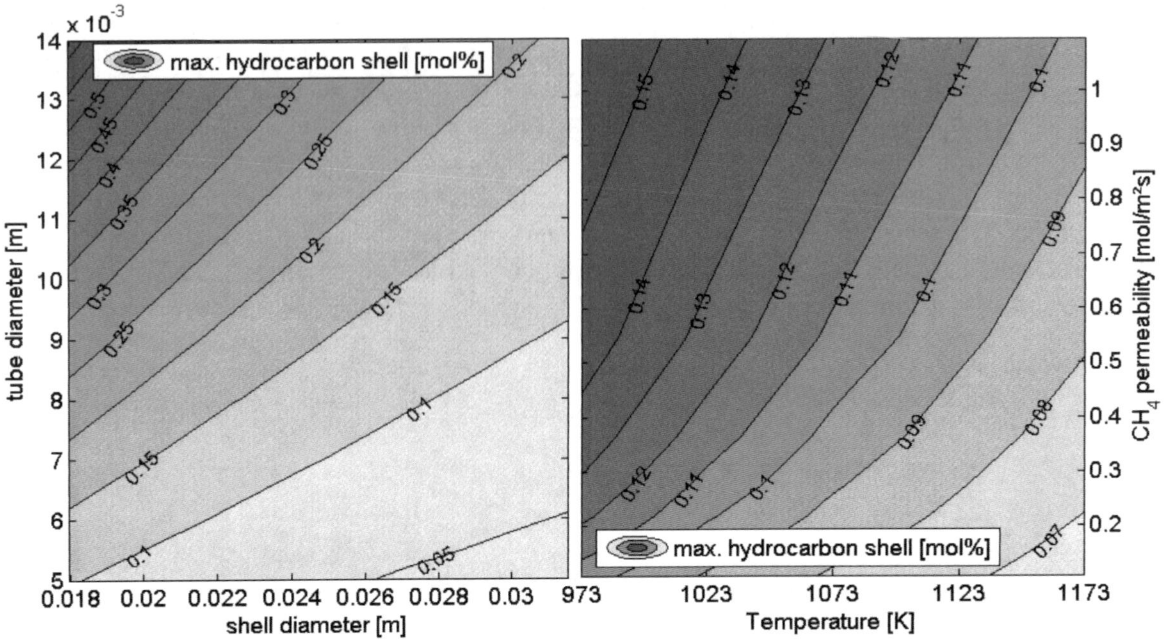

Figure 7: Influence of tube and shell diameter (a), CH_4 permeability and temperature (b)

A larger reactor shell results in an increase of the gas phase reactions and a higher dilution of the permeating hydrocarbons. This goes along with an increase in the free gas volume. In case of an ignition inside the reactor, there would be more space left for flame propagation with a possibly higher resulting explosion pressure. A possible measure is the filling of the shell with inert particles that can reduce the probability of flame propagation.

The effect of operating temperature and methane permeability, the latter estimated at inlet conditions, is shown in figure 7b. The permeability has a strong effect on the maximum, as it is a direct measure for the mass transport through the membrane. High values lead to increased accumulation.

The temperature affects various processes that occur in the reactor, e.g. the flow velocities in the tube and the shell, the reaction rate as well as the mass transport across the membrane. Higher temperatures lead to lower accumulation, which shows that the reducing effects prevail. Depending on the choice of operating conditions, a wide range of gas mixtures can form in the reactor shell. It is shown that the lower flammability limit is passed in many cases leading to situations where an explosion can occur on the membrane reactor in normal operation.

3.3 Maximum reactor length

A different strategy, than manipulating reaction conditions to avoid flammable mixtures, is to build a short reactor in a way that a critical accumulation is not exceeded before the outlet is reached. To gain a maximum degree of freedom concerning the choice of operating conditions, a setup is needed, where inlet concentrations and flow rates can be varied over the entire range of conditions without reaching the flammability limit in the reactor shell.

The worst conditions have been shown to occur at high tube flow rates, low shell flow rates, high methane and low oxygen feed content. These values are thus set to their minimum respectively maximum value in the following section. The reactor temperature has been shown to have a reducing effect on the shell side hydrocarbon concentration. It is fixed at the minimum value of

973K. Now, reactor inner and outer diameter and maximal methane permeability are varied to estimate, at which reactor length the threshold value is reached. Figure 8 shows contour plots of reactors lengths at which the threshold value is reached.

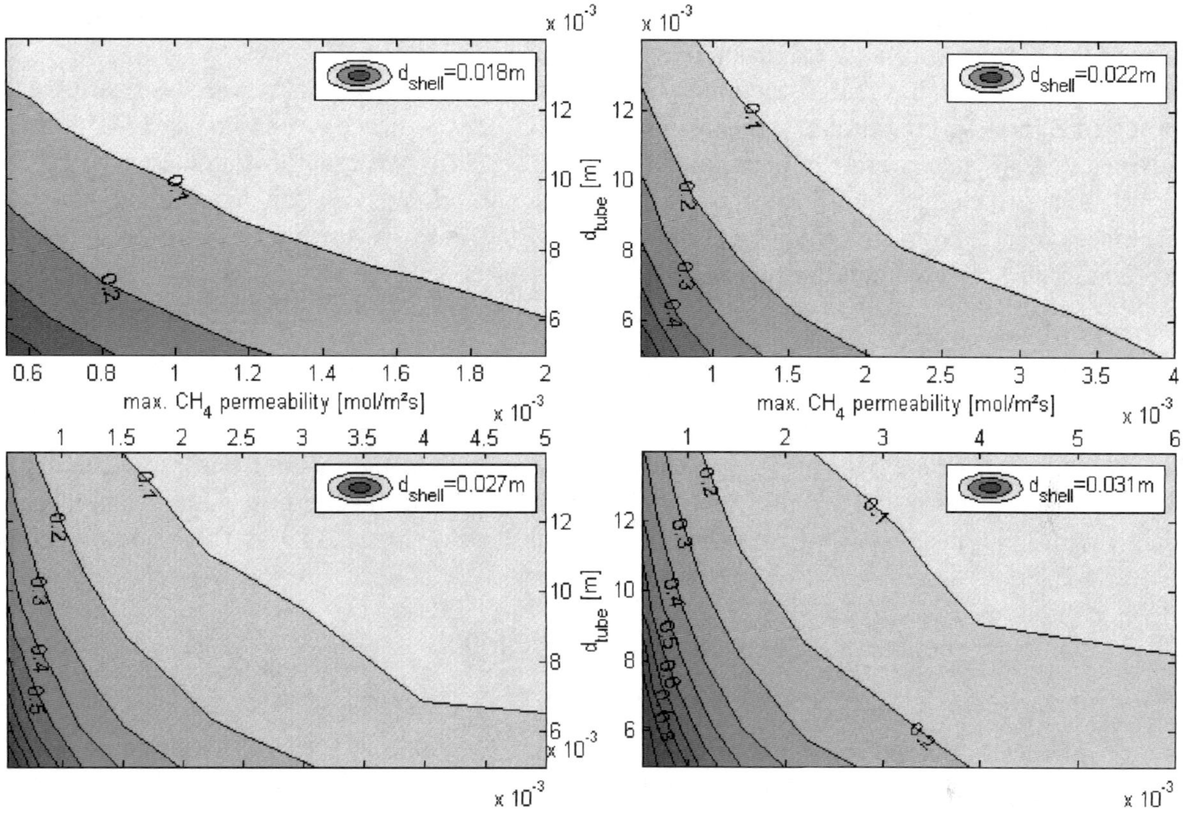

Figure 8: Maximum reactor length [m] until $x_{hydrocarbon}$=2mol-% is reached

The four graphs allow to determine a combination of four variables of the reactor, the inner and the outer diameter, the reactor length and the maximal methane permeability. Any reactor length, from very short beds up to 1m, can be realized when the three other parameters are chosen appropriately. The indicated reactor lengths are maximum values and a safety margin is not incorporated.

4. Conclusion

Membrane reactors are often describes as inherently safer reaction devices, concerning explosion hazards, in comparison to packed-bed reactors with co-feed operation. This is attributed to the fact, that the feed gases are not premixed before they enter the reactor and are contacted only by transport across the membrane.

Nevertheless, our analysis shows that some constraints exist for membrane reactor operation. Leakages and membrane rupture can lead to direct gas contact in the hot part of the reactor. Explosion-proof design is therefore indispensable.

Moreover, it is shown that permeation of hydrocarbons through the non-selective porous membrane can lead to the formation of flammable gas mixtures. This effect can be controlled when operating conditions and reactor geometry are chosen appropriately.

High shell flow rates and high oxygen feed contents lead to a reduced hydrocarbon accumulation. On the other hand, high methane contents and high tube flow rates lead to an increase in permeation driving force leading to higher shell side concentrations. An increase of the reactor temperature appears to be beneficial, since it favors shell gas phase reactions that convert the permeating gases to carbon oxides.
To determine a suitable reactor geometry with a maximum degree of freedom concerning the choice of operating conditions, a scenario is defined under which hydrocarbon concentration is maximal. These values are the base for the variation of the two reactor diameters, the reactor length and the methane permeability. A graphical approach is chosen to combine the four quantities for OCM membrane reactors in a way that the lower flammability limit is not exceeded in the reactor shell during normal operation.

5. References

This work is part of the Cluster of Excellence "Unifying Concepts in Catalysis" coordinated by the Technische Universität Berlin. Financial support by the Deutsche Forschungsgemeinschaft (DFG) within the framework of the German Initiative for Excellence is gratefully acknowledged (EXC 314).

[1] Zaman, J. Fuel Process. Technol., 58, pp61-81, 1999.

[2] Stansch, Z; Mleczko, L; Baerns, M, Ind. Eng. Chem. Res. 36, 2568-2579, 1997.

[3] J. Coronas, J. Santamaria, M. Menendez, Chem. Eng. Sci. 49, 2015, 1994.

[4] Holst N; Khakpour T; Holtappels K; Steinbach J: 2011 AIChE Spring Meeting & 7th Global Congress on Process Safety Chicago, IL March 13-16, 2011.

[5] Stuenkel S; Holst N; Repke J; Steinbach J; Schomäcker R; Wozny G. PSE'09 August 16-20, Salvador-Bahia, 2009.

[6] S Liu, X Tan, and R Hughes. Catalysis Reviews, 43((1&2)): pp147–198, 2001.

[7] Lane, G; Wolf, E. J. Catalysis, 113, pp144-163, 1988.

[8] Zabetakis, MG, Bureau of mines, Bulletin 627, 1965.

[9] Bartknecht, W. Springer ISBN 978-0387554648, 1993.

[10] Lafarga D; Santamaria J; Menendez M, Chem. Eng. Sci. 49, 2005, 1994.

[11] Kao YK; Lei L; Lin YS, Catalysis Today 82 pp255–273, 2003.

Synthesis of Nonylphenol Propoxylate Sulfates and Nonylphenol Propoxylate *Φ*-Sulfonates

Meijun Zhang, Zhiqiang He, Yun Fang*

School of Chemical & Material Engineering, Jiangnan University, Wuxi, Jiangsu 214122, China

Abstract

A series of nonylphenol propoxylate sulfates (NPS) and nonylphenol propoxylate *Φ*-sulfonates (NPSO) was synthesized from nonylphenol by propoxylation using propylene oxide (PO) and followed by sulfation in the end hydroxyl or sulfonation at the phenyl ring just using chlorosulfonic acid as sulfonating agent in different media and different temperatures. The propoxylation of nonylphenol was performed using $Ba(OH)_2$ as catalyst and CaO as co-catalyst, and the propoxylates with narrow distribution of PO adduct number 3, 5, 7 and 9 were obtained. When $ClSO_3H{\cdot}DMF$ complex used as mild sulfonating agent under lower temperature the reaction would only be seen in the end hydroxyl of propoxylates to form target product NPS instead of conventional sulfamic acid route. While $ClSO_3H$ used as sulfonating agent in 1, 2-dichloroethane under higher temperature the reaction could be found both in the end hydroxyl of propoxylates and at the benzyl ring. Then the reaction mixture of sulfate and *Φ*-sulfonate was hydrolyzed in acidic condition and the stable target product NPSO was gotten. The relationship of structure-performance of NPS and NPSO was measured and compared with each other.

Keywords: Nonylphenol, Nonylphenol propoxylate, Nonylphenol propoxylate sulfate, Nonylphenol propoxylate *Φ*-sulfonate

----

*Corresponding author: yunfang@126.com.

(75a) PSA In Biomass to Fischer-Tropsch Fuels Processes with CO_2 Capture

Ana Mafalda Ribeiro, Jo„o Carlos Santos and Alírio Egídio Rodrigues, Laboratory of Separation and Reaction Engineering, Department of Chemical Engineering, Faculty of Engineering, University of Porto, 4200-465 Porto, Portugal

The use of biomass as a source of renewable energy has received much attention lately. The conversion of biomass into valuable energy can be accomplished through a variety of technologies from which gasification appears to be the most promising and efficient one. Gasification converts biomass into synthesis gas that can be further processed into a wide range of high value products such as methanol, hydrogen and Fischer-Tropsch products. However, prior to the synthesis process, the bio-syngas must be conditioned, which includes the removal of carbon dioxide for subsequent sequestration and capture. The most common technologies used for CO_2 removal from syngas are absorbent-based systems such as the Rectisol and Selexol processes. In this study a PSA process is suggested for the replacement of the absorbent-based CO_2 removal process. A pressure swing adsorption cycle was developed for the simultaneous production of a carbon dioxide enriched stream and a (H2+CO) stream with the required specifications of the Fischer-Tropsch reactor feed. An activated carbon was selected as adsorbent. Simulation results showed that it was possible to obtain a CO_2 product with purity above 95% and with 90% recovery. Additionally a H_2/CO stoichiometric ratio of 2.14 was obtained for the produced syngas stream which is adequate for further processing in the Fischer-Tropsch reactor. The power consumption of the CO_2 capture was found to be of 3.36 MW. This value represents a reduction of about 28 % when compared to a Rectisol process with the same CO_2 recovery.

See more of this Session: Gas Separation Technologies
See more of this Group/Topical: Topical 6: 11th Topical Conference on Gas Utilization

(75b) Modified 'Molecular-Gate' PSA for CO_2 Capture From Flue Gas

Raghwendra Singh Thakur[1], Nitin Kaistha[1] and D. P. Rao[2], (1)Department of Chemical Engineering, Indian Institute of Technology Kanpur, Kanpur, India, (2)Process Intensification Consultant, Hyderabad, India

Keller II and Kuo in 1982 proposed a dual-piston Pressure Swing Adsorption (PSA) for separation of mixtures and termed it as *molecular gate* PSA. It consisted of a single adsorbent bed connected with two piston-cylinder assemblies at the ends of the bed to impose cyclic flow of gas with feed entering at an intermediate location along the bed. High and low pressures in the bed are varied by manipulating the displacement volume of pistons. They claim that high purities of both products in the range 99 mol% and high productivity can be achieved at moderate vacuum. However, its application is limited by the size of piston-cylinder assembly that could be used in practice, for example, for CO_2 capture from flue gas. To overcome this limitation, we propose a modified molecular gate PSA replacing the piston-cylinders with two surge tanks, a compressor, a vacuum pump and mass flow controllers. The cycle essentially has only two steps, pressurization and depressurization in which cyclic flow of gas is maintained through the bed. The lead/lag functionality of pistons in molecular gate PSA has been realized using the mass flow controllers. An additional modification of molecular gate has been proposed in timing products withdrawal by passing tanks during the steps to enhance purity and productivity. We have carried out simulation studies on separation of synthetic flue gas containing 20 mol% CO_2 and 80 mol% N_2, using Zeolite 13x. The results indicate that it is possible to get CO_2 purity and recovery in excess of 99% with high pressure of 1 bar and low pressure 0.08 bar with productivity of 43.9 LSTP/h/kg a performance superior to the performance of conventional PSA. The productivity can be further enhanced by 3.5 times upon relaxing the purity specification from 99 mol % to 95 mol%.

Keywords: modified molecular gate, CO_2 capture, flue gas

Reference:

Keller II, G.E. and Kuo, C.H.A.: Enhanced Gas Separation by Selective Adsorption, US Patent 4,354,859 (1982).

See more of this Session: Gas Separation Technologies

See more of this Group/Topical: Topical 6: 11th Topical Conference on Gas Utilization

PSA SEPARATION OF CH4/N2 MIXTURE WITH CARBON MOLECULAR SIEVE

Fang Xi, Wensheng Lin, Anzhong Gu*

*(*Corresponding author: linwsh@sjtu.edu.cn)*

Institute of Refrigeration and Cryogenics, Shanghai Jiao Tong University, Shanghai, China

Abstract

Coalbed methane (CBM) is a kind of unconventional natural gas, and liquefaction is a good alternative to make use of it. In China, CBM often contains air when it comes out of the well. Considering issues regarding safety, it is natural that oxygen must be removed firstly. Then, the effective enrichment of methane from methane-nitrogen mixture is one of the most important issues in the liquefaction of CBM. In this paper, a two-bed pressure swing adsorption (PSA) experimental apparatus was used to separate CH_4/N_2 mixture on a selected type of carbon molecular sieve (CMS) adsorbent. Owing to the kinetic separation mechanism on CMS, nitrogen is adsorbed on CMS, while methane with certain pressure is delivered continuously to the next process of liquefaction. The influence of adsorption pressure, feed gas velocity and cycle time on the overall performance of the adsorption process was discussed. Feeding a mixture of 50% CH_4/50% N_2, methane purity of 95.45% was obtained; and 94.89% purity was obtained while feeding a mixture of 30% CH_4/70% N_2. The results show that all three parameters may affect the separation performance, and the later two have greater influences. It is easier to achieve methane purity above 90% at low adsorption pressure and low feed gas velocity. And the shorter the cycle time, the higher the methane purity is.

Key words: methane-nitrogen mixture; pressure swing adsorption; carbon molecular sieve; coal-bed methane

Introduction

Methane contained in the coal bed methane (CBM) is a kind of clean fuel, but its emission to atmosphere would contribute to the greenhouse effect. The methane content in CBM is usually only 20% to 40% due to the mixing of air during the exploitation process[1]. In spite of the large coal production, CBM is not well utilized in China. Besides its low methane content, CBM wells are often remote and the gas production of a single well is so small that delivering it through pipeline is economically unpractical. An effective way of utilizing CBM is to deliver it to local natural gas supply systems after liquefaction[2].

For oxygen mixed in CBM, it must be removed firstly considering the safety issue. So the major and difficult task of utilization CBM is to separate methane from nitrogen economically. Although cryogenic distillation is a mature application, it usually needs higher energy consumption for the undesired liquefaction of nitrogen. An alternative way of separating the CH_4/N_2 mixture is pressure swing adsorption (PSA) process by using the mechanism of kinetic selectivity to produce enriched methane with certain pressure.

Principally, a kinetic separation is achieved by virtue of the difference in diffusion rates which is higher for the smaller nitrogen molecular compared to methane. Although simple in principle, it is a relatively difficult separation compared to an equilibrium separation because of the low values of the diffusion time constants and the adverse equilibrium selectivity towards methane and nitrogen. Ackley [3] enriched methane from 50% to 80% by vacuum pressure swing adsorption (VPSA). However, it is hard to enrich methane up to 90% through kinetic PSA while the methane content is lower than 50% in the feed gas until now[4]. The selection of appropriate adsorbent is crucial to the successful separation of methane and nitrogen.

In our recent study[5], single bed kinetic PSA separation experiments were carried out under a wide range of adsorption pressures(P), feed gas velocities(v) and different methane contents in feed gas with a selected carbon molecular sieve (CMS). The breakthrough curves showed that at the initial stage of adsorption there was only methane at the outlet of adsorption bed, indicating that methane could be enriched up to 90% with a single PSA unit.

In this paper, the same two simulated methane-nitrogen mixtures (30% and 50% methane, respectively) were separated using a two-bed PSA unit with the same CMS adsorbent. The experimental studies under various conditions of adsorption pressures, feed gas velocities and cycle times were investigated.

Experimental

A simple two-bed PSA process operated in a cyclic manner as shown in *Figure 1* comprises four basic steps: repressurization, feed, counter-current blowdown and counter-current vacuumization. High pressure feed flows through bed B where the faster diffusing nitrogen is preferentially adsorbed, leaving a stream richer in methane at the outlet of the bed after the regenerated bed repressurized with product. Vacuum regeneration is used to remove most of the nitrogen which has been adsorbed in order to avoid slow uptake of methane which would saturate the adsorbent and reduce its effectiveness for taking up nitrogen during the subsequent adsorption step.

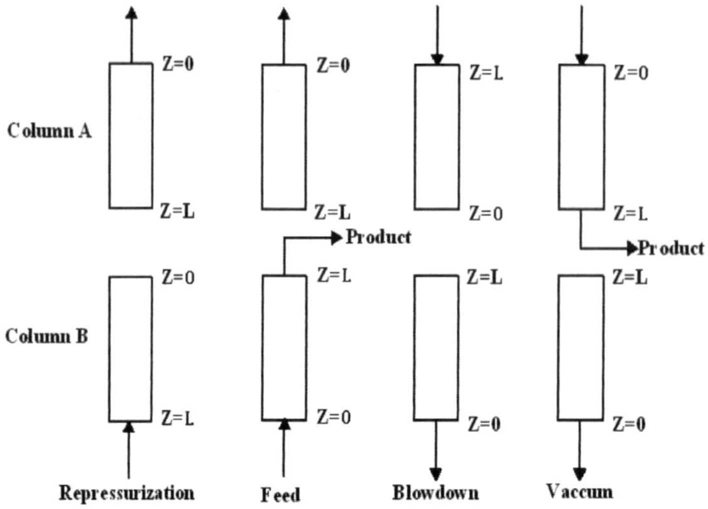

Figure 1 Schematic diagram of steps involved in PSA cycle

Table 1 Experimental conditions of the two-bed PSA unit

Feed gas composition	50%CH_4, 50%N_2
	or 30%CH_4, 70%N_2
Column length	30cm
Column diameter	3cm
Adsorbent	CMS
Mass of adsorbent in each column	136g
Particle size	φ1.5×3 cylinder
High pressure	0.3 or 0.6MPa(Abs.)
Velocity of feed gas	0.2 or 0.4 cm/s

The experimental unit is shown in *Figure 2*. The two stainless steel columns were packed with a known amout of CMS(Table 1), which was produced by Changxing Shanli Chemical Material Technology Co.,Ltd. Activation of CMS was done at 423K overnight under high vacuum conditon. Each column was connected to three pneumatic valves: one for the inlet of feed gas, one for the outlet of product in feed step and the inlet of product gas in repressurization step, and the other for the outlet of absorbed nitrogen in blowdown and vacuumization steps. The pneumatic valve, which was used to prestect the experimental apparatus from possible explosion, was driven by pressurized air. The circulation of the pressurized air was controlled by two-way solenoid valve. Sequential switching of solenoid valves was controlled by relays. An electronic mass flow controller (MFC)(±0.3%) was used to controll the feed gas flow. The product and feed stream are analysed by withdrawing gas samples through a manual sampling probe and submitting them to a GC900A chromatograph. The experimental unit was evacuated for 12 hours and purged with pure helium before each experiment. The pressure of the unit was controled by a back pressure valve.

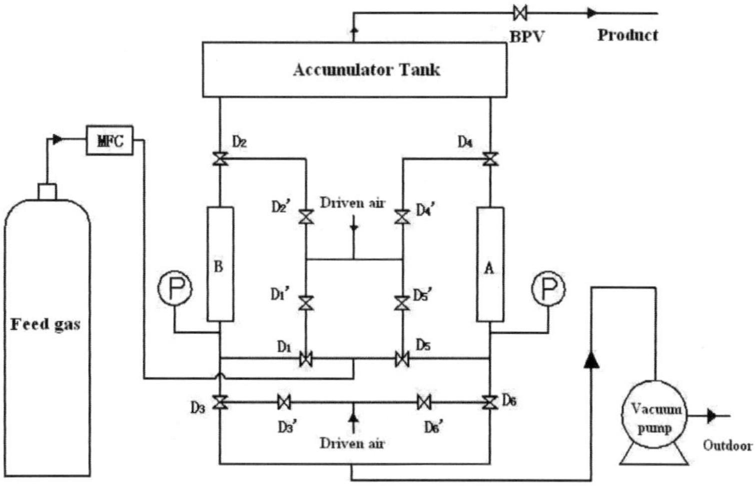

D_1, D_2, D_3, D_4, D_5, D_6: solenoid valve; D_1', D_2', D_3', D_4', D_5', D_6': pneumatic valve;

MFC: mass flow controller; A, B: column; BPV: back pressure valve; P: pressure gauge.

Figure 2 Schematic diagram of the two-bed PSA experimental unit

Results and discussion

It has been proved that velocity of feed gas, adsorption pressure, and cycle time can affect the performance of a PSA unit[6]. Our recent study[5] showed that the greater the velocity of feed gas was, the lower the methane purity was in product while the velocity was between 0.2-1.0cm/s. In this two-bed PSA research, velocities of 0.2 and 0.4cm/s were chosen according to the single-bed experiments. The adsorption pressures studied were 0.2 and 0.5MPa.

The unit was purged and pressurized by helium at the beginning of experiments, so there are two kinds of curves in the methane concentration plots as showed in *Figure 3*. The solid ones were the methane concentration in methane and nitrogen binary mixture(*CH4%2*) in buffer tank, manually eliminated helium. This profile represented the separation effect of the CMS directly. The hollow ones were the methane concentration in methane, nitrogen and helium ternary mixture($CH_4\%3$), which were the true methane concentration of the product in buffer tank, representing the productivity of adsorbent. In these two figures, *t* was the adsorption time of a single bed. All the concentration points correspond to those measured 2s prior to the end of each cycle.

The results show that the methane concentration is enriched to a certain extent under each condition. While *v* is 0.2cm/s, $CH_4\%2$ is up to 90% under all experimental conditions, and the highest one is 95.45%. But it is different while *v* is 0.4cm/s. $CH_4\%2$ is up to 90% only when the methane concentration in feed gas($CH_4\%1$) is 50%, and the highest one is 94.87%. Table 2 is the summary of experiments.

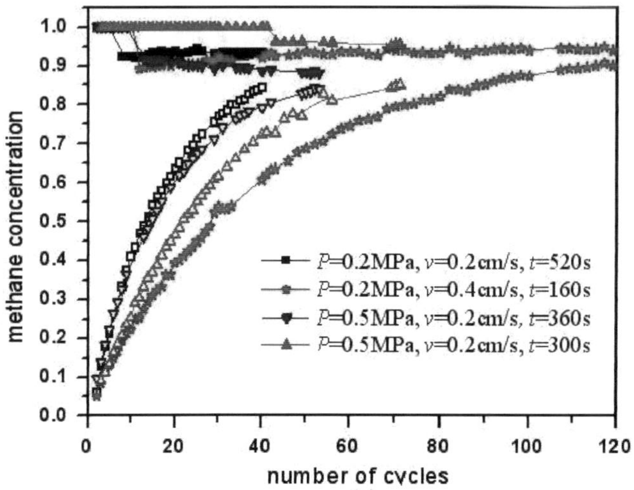

(a)

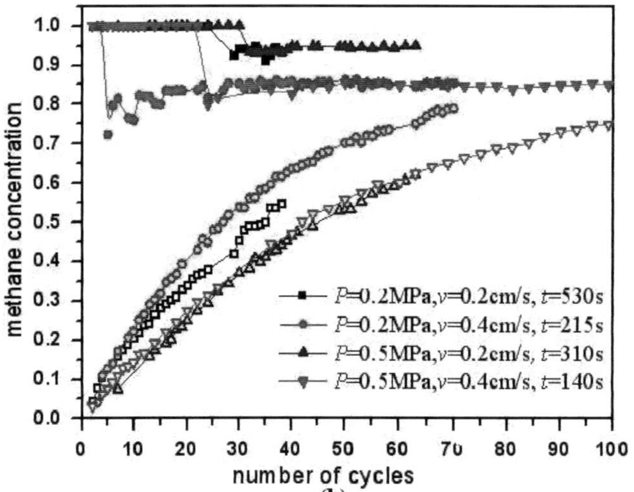

(b)

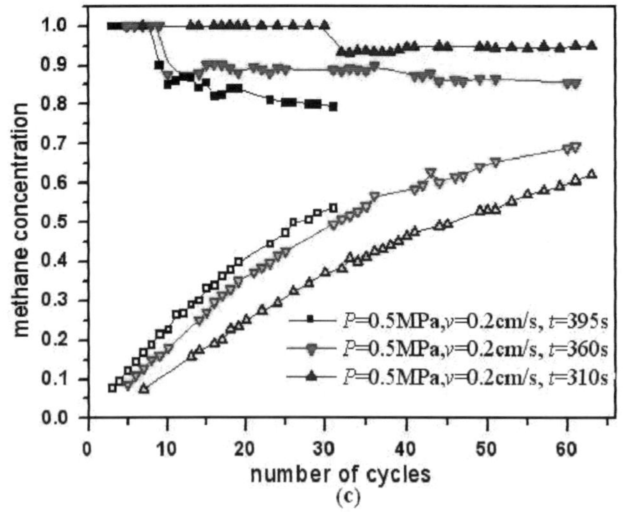

(c)

Figure 3 Methane concentration profiles while the methane concentration was 50%(a) and 30%(b, c) in feed gas balanced with nitrogen.

Table 2 Summary of two-bed PSA experiments with CMS (temperature: 300K)

P (MPa)	*v* (cm/s)	$CH_4\%1$ (V%)	*t* (s)	$CH_4\%2$ (V%)	$CH_4\%3$ (V%)
0.2	0.2	50	520	93.64	84.54
		30	530	93.23	54.76
	0.4	50	160	94.37	90.78
		30	215	85.67	79.05
0.5	0.2	50	360	88.04	84.21
			300	95.45	85.01
		30	395	79.53	53.62
			360	85.96	69.44
			310	94.89	62.18
	[illegible]	[illegible]	[illegible]	[illegible]	[illegible]

It has to be noted that the reason of the slow increase of $CH_4\%3$ is the much greater volume of buffer tank compared with the volume of column, in order to make sure that the pressure decline of the system after repressurization is small enough. It is interesting that $CH_4\%2$ is 100% at the beginning of the adsorption. In fact, the nitrogen extent is too small to be analyzed though GC900A. In the enlarged profiles from the chromatograph, there is a chromatographic peak of nitrogen.

It can be seen from the results that cycle time has a great influence on methane purity. Methane purity in the product is higher while the cycle time is shorter no matter what other experiment conditions are. The influences of adsorption pressure and feed gas velocity are the same as that in the single-bed study: the increases of both parameters have negative effects on the methane purity. As can be seen from the profiles, while the methane concentration in feed gas is 30%, the accumulation of methane in buffer tank is slower than that when methane concentration is 50%.

Conclusions

Experimental PSA studies were performed with two methane-nitrogen feeds (50% and 30% methane in nitrogen) using a two-bed unit packed with a selected CMS. A four-step PSA cycle was used. The exit methane concentration ranged from 95.45% to 88.04% for the 50% feed and from 94.89% to 79.53% for the 30% feed.

The cycle time, feed gas velocity and adsorption pressure could affect the methane purity, and the methane concentration in feed gas significantly influences the accumulation speed of methane in buffer tank. It is easier to achieve methane purity above 90% at low pressure and low velocity. And cycle time has a negative effect on methane purity.

The study shows that it is possible to enrich methane concentration from a relatively low level to a commercially acceptable one, implying potential application in CBM utilization practices.

References

[1] Nie LH , Xu SP , Su YM , Liu SQ. Progress of recovery of low concentration coal bed methane. Chemical Industry and Engineering Progress, 2008,27(10):1505-1511.

[2] Lin WS, GU M, GU AZ, et al. Analysis of coal bed methane enrichment and liquefaction processes in China. In: Proceedings of the 15th International Conference & Exhibition on Liquefied Natural Gas. Barcelona, Spain, 2007.

[3] Ackley M W, Yang R T. Kinetic separation by pressure swing adsorption: Method of characteristics model. AIChE Journal, 1990, 36 (8):1229-1238

[4] Guo P, Li M. Research progress of separation of CH_4/N_2 in coal-bed methane. Chemical Industry and Engineering Progress, 2008, 27(7):963-967.

[5] Xi F, Lin WS, Gu AZ. Separation of methane/nitrogen mixture by pressure swing adsorption on carbon molecular sieve. Journal of the China Coal Society, in press.

[6] Fatehi A. Separation of methane nitrogen mixtures by pressure swing adsorption using a carbon molecular sieve. Gas Separation & Purification, 1995,31(9):199-204.

(75d) Two-Dimensional Modeling the Transfer Phenomena In the Radial Adsorber

Xingang Zheng, **Yingshu Liu**, Wenhai Liu, Yongling Li and Hui Zhang, School of mechanical engineering, University of science and technology Beijing, Beijing, China

In this study, a two-dimensional model is used to study the transfer phenomena occurred in the radial adsorber with consideration of heat effect, gas compressibility and dead volumes. The distributions of concentration, temperature and velocity in four possible configurations, CP-z, Cp-Π, CF-z, and CF-Π of a radial flow adsorber are investigated. As expected, the results visually show the evolution of process variables and their distribution. Some discussions of the effect of different configurations on variables distribution are made. The two-dimensional simulation is found to be very useful for a better understanding of the dynamic sorption and separation process in the radial adsorber.

See more of this Session: Gas Separation Technologies
See more of this Group/Topical: Topical 6: 11th Topical Conference on Gas Utilization

EFFECTS OF HYDROGEN CONTENT ON NITROGEN EXPANSION LIQUEFACTION PROCESS OF COKE OVEN GAS

Lin Zhang, Wensheng Lin, Anzhong Gu*

*(*Corresponding author: linwsh@sjtu.edu.cn)*

Institute of Refrigeration and Cryogenics, Shanghai Jiao Tong University, Shanghai, China

Abstract

Coke oven gas (COG) is a by-product when producing coke from coal. It is not always easy to make use of COG, causing waste of energy. Producing liquefied natural gas (LNG) is an efficient way of utilizing COG. The amount of hydrogen affects liquefaction process significantly, because its thermal properties are quite different from the other compositions (methane, carbon monoxide, etc.) of COG. Based on nitrogen expansion liquefaction process, this paper simulates a series of liquefaction processes of COG containing different amount of hydrogen. After optimizing these processes and comparing their parameters, it turns out that the hydrogen content exerts a great influence on the unit power consumption and the liquefaction rate of the processes; however, its effect on the product purity is not distinct. Furthermore, the liquefaction process is integrated with distillation separation of hydrogen to upgrade the quality of LNG. The ways of achieving energy integration between liquefaction and distillation are suggested, and the influences of some important parameters, such as the highest pressure of nitrogen cycle, are analyzed.

Key words: liquefied natural gas (LNG); coke oven gas; nitrogen expansion liquefaction process; unit power consumption; distillation

Introduction

The coke oven gas (COG), a by-product from coking plants, is emitted at a level of 300~350Nm^3/t of coal [1]. Along with the development of the coke industry, a large amount of COG has not been utilized properly[2]. In China, for example, there is about 70×10^9 Nm^3 available COG each year according to the overall coke output in 2006[3]. But in fact, the utilization amount of COG is very small and at a very low level, this leads to serious environmental pollution and resource waste.

Recent years, many ways have been developed for the utilization of COG[4,5], but they are usually not efficient and not economical for utilizing small or medium scale of COG. In order to use this kind of COG, Yao[6] and Zhang[7] have developed a process for producing LNG from COG, and this process produce hydrogen synchronously.

When liquefying COG with the liquefaction process of natural gas, the difference between their compositions has an influence on the liquefaction process. Gao et al [8~10] have studied the effects of nitrogen on the liquefaction processes. The effects of CO are similar those of nitrogen, because their thermal properties are similar in many aspects. In a previous study, Zhang et al[11] have studied the effects of hydrogen on the liquefaction process of COG without distillation. Because of the huge difference of thermal properties between hydrogen and other components of COG, distillation may have some impact on the product quality and the unit power consumption of the process. In order to study the effects, a series of nitrogen expansion liquefaction processes of COG containing different amount of hydrogen at different methane recovery rates are simulated in this paper with HYSYS software.

Pre-Simulation

Before liquefaction, CO_2 and other impurities will be removed, and the remaining components are H_2, CH_4, CO and N_2. And H_2 and CH_4, the chief components of COG, account for about 55% and 28% of the total volume respectively [12].

As shown in Table 1, the thermal properties of hydrogen are very different from that of other components of COG, so, in principle, it is possible to separate CH_4 with hydrogen with liquefaction process only. However, very little hydrogen contained in product has great impact on the temperature of product. Fig. 1 shows the bubble point temperature of CH_4/H_2 mixture. When the hydrogen content goes up to 300ppm from 0ppm, the bubble point temperature decreases about 10°C. So removing hydrogen from product may help to increase the liquefaction temperature, and thus, to reduce the unit power consumption. Considering this, distillation is used together with the liquefaction process.

Because of the extremely low boiling point temperature of hydrogen, lower temperature is needed to liquefy COG than that of natural gas. As shown in Fig. 2, under different methane recovery rate (the molar flow rate ratio of methane in LNG product to methane in CBM feed gas, x), the temperature of COG before throttling decrease with the increase of hydrogen content. As a result, the nitrogen expansion liquefaction process is used in this research, because it can provide very low temperature.

Table 1 Thermal properties of the main components of COG

	H_2(n-H_2)	CH_4	CO	N_2
Critical point temperature(℃)	-240.0	-82.4	-140.2	-146.9
Critical point pressure(MPa)	1.315	4.64	3.499	3.398
Normal boiling point temperature(℃)	-252.8	-161.4	-191.5	-195.8

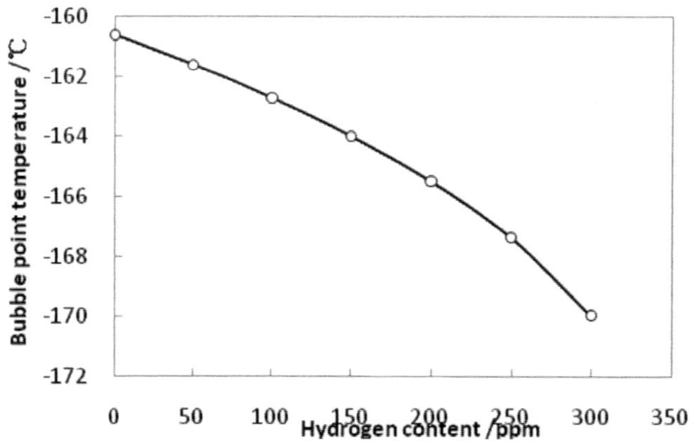

Fig.1 Bubble point temperature of CH_4/H_2 mixture

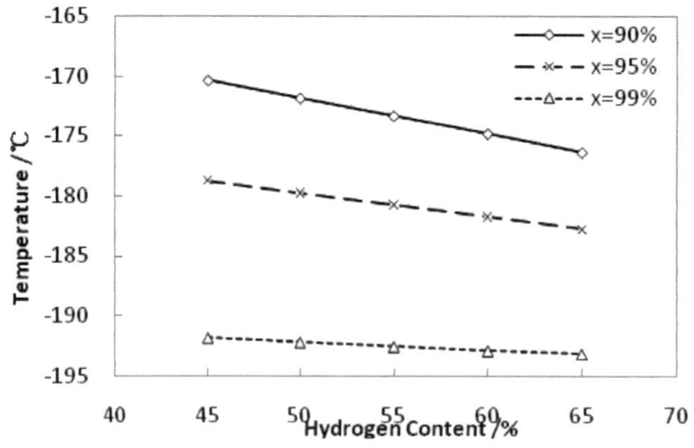

Fig. 2 Temperature of COG before throttling

SimulationDetails

The Liquefaction Process

The nitrogen liquefaction process is shown in Fig. 3. After two stages of compression and cooling by a water-cooler, feed gas is gradually cooled with nitrogen in two heat exchangers. After that, the pressure is reduced by throttling, and finally the LNG product is separated out in distillation tower. The gas from the distillation tower is diverted back into these two heat exchangers to recover its cold energy.

For the nitrogen cycle, nitrogen firstly undergoes two stages of compression and cooling by a water cooler, and then it is cooled in heat exchanger HEX-101. And after expansion and cooling in heat exchanger HEX-102, it provides heat energy for the reboiler. Finally, it supplies cold energy for the two heat exchangers after the second stage expansion.

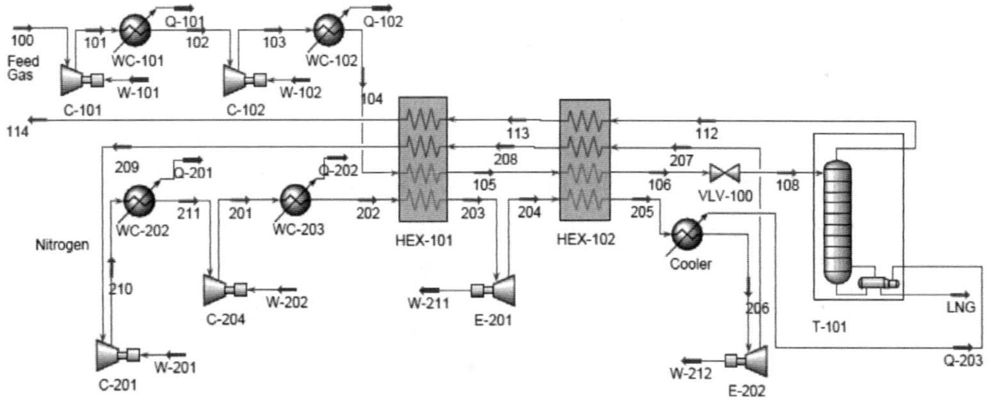

C – Compressor, E – Expander, W – Work of compressor or expander,

HEX – HeatExchanger, WC – Water Cooler, T – Distillation Tower

Fig. 3 The nitrogen expansion liquefaction process of COG

Parameter Setting

Some parameters in the process are firstly set for the simulation.

The initial temperature T_{100} and pressure p_{100} of feed gas are 35°C and 101.3kPa, respectively, and after two stages of compression its pressure p_{104} is 5.0MPa. The pressure of the product LNG p_{LNG} is 110kPa, and its temperature is decided by the composition, which is decided by the demanded methane recovery rate and the boil-up ratio of distillation.

Additionally, the adiabatic efficiency of compressor and expander are 0.85 and 0.8 respectively [13]. After cooling by a water cooler, the temperature of all gases goes down to 35°C. The pressure drop of heat exchangers are taken as 0kPa to simplify the simulation, and the minimum temperature difference is 5°C.

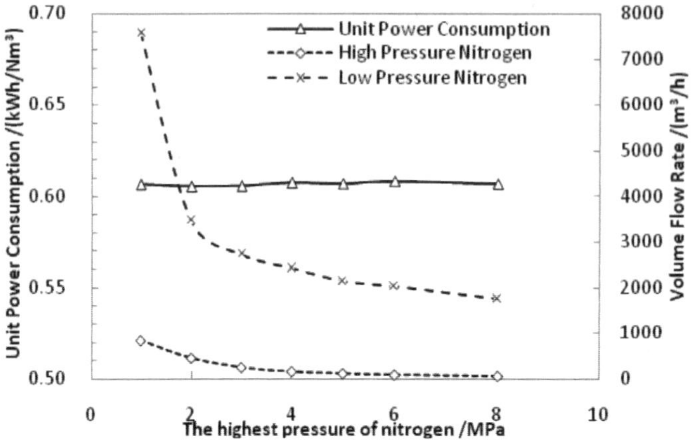

Fig. 4 Effects of the highest pressure of nitrogen on the process

Finally, the highest pressure of nitrogen cycle has a great influence on the size of devices. As shown in Fig. 4, the highest pressure of nitrogen has little effect on unit power consumption, but it has great influence on the volume flow rate of nitrogen, which dramatically affect the size of devices. To balance the effects of the pressure and the volume flow rate, the pressure $p_{202}$is set at 3.0MPa.

Simulation Range

After pretreatment, COG still contains some CO and N_2, which cannot be removed easily. Thus the feed gases in these simulations contain 3% (molar fraction, the same below) CO and 3% N_2.

This paper has simulated COG liquefaction processes with hydrogen content from 45% to 65% by the step of 5%, when the methane recovery rate is fixed at 92%, 95% and 98%, respectively.

Additionally, processes without distillation have also been simulated to demonstrate the effects of distillation on the processes, in the case that the methane recovery rate is 95%.

Results and Discussions

Unit Power Consumption (UPC)

As one of the most important factor of the process, the UPC is defined as follow:

$$UPC = \frac{W_{201} + W_{202} - W_{211} - W_{212}}{q_{v,LNG}}$$

Where, W_{201} and W_{202} are the input work of compressor C201 and C202, respectively, W_{211} and W_{212} are the output work of expander E201 and E202, respectively, and $q_{v,\mathrm{LNG}}$ is the gas flow rate of product at standard state.

All the processes are optimized to minimize the UPC, and the results are shown in Fig. 5. As shown in figure 5(a), when the methane recovery rate is fixed, the UPC increases with the increase of hydrogen content, and by comparing the distances of these three curves, it turns out that the UPC increases more rapidly with the rise of methane recovery rate. From the figure 5(b), we can obtain that the UPC of the process with distillation is about 10% lower than that of the process without distillation.

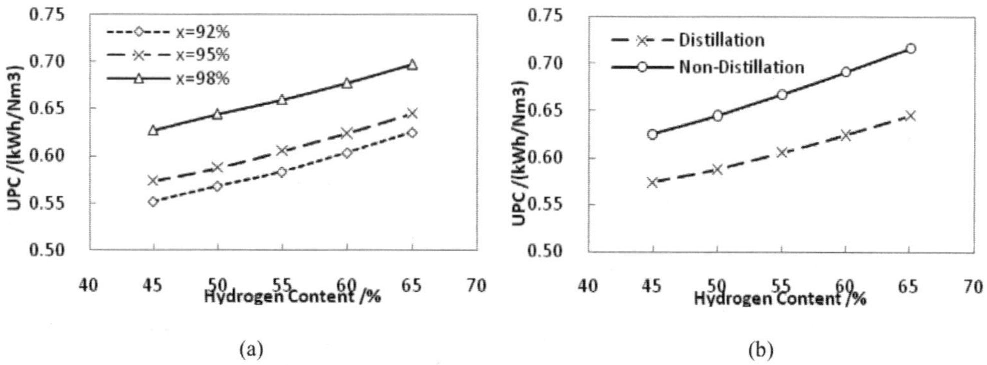

(a) TheUPC curves for different methane recovery rates

(b) The UPCsfor different separation ways when the methane recovery rate is 95%

Fig.5 Unit power consumption at different conditions

Product Quality

As the most important factor of the product, methane content is determined by the methane recovery rate, the methane content of feed gas and the inlet temperature of distillation column. Fig. 6 shows the calculation results. It indicates that the hydrogen content has no certain effects on the methane content of product when the processes are optimized. And on the other hand, the methane content generally decreases with the increase of methane recovery rate. Additionally, Fig. 6(b) shows that distillation improves the methane content of product slightly.

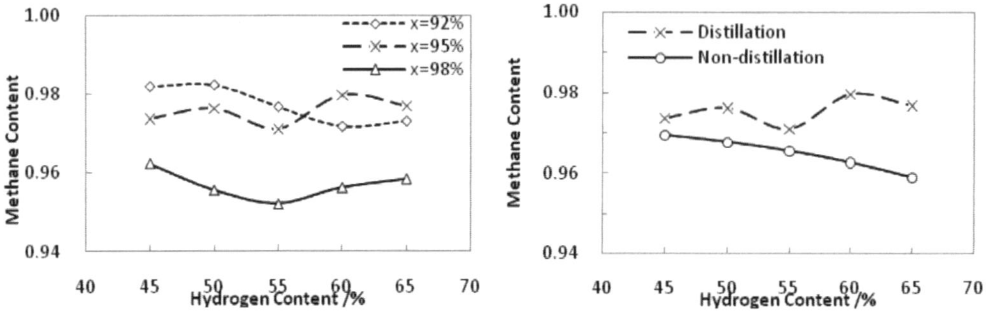

(a) (b)

(a) Methane content at different methane recovery rate

(b) Methane content at different separation ways when the methane recovery rate is 95%

Fig. 6 Methane content of product at different conditions

Furthermore, for the optimized processes, hydrogen content of the product decreases from 295ppm to 265ppm linearly, when the hydrogen content of feed gas increase from 45% to 65%. Thus, hydrogen is almost entirely separated out from product by distillation. The unit power consumption of the process with distillation is lower than that of the process without distillation.

Liquefaction Rate

Fig. 7 shows the liquefaction rate of the proceses. Although the liquefaction rate increases with the increase of methane recovery rate, the most influencing factor is hydrogen content. The effects of distillation and the methane recovery rate are negligible.

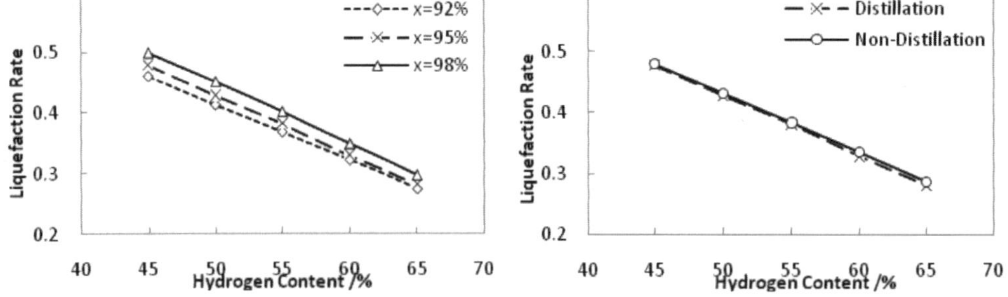

(a) Liquefaction rate at different methane recovery rate

(b) Liquefaction rate at different separation ways when the methane recovery rate is 95%

Fig. 7 Liquefaction rate at different conditions

Conclusions

The following conclusions are drawn from the simulations.

The highest pressure of nitrogen cycle has little effect on the unit power consumption, though it may strongly affects the volume flow rate of nitrogen, which determines the size of devices.

The unit power consumption increases with the increase of hydrogen content of COG and the increase of the methane recovery rate. And the unit power consumption of the process with distillation is about 10% lower than that of process without distillation, when the methane recovery rate is fixed.

Finally, when the methane recovery rate is appropriate, the unit power consumption of the process with distillation is about 0.6kWh/Nm^3, and the methane content of product is more than 95%. These figures are commercially acceptable. So it is feasible to use the process for producing LNG from COG.

References

[1] Koyo Norinaga, Hiroshi Yatabe, Masahiro Matsuoka, Jun-ichiro Hayashi.Application of an Existing Detailed Chemical Kinetic Model to a Practical System of Hot Coke Oven Gas Reforming by Noncatalytic Partial Oxidation. Ind. Eng. Chem. Res.2010;49 (21):10565–10571

[2] Zhang Jiayuan, Zhang Xiaohui, Chen Zhuo, Li Li. Thermodynamic and kinetic model of reforming coke-oven gas with steam. Energy2010;35:3103-3108

[3] Shi Weizhu, An Liansuo, Chen Haiping, Zhang Xuelei. Performance Simulation of Gas Turbine Combined.

[4] Yang Wei-hua, Xu Tao, Li Wei, Chen Guang, Jia Li-yue, GuoYue-jiao. Waste Gases Utilization and Power Generation in Iron and Steel Works. Third International Symposium on Intelligent Information Technology Application Workshops,2009; 246-249

[5] J.M. Bermúdez, B. Fidalgo, A. Arenillas, J.A. Menéndez. Dry reforming of coke oven gases over activated carbon to produce syngas formethanol synthesis. Fuel 2010; 89:2897~2902

[6] Yao Zhanqiang, Ren Xiaokun, Sun Yu, He Qiuju, Zhang Wu. Production of liquefied natural gas from coke oven gas by comprehensive utilization. Fuel & Chemical Processes, 2009, 40 (4): 44-46

[7] Zhang Wu, Ren Xiaokun, Wang Wenchuan, Lu Wenjun, Gao Yuanjing, Sun Yu, Wang Xiaoye. Separation technology and equipment of producing liquefied natural gas from coke oven gas : CN, 101463281A.2009-06-24

[8] Gao Ting, Lin Wensheng, Gu Anzhong, et al. Effects of nitrogen content in coalbed methane on nitrogen expansion liquefaction process. Proceedings of International Conference on Cryogenics and Refrigeration. Shanghai, China, 2008: 309-312

[9] Gao Ting, Lin Wensheng, GuAnzhong, et al. Optimization of coalbed methane liquefaction process adopting mixed refrigerant cycle with propane pre-cooling. Journal of Chemical Engineering of Japan, 2009, 42 (12): 893-901

[10] Gao Ting, Lin Wensheng, GuAnzhong, et al. Coalbed methane liquefaction adopting a nitrogen expansion process with propane pre-cooling. Applied Energy, 2010,87(7):2142-2147

[11] Zhang Lin, Gao Ting, Lin Wensheng, GuAnzhong. Effects of hydrogen content on nitrogen expansion liquefaction process of hydrogen-containing methane mixture. CIESC Journal, 2010; 61(S2): 21-24

[12] Cheng Hongwei, Lu Xionggang, Liu Xu, Zhang Yuwen, Ding Weizhong. Partial oxidation of simulated hot coke oven gas to syngas over Ru-Ni/Mg(Al)O catalyst in a ceramic membrane reactor. Journal of Natural Gas Chemistry, 2009; 18:467-473

[13] Remeljej C W, Hoadley A F A. An exergy analysis of small-scale liquefied natural gas (LNG) liquefaction processes. Energy, 2006, 31: 2005-2019

HANDLING LPG COMPONENTS IN LNG PLANTS

Stanley Huang[a], Robert Saunderson[b], Chittarajan Pathak[c], Bobby Martinez[a]

a: Chevron Energy Technology Company, Houston, Texas
b: Air Products and Chemicals Inc., Allentown, Pennsylvania.
c: Chevron Australia Pty Ltd., Perth, Western Australia

ABSTRACT

Natural Gas Liquid (NGL) recovery is an inherent part of most Liquefied Natural Gas (LNG) production trains. The requirement serves many purposes, such as adjusting the heating value of the product LNG, mitigating the freezing risks caused by specific intermediate and heavy components, and enhancing economic returns by selling NGL separately. In recent years attention has been paid to enhance the overall plant efficiency by integrating the refrigeration requirements of LNG and NGL sections.

This paper shows different heating value specifications in different regions of the world. Therefore, an LNG plant may be required to produce different grades of LNG products to meet the market demands. An ideal design of NGL facilities should be sufficiently flexible to allow operators to adjust the level of NGL recovery. In other words, operators can recover the maximum amount of NGL for sales or leave the LPG components (propane and butane) in the LNG to maximize its heating value.

This paper also presents these NGL technologies which have been proven and implemented in existing LNG plants. The NGL process configurations are classified into several categories and discussed. Additionally, the paper uses a traditional scrub column as an example to highlight the process parameters available to operators for regulating the levels of NGL recovery.

1.0 INTRODUCTION

The components of a produced gas can be broadly classified into the following categories: light, intermediate, heavy components, and impurities [GPSA, 2004]. There may be overlap between adjacent categories; hence, ambiguities may arise in the use of these broad terminologies. To avoid confusion in the following discussions, the chemical species in each category are explained below. The intermediate components are designated as LPG below.

Lights:	mainly methane (C1), ethane (C2), and nitrogen (N2)
LPG:	mainly propane (C3) and butanes (C4); may include some C2 or C5; also known as intermediate components
Heavies:	mainly C5+; may include some C4; may be sold as natural gasoline after stabilization. NGL is LPG plus heavies.
Impurities:	acid gases (CO2 and H2S), and mercaptans

In LNG industry, the light components are the major species in the delivered product. The heavy components are subjected to strict specifications due to freezing-out concerns. Similarly, the impurities are removed to very low levels due to environmental or freezing concerns. Only the LPG components are subjected to negotiations between vendors and purchasers.

The following two terms are borrowed from gas processing industry, in which the desirable LPG components are "recovered" from the gas stream and the rest of the components are "rejected" into the pipeline gas. Hence, the lean gas in the pipeline is referred to as "residue gas". In the same token, when LPG components are to be sold as separate commodities from LNG, the process terminology is "LPG recovery" because the components are recovered from the gas stream. In contrast, if LPG components are to be retained in the product LNG, the process terminology is "LPG rejection".

This paper is divided into several sections. Section 2, Technical Background, discusses the impacts of LPG components on the High Heating Value (HHV) of the LNG and the NGL facilities used to perform the compositional adjustments. Section 3, Process Configurations for NGL Separation, discusses different configurations of NGL handling systems. Section 4, Scrub Column Design and Operations, uses an example to illustrate important features of scrubber operations. Finally, the conclusions and references are provided at the end.

Aside from offering insightful analysis for traditional LNG plants, the information presented is applicable to Floating LNG (FLNG) processing facilities. An FLNG facility is expected to process gases from several production fields throughout its designed life. Since different gas fields have varied gas compositions, the NGL section of an FLNG design should provide adequate flexibility for operators to shift from LPG Recovery to LPG Rejection, and vice versa.

2.0 TECHNICAL BACKGROUND

The HHV specification for LNG varies in different parts of the world. Therefore, an LNG plant will be designed to produce a product specifically for the market it serves. Depending on the gas supply, the LNG plant could be equipped with an NGL removal and fractionation section and/or a nitrogen rejection. The following subsections discuss LNG HHV specifications in world market. The discussion also covers the NGL processing units inside an LNG plant and their relations with other functional units.

2.1 LNG heating values

Figure 1 shows variations in HHV specifications in different regions of the world [Bramoulle et al., 2004]. Immediately obvious is that East Asian countries prefer relatively high HHV specifications, while the U. S. markets prefer relatively low HHV specifications. European countries are fairly tolerant in that their specifications are rather broad. The reasons behind the observed variations are mostly historical, based on locally produced gases. Of course, the willingness of local residents to adapt to changes also plays an important role.

The circled area in Figure 1 indicates potential opportunities of using LNG to carry LPG components. As will be explained later, high HHV implies high contents of LPG components in the product LNG which is referred to as "hot LNG". For LNG plant operators, this means LPG components are rejected to the overhead residue gas stream of the scrub column..

Figure 2 plots the HHV versus the methane content of selected LNG sources [McCartney, 2003; Huang et al, 2007]. Two convergent lines of (C1 + C2) and (C1 + C3) are used to highlight the convergence. The majority of LNG sources fall within the boundaries. The relative location of each point between the two lines reflects the richness of its C3 or C3+ components. Two sources fall below the lower boundary because these two sources are relatively lean in C3+ components and have high nitrogen contents (about 1%).

About half of the 20 LNG sources require heating value adjustments before the re-gasified LNG can be delivered into U.S. domestic pipelines. Eleven LNG sources meet the U.S. pipeline specifications without any heating value adjustments. In other words, products from about half of the global LNG sources require heating value adjustments to meet U.S. market specifications.

Adjusting LPG components is the most efficient way to control HHV. There are two approaches: at the LNG plant or at the receiving terminal. Between the two approaches, adjusting at the LNG plant is easier and more straightforward technically. Another advantage is that LNG product can be readily accepted in any receiving terminals at target market.

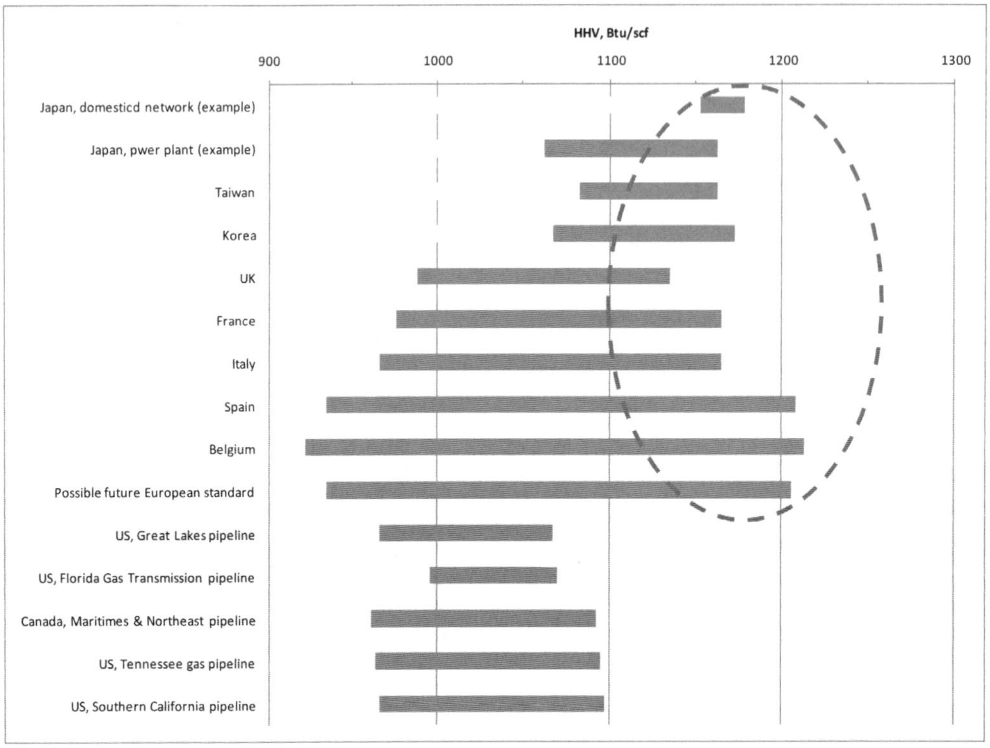

Figure 1 Global markets with potential for accepting rich LNG

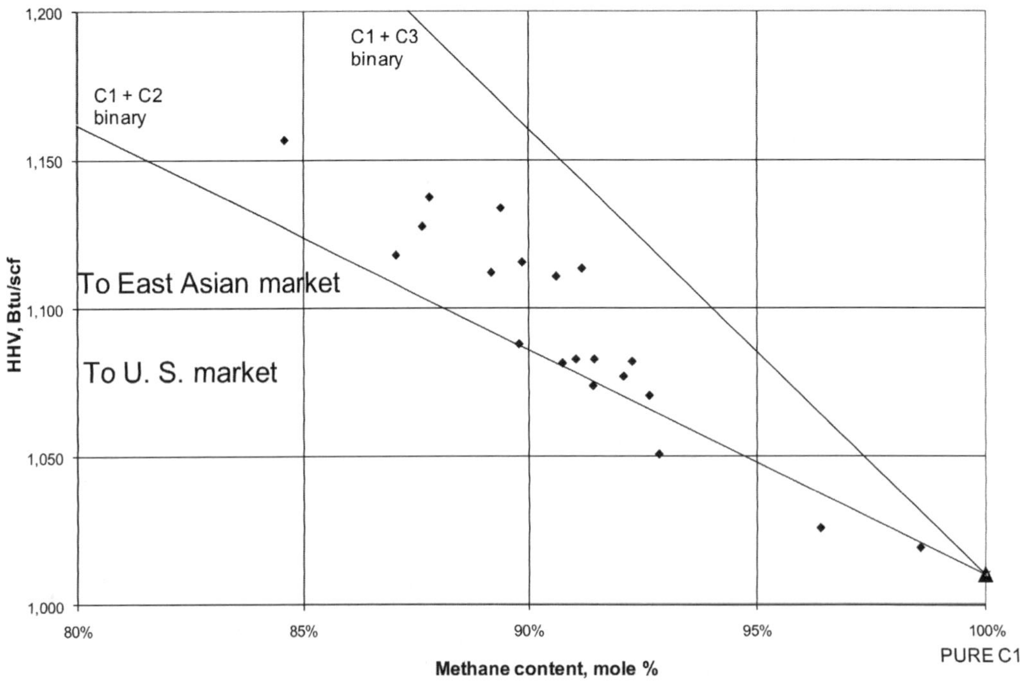

Figure 2 Different LNG specifications in global market

2.2 NGL section in an LNG plant

A typical natural gas liquefaction process is depicted in Figure 3. Sweet, dry, and pre-chilled natural gas goes into the cryogenic section of the LNG plant. The heavies are removed to meet LNG heating value specifications and to prevent possible freeze-out of components, such as benzene at cryogenic temperatures. The lean, high-pressure gas is then liquefied and sub-cooled. Typically, there is an End-flash section or Nitrogen Rejection Unit (NRU) to adjust the N2 composition before the LNG enters the storage tanks.

During the plant holding mode, the flashed gases from NRU and Boil-Off-Gas (BOG) from LNG tanks are collected and compressed to the plant fuel gas system. During the loading mode, the ship BOG is shown totally recovered, re-compresssed, and recycled to the LNG plant area.

Figure 4 shows an example of the entire NGL handling section [Chiu, 1997]. The key equipment in the NGL handling section is the scrubber. This is the most expensive equipment item in the entire section due to its cryogenic operating conditions and relatively large flows. Although the refrigeration shown in the figure comes from the propane (C3) loop, it can be from the low-temperature refrigeration loop of the LNG plant. The choice depends on the temperature requirement.

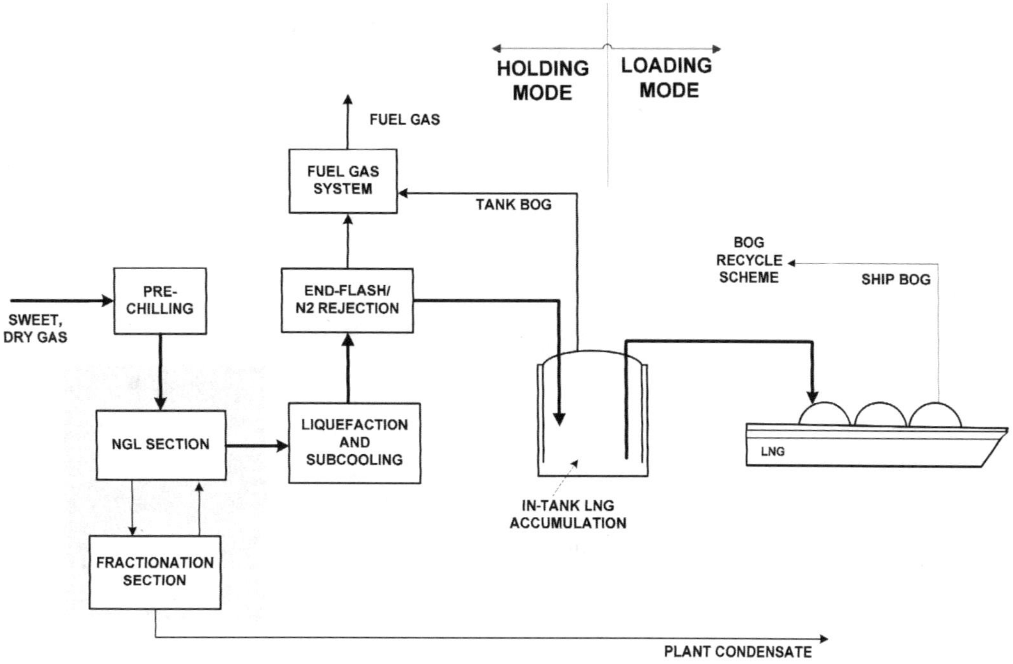

Figure 3 SCHEMETIC DIAGRAM OF AN LNG PLANT

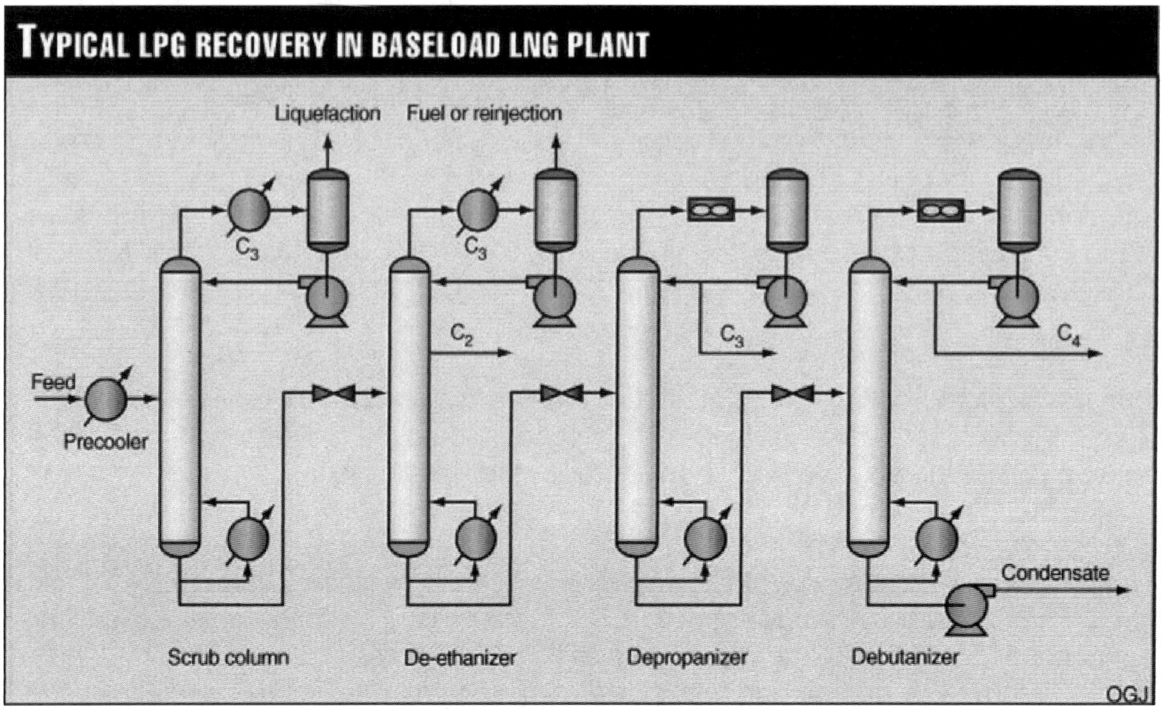

Figure 4 Typical process scheme for NGL recovery in LNG plant (Courtesy of Oil & Gas Journal)

3.0 PROCESS CONFIGURATIONS FOR NGL SEPARATION

The integration of NGL and LNG sections has been a topic being rigorously discussed. Depending on process configurations, equipment efficiencies, and evaluation criteria, different authors made various observations and claims [Hudson, et al., 2004; Attaway et al., 2005; Hagyard, et al., 2005; Qualls, et al., 2005]. It is not the purpose of the paper to distinguish the advantages of various configurations. Rather, this paper would describe various process configurations which are technologically proven and have been implemented in LNG plants.

The following discussions will use the Air Products' C3MR process as an example to facilitate discussions. This example process is the most utilized process in the LNG industry. The configurations of NGL section can be classified into several categories. Each category is shown in its respective illustration and discussed. The materials of this section are mostly based on a previous presentation by Spilsbury, et al. [2006].

3.1 HP scrub column designs

Figure 5 depicts an LPG extraction scheme practiced in the LNG industry using a scrub column in the C3MR process with a two-bundle main cryogenic heat exchanger. In this scheme, the Main Cryogenic Heat Exchanger (MCHE) contains two bundles – a warm bundle which liquefies the feed gas and a cold bundle which subcools it. The refrigerant used to provide overhead condensation of the scrub column is propane coolant. In order to avoid vacuum conditions in the propane refrigeration system, the minimum achievable scrub column condensing temperature is about -35 °C. For rich feed gases, sufficient condensation occurs at this temperature to provide enough reflux to remove LPG's and heavier components from the LNG.

Figure 6 shows another common LPG extraction scheme practiced in the LNG industry. The scheme uses a scrub column with a three-bundle MCHE. In this process the scrub column condenser duty is provided by mixed refrigerant in the warm bundle of the main cryogenic heat exchanger. The configuration can achieve an overhead temperature as cold as about -60 °C, subject to phase density difference criteria. The middle and cold bundles of the MCHE liquefy and subcool the LNG. The three bundle process is typically implemented with medium to lean feed gas compositions. Depending on how cold the scrub overhead becomes, some C2 recovery, up to about 20%, is possible.

3.2 Recycle streams

Figure 7 shows a process scheme utilizing recycle streams to enhance LPG recovery. High LPG extraction can be achieved without utilizing expanders. Recycling LPG and heavier components which act as lean oil absorbent can extract over 95% of LPG components.

A sharper separation of pentanes and heavier, including benzene, in the scrub column, can be achieved by recycling some of the butane and lighter hydrocarbon liquids produced in the plant fractionation unit. If a depentanizer is present in the fractionation train, C5 can be added to the butane recycle. C5+ can be recycled to stages below the top of the column above which a lighter recycle or reflux stream ensures no freezing components can leave the top of the column.

This increased LPG extraction makes the scrub overhead product increasingly lean. Making this overhead too lean limits the distillation pressure of the scrub column as the mixture critical pressure is lowered.

It can be shown that recycling ethane to the scrub overhead compensates for this effect, raising the critical pressure and allowing normal scrub column operating at high LPG recovery. The ethane reinjection is shown here as a separate circuit in the warm bundle of the main cryogenic heat exchanger. It could alternatively be reinjected into the feed circuit to the warm bundle.

3.3 Front-End Turbo-Expander Process

The turbo expander process is the most sophisticated LPG recovery system that can be integrated with a liquefaction process. Figure 8 shows a schematic for Front-end Turbo-expander Process.

A front end expander plant minimizes the variation in the feed composition entering the liquefaction unit. In this process, feed gas is partially condensed against the lean gas from the separation column overhead. The vapor is expanded through a turbo-expander before entering the separation column. After cooling the feed gas, the column overhead is re-compressed to liquefaction pressure. If ethane extraction is desired, a portion of this high pressure gas is fully condensed against the column overhead stream and is used as reflux to the column.

3.4 Integrated Turbo-Expander Process

Figure 9 shows the schematic of an integrated expander plant. The NGL section is similar to a front-end expander plant, except that the upstream propane precooling decreases the load on the recompression resulting in a higher efficiency for the extraction process. However, the propane refrigeration system load is increased.

It should be noted that the introduction of the feed gas compressor and expander in series with the liquefaction process could have some impacts on the overall plant availability.

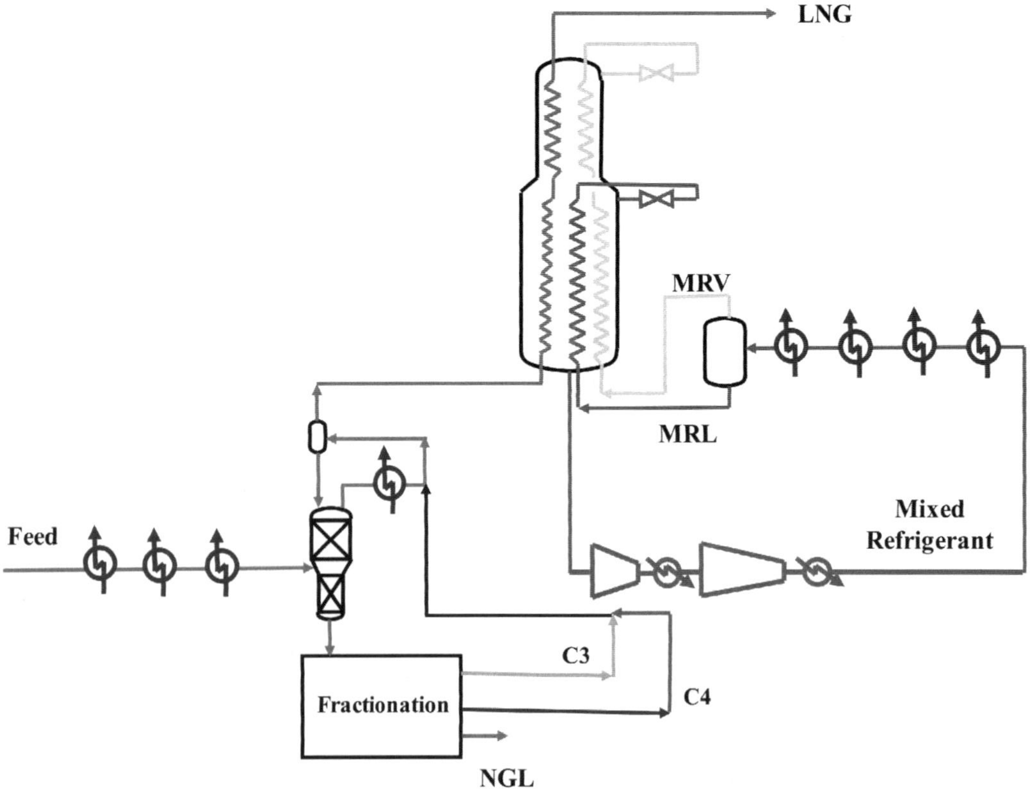

Figure 5 Two Bundle design of APCI C3MR Process

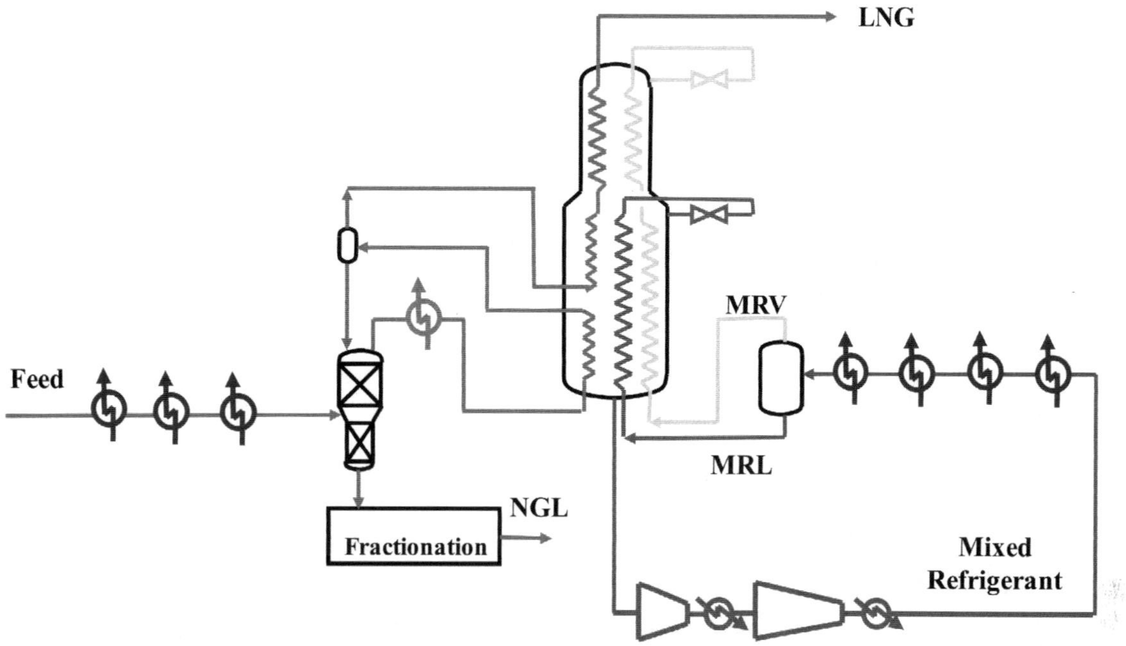

Figure 6 Three Bundle design of APCI C3MR Process

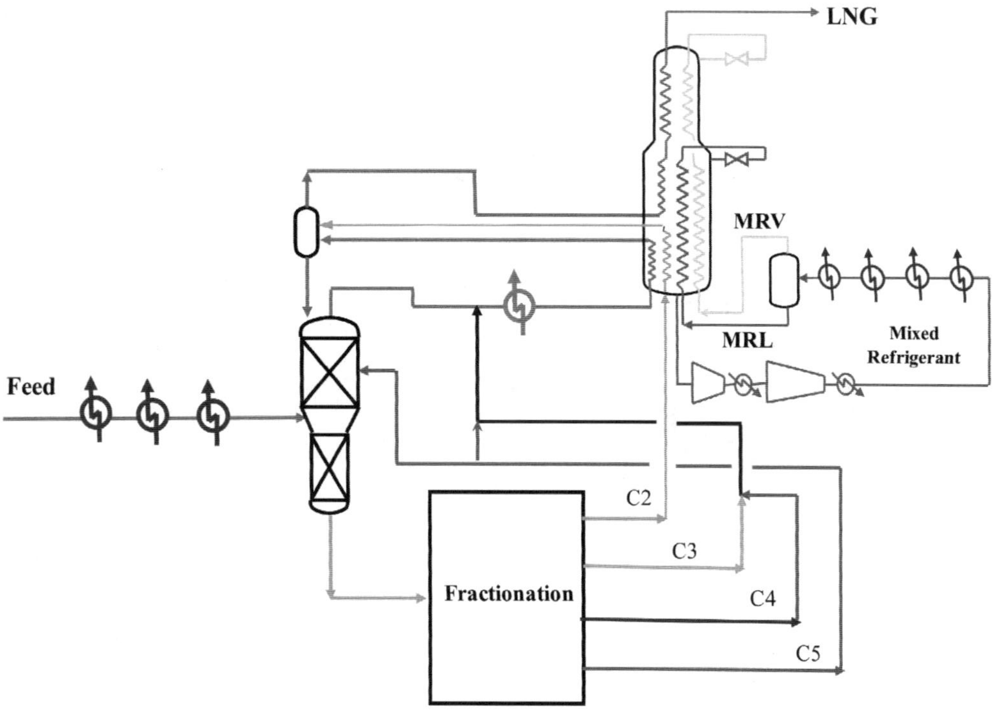

Figure 7 Enhancing LPG recovery using recycle streams

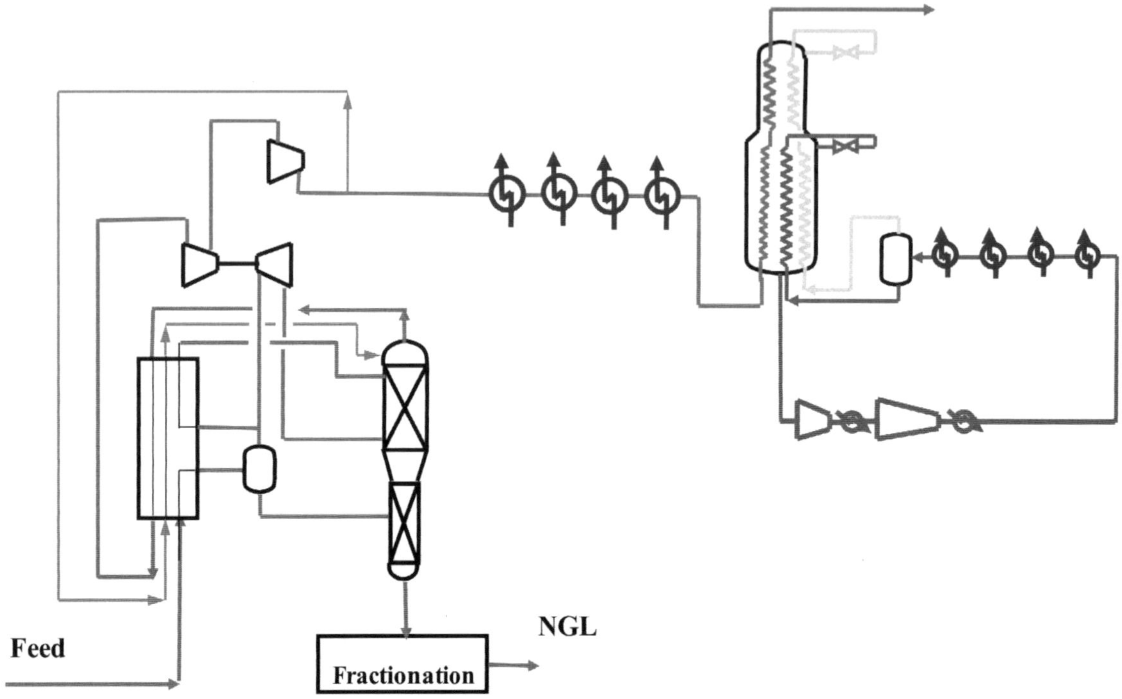

Figure 8 Front-end design for LPG recovery

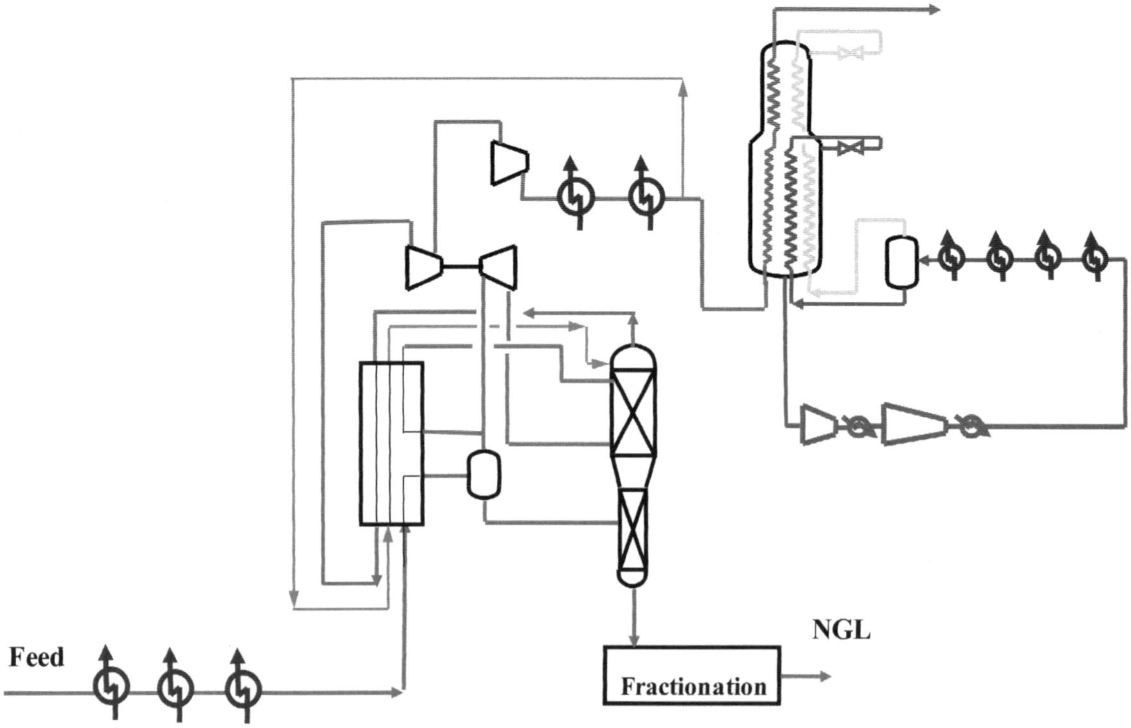

Figure 9 Integrated Turbo-Expander Process

4.0 SCRUB COLUMN DESIGN AND OPERATIONS

This section presents operations of a typical scrub column as illustration. Of particular interest is the adoption of a pump-around (PA) line for rich gases. There are two subsections. The first one discusses the PA line and necessary conditions under which the PA line can be applied. The second subsection presents semi-quantitatively the simulated results based on actual scenarios.

4.1 Reflux with pump-around provision

Figure 10 shows a scrub column with reflux drum, the liquid from the reflux drum can return to the column top as the reflux stream, or can be taken out of the column system. A PA line is shown to channel the outgoing liquid to the vapor stream, which enters the MCHE. The PA line serves as an expedient bypass to avoid sending LPG components through various fractionators shown in Figure 4.

Figure 10 also indicates the condenser duty. As explained in Figure 4, the condenser can be a C3 exchanger or the warm bundle in an MCHE, depending on actual condensing temperatures. The LPG components can leave the reflux drum in vapor or liquid phases. The difference reflects in the stream enthalpy. Fine-tuning the split between the two phases should consider the distribution of chilling duties among various exchangers and the reflux requirements in the scrubber. Thus, the PA line provides flexibility for LNG plant operators.

If significant amount of benzene exists in the feed gas, the use of PA line may be constrained. This is because benzene tends to freeze-out in cold sections of the LNG plant and should be removed to within specifications. The benzene removal is case-specific and the project team is expected to perform careful evaluations.

4.2 Illustration using a realistic case

This subsection presents semi-quantitatively the simulated results based on realistic cases where the feed gas compositions become increasingly richer with time.

Figure 11 plots the reflux rate versus the molecular weight (MW) of inlet gas to the scrubber. There is a minimum reflux rate for the column calculations to converge. Common industrial practices adopt 20% design margin for safe operations. High reflux rate means heavy traffic inside the column. This would lead to increased reboiler and condenser duties. Consequently, the operating efficiency may be compromised.

As the gas becomes richer, condensation inside the column occurs more readily. Therefore, smaller reflux rate from the top is required to maintain the liquid traffic inside the column. The minimum reflux rate shows a downward slope.

Figure 12 plots the vapor fraction versus the MW of inlet gas to the scrubber. The amount of condensed liquid in the reflux drum is inversely related to the vapor fraction.

When the gas is lean, relatively high reflux rates are required for column operations. Therefore, vapor fraction should be low and no PA is desirable. When the gas becomes rich, smaller rate in reflux liquid to column top is required. Therefore, the vapor fraction increases with MW. Two trend lines of minimum reflux and low condensing are indicated in the figure. Also shown is a region marked as High Condensing.

In the case where MCHE also serves as the condenser, high vapor fraction may lead to under-utilization of the lower bundle. Therefore, it may become desirable to operate in the High-Condensing region to improve the equipment utilization, and then use PA to reduce the reflux rate to the column. This practice will be elaborated again below.

The issue of controlling the condensing levels is particularly important when feed gas compositions may change over time. Situations where changing feed compositions are incurred include: (1) gas supplies coming from multi production basins, or (2) FLNG which is designed to handle small gas fields and be moved away when fields are exhausted.

Figure 13 plots the scrubber bottom flow rate versus the MW of inlet gas to the scrubber. The scrub column bottom flow consists of intermediate or heavy components from the inlet gas. Accordingly, as the gas becomes richer the bottom flow is anticipated to increase. This trend is indicated by the upward diagonal line in the figure.

The NGL recovery level is strongly influenced by the reflux rate at the column top. The higher the reflux rate, the more NGL will be recovered at column bottom. When reduced NGL recovery is desired, PA provides an expedient means to reduce the reflux rate without impacting the distribution of cooling duties among chillers in the LNG plant. The arrow in the figure shows the idea of using PA to reduce the bottom flow rate to within designed capacity of fractionators.

Figure 13 shows a horizontal line designated as Design Capacity of Fractionators. When the feed gas becomes richer than the original design cases, the PA line may be properly utilized to mitigate the impacts on equipment.

5.0 CONCLUSIONS

The conclusions of this work are listed as follows:

- Incentives to recover or reject LPG components from the gas streams are explained.
- Design features of NGL section in an LNG plant are elaborated.
- Illustrations for scrubber operations are provided using a realistic case. It is demonstrated that PA line provides flexibility for plant operators when the inlet gas becomes richer than originally designed conditions.

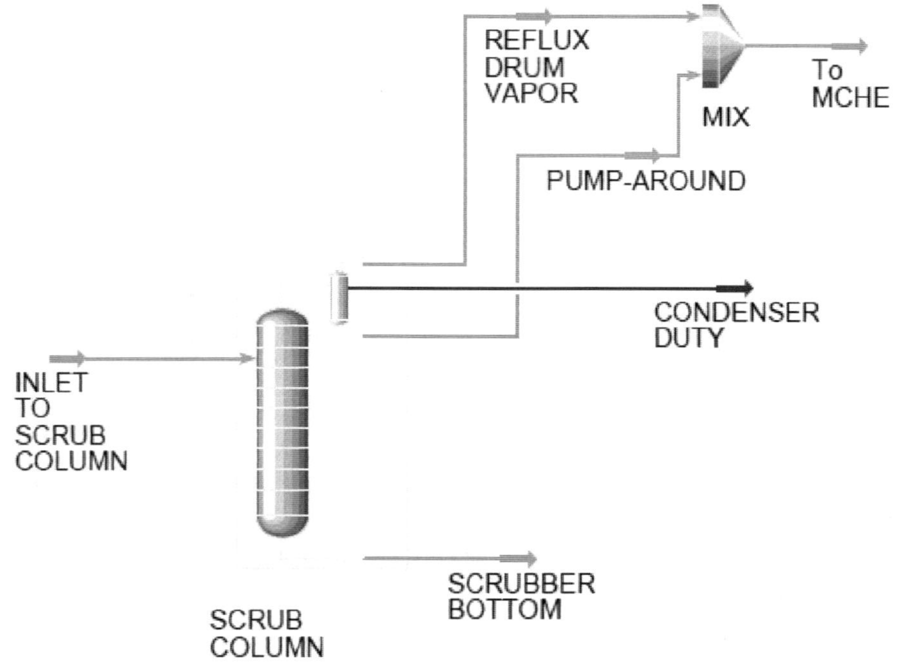

Figure 10 Simulating typical scrub column in LNG plant

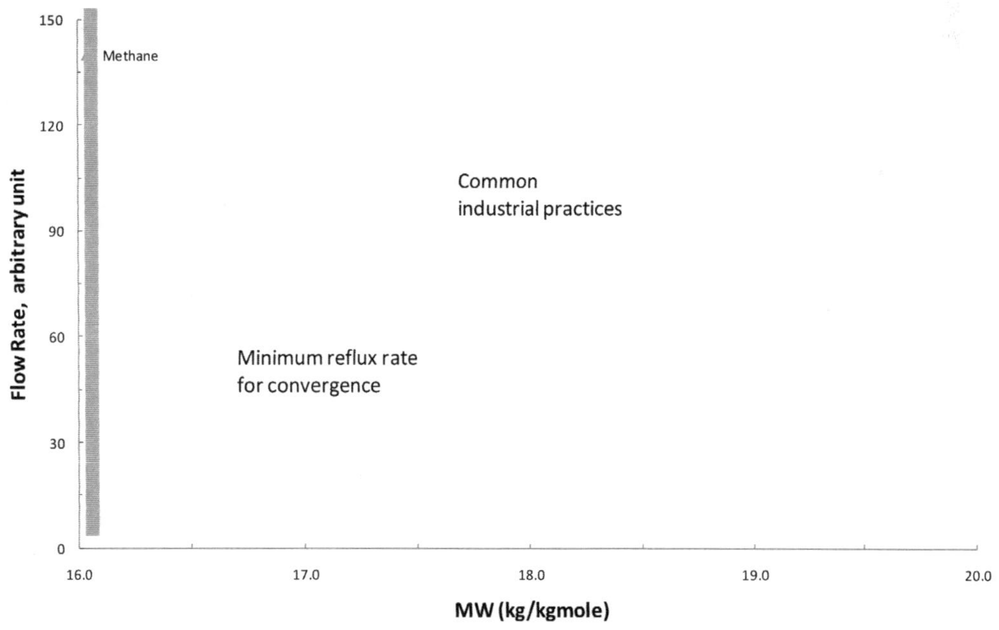

Figure 11 Reflux rates to scrub columns

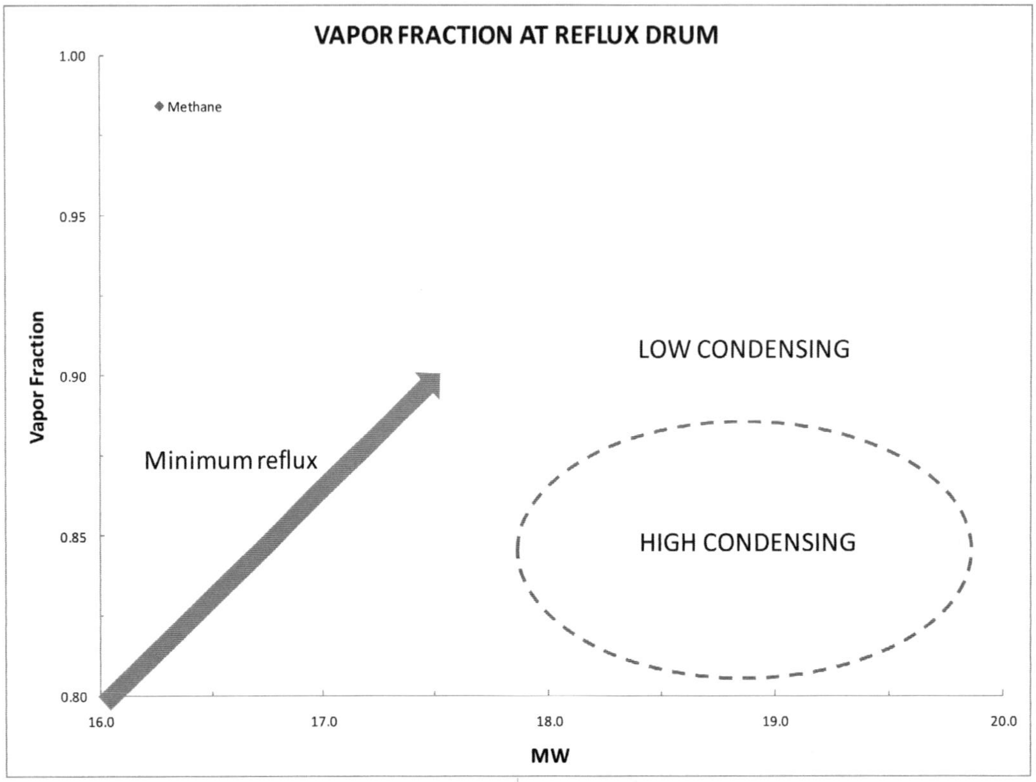

Figure 12 Vapor fractions at reflux drum

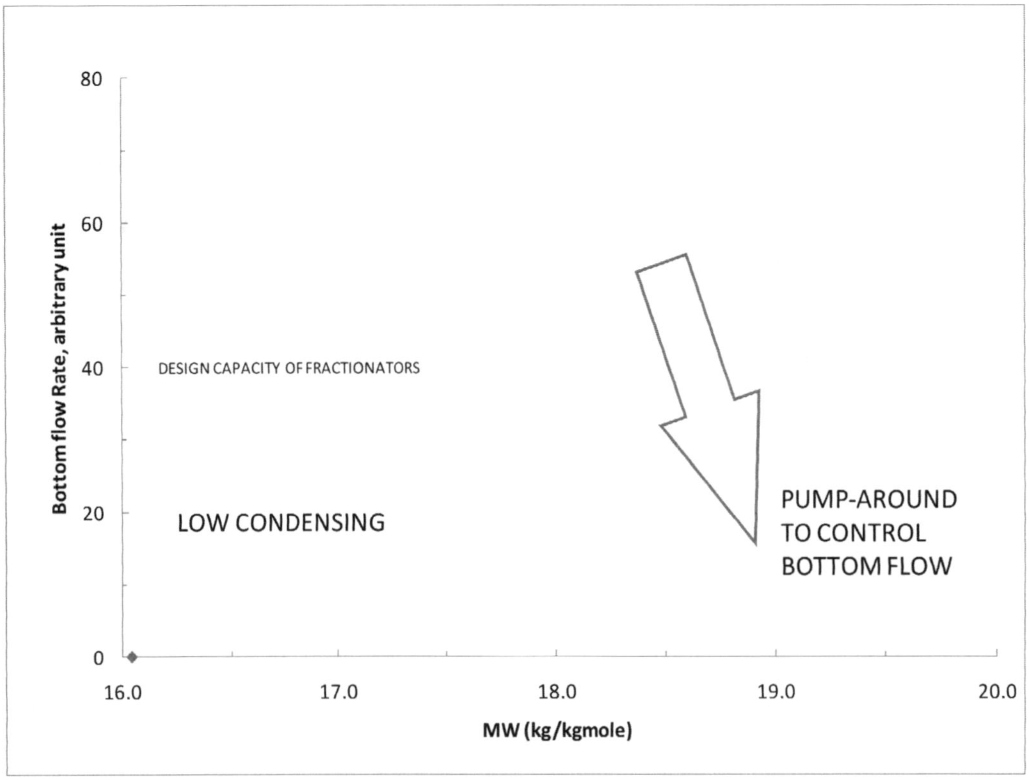

Figure 13 Scrubber bottom rates

ACKNOWLEDGEMENTS

The authors would gratefully acknowledge the contributions by Ida K. Chan of Chevron. Her persistent efforts and masterful process simulations made this work possible.

REFERENCES

- Attaway, D. A., Huang, S., Kotzot, H., Durr, C. A., "Optimal Process Locations for NGL Recovery in LNG Plant", paper presented in AIChE Spring National Meeting, Atlanta, GA (2005).
- Bramoulle, Y., Morin, P., Capelle, J.-Y., "Differing Market Quality Specs Challenge LNG Producers", Oil & Gas Journal, issue of October 11 (2004).
- Chiu, C.-H., "LPG-Recovery Processes for Baseload LNG Plants Examined", Oil & Gas Journal, November 24 (1997).
- GPSA Engineering Data Book, 12th Edition, Gas Processors Suppliers Association, Tusa, OK (2004).
- Hagyard, P., Paradowski, H., Gadelle, D., "Simultaneous Production of LNG and NGL", paper presented at LNG 14 Conference, Doha, Qatar (2004).
- Huang, S., Chiu, C.-H., Elliot, D., "LNG: Basics of Liquefied Natural Gas", book published by The University of Texas, Continuing Education, Petroleum Extension Services (PETEX), Austin, Texas (2007).
- Hudson, H. M., Wilkinson, J. D., Cuellar, K. G., and Pierce, M. C., "Integrated Liquids Recovery Technology Improves LNG Production Efficiency", paper presented at GPA 2004 Spring National Meeting, New Orleans, LA (2004).
- McCartney, D.; "Gas Conditioning for Imported LNG", paper presented in 82nd GPA Annual Convention in San Antonio, TX (2003).
- Qualls, W. R., Huang, S., Elliot, D., Chen, J. J. R., Lee, R. J., Yao, J., Zhang, Y. I., "Benefits of Integrating NGL Extraction and LNG Liquefaction Technology", paper presented in AIChE Spring National Meeting, Atlanta, GA (2005).
- Spilsbury, C., Roberts, M. J., Brostow, A., Liu, Y. N., "Meeting LNG Heating Value Requirements in a Pre-cooled Mixed Refrigerant Process", paper presented in Air Products Owners' Seminar, Allentown, PA, September (2006).

Biographies

Dr. Stanley Huang has a specialty area in cryogenic applications, particularly in LNG and gas processing. He graduated from National Taiwan University with a B.S. degree and attended Purdue University in 1981. He earned his Master and Ph.D. degrees in Chemical Engineering and a Master of Science in physics there. Stanley has 20 years of industrial experiences and is a Registered PE in Texas. Currently he is a Staff LNG Process Engineer of Chevron. E-mail: Shhuang@Chevron.com

Robert Saunderson joined Air Products and Chemicals in 1990, and has more than 17 years experience as a process engineer in the baseload LNG industry. He is responsible for process designs of the LNG liquefaction unit. He was APCI's lead process engineer for the Ras Laffan Expansion Project: Trains 3, 4, and 5, and is currently APCI's lead process engineer for the Gorgon, GNL3Z, and Papua New Guinea LNG projects. E-mail: saunderp@airproducts.com

Chittaranjan Pathak is a process engineer for Chevron Australia. He graduated in 19991 from Assam Engineering College India with Bachelor degrees in Chemical Engineering. He earned his Master Degree in Advanced Process Design from Monash University. Chitta has 19 years of experience in process design having worked in companies like Kellogg Joint Venture, Technip, and Engineers India before joining Chevron.

Mr. Bobby Martinez has over 20 plus years in the oil and gas industry. He earned a bachelor degree in Chemical Engineering and Chemistry from New Mexico State University. He is a Texas registered professional engineer and holds several co-authored patents, and industry publications in the art of LNG. Mr. Martinez is currently Chevron's LNG Major Capital Projects Technical Team Leader.

(76c) What We Learned by Using Dynamic Models for LNG Work

David Hill, CHEMCAD Technical Support, Chemstations, inc., Houston, TX, Aaron Herrick, Development, Chemstations, inc., Houston, TX and Kenichi Matsuoka, Inpex Corporation, Tokyo, Japan

Chemstations has advised on several modeling projects for LNG processes. Dynamic simulation helped the engineers gather insight into their process. We will provide a discussion of scenarios, and what was achieved by the engineers modeling them.

See more of this Session: LNG Simulations and Controls

See more of this Group/Topical: Topical 6: 11th Topical Conference on Gas Utilization

(76d) Study on Energy Consumption for Natural Gas Liquefaction

Meiqian Wang, Jian Zhang, **Qiang Xu** and Kuyen Li, Dan F. Smith Department of Chemical Engineering, Lamar University, Beaumont, TX

The main purpose of this work is to optimize natural gas liquefaction process by reducing energy consumption. At first, a rigorous simulation of natural gas liquefaction is conducted. Based on the simulation results, the refrigerants and natural gas thermodynamic states along with whole process are analyzed in order to find out the direction in which the energy consumption could be reduced and what parameters could to be adjusted. Using thermodynamic analysis results, an optimized simulation is successfully undertaken, and the results before and after optimization are compared to validate previous analysis. Moreover, traditional natural gas liquefaction process performance analysis methods, like cooling curve and Coefficient of Performance (COP), are also used to confirm this thermodynamic analysis methodology is practical and useful.

See more of this Session: LNG Simulations and Controls
See more of this Group/Topical: Topical 6: 11th Topical Conference on Gas Utilization

(76e) Improved LNG Plant Performance Through Higher Thermodynamic Efficiency

Christopher M. Ott, Yu-Nan Liu and Xukun Luo, Air Products & Chemicals, Allentown, PA

For many years, the LNG industry has focused on large baseload plants. Recently, more attention is being given to smaller LNG facility sizes and alternative liquefaction strategies. This presentation will examine the impact that the thermodynamic efficiency of a particular cycle can have on the overall plant performance.

See more of this Session: LNG Simulations and Controls
See more of this Group/Topical: Topical 6: 11th Topical Conference on Gas Utilization

Improving Refinery Unit Transitions Using Process Automation Technology in a Base Oil Hydroprocessing Facility

Robert M. Tsai, Chevron, Richmond, CA

Introduction

Automation in Refining

The application of process automation technologies in US refineries are still relatively rare and limited, outside of safety systems and simple sequential automation systems for switching parallel unit operations. Many tools exist in the industry to optimize a unit once it is at steady state using simple regulatory or multivariable control techniques. However, few applications exist where a refinery unit is completely transitioned from one process state to another, such as adjusting unit operation for product grade or feedstock quality changes, let alone starting up or shutting down a unit. Several factors are often cited in preventing automation technologies from taking root in the refining industry. Some of these include: lack of resources to maintain these technologies, unit and operational complexity, difficulty estimating tangible benefits and difficulty changing the status quo. Chevron has had success overcoming these hurdles and proving hard dollar benefits in a base oils hydroprocessing facility.

Base Oils Hydroprocessing

In lube oil hydroprocessing, heavy vacuum gas oil (VGO) is upgraded into high viscosity index (VI), low viscosity lube oil using a combination of hydrocracking and isomerization reactions to transform undesirable aromatics, olefins, and asphaltenes into desirable n-paraffin and iso-paraffin and components. In severe hydrocracking applications, undesirable sulfur, nitrogen and oxygen containing compounds are also hydrotreated and removed. Lastly, the hydrocracked lube oil stock is hydrofinished to improve color, thermal and oxidative stability, and reduce sulfur and nitrogen levels further.

Chevron uses a Chevron Lummus Global (CLG) All-Hydroprocessing technology that combines ISOCRACKING™, ISODEWAXING™, and ISOFINISHING™ in series to produce finished base oil stock. The VGO feedstock is heated by a furnace then passed through a series of catalyst beds in multiple reactors. The reactions over the catalyst beds are generally exothermic, and hydrogen quench is used to control catalyst bed temperatures to targets and to prevent reactor runaway. Figure 1 shows a basic diagram of this process.

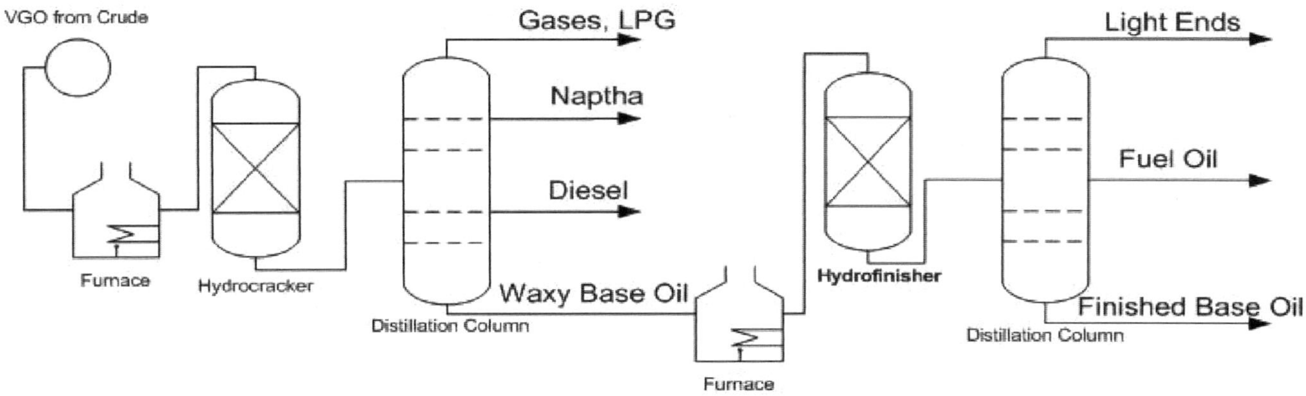

Figure 1 – Typical Chevron Base Oil Hydroprocessing Unit

Unit Transitions

Various grades of base oils are produced from a single unit over the course of a month depending on market conditions. This causes the unit to operate on a feed wheel with frequent feed and product changes that require the plants to be adjusted several times a month. Multivariable control applications are in place to optimize each product run, but a large optimization opportunity existed in the eight-hour transitions between product runs. This transition was done "on the fly" without routing product off-test during the transition; therefore, having a seamless and reproducible transition would maximize throughput of salable product, greatly improving the overall profitability of the unit. Manually executing these transitions resulted in large variations in transition time and product yield. An automation solution was implemented to capture this optimization opportunity using an application that employed simple, graphically-based instruction logic to execute complex programs over an OPC connection. This automation solution is available to the operator though his or her existing control console. Since some of the tasks in the transition needed to be executed by an operator in the field, such as routing tanks or stopping and starting non-automated pumps, it was important for the software to directly interface with the console operator, who was responsible for monitoring the automation solution that was coordinating the automated tasks with field tasks in the transition.

This particular transition, known as a "feed switch" in the unit, involves two major tasks: adjusting reactor conditions and feed rate as fast as possible to maximize new product yield and adjusting the distillation train operation as the new product progresses through the unit to maximize product recovery. Figure 2 shows a graphical representation of a typical feed rate and reactor temperature curve for the reactor portion of the feed switch transition. Because reactor temperature targets are dependent on both feed type and rate, the curves are vastly different between product combinations.

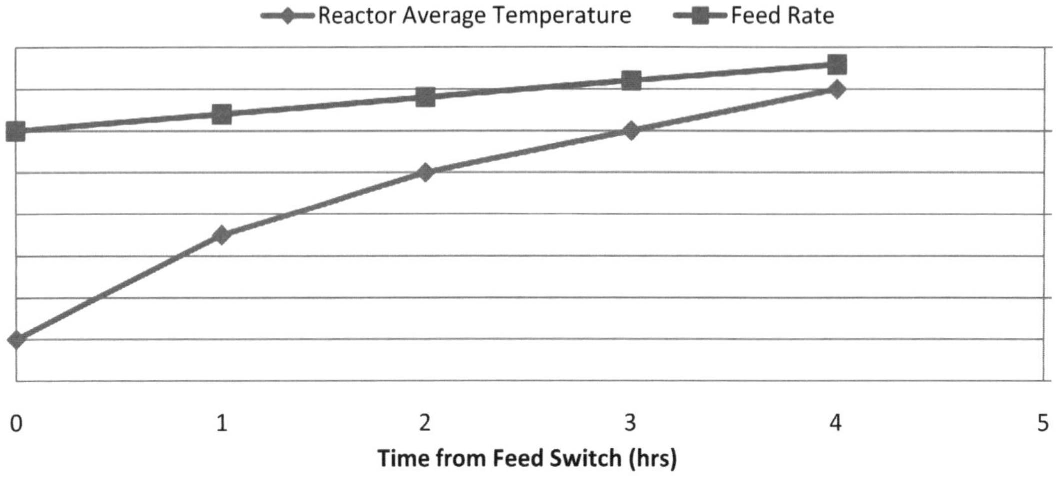

Figure 2: Example Feed Switch Reactor Temperature and Feed Rate Curve

After the reactor temperatures and feed rates are moved to targets, distillation targets are set with various timings based on the expected time it takes the new product composition change to reach the distillation portion of the unit. An additional layer of automation coordinates feed tank and product tank switching as to maximize product recovery and minimize giveaway. Before the implementation of this multi-layered automation solution, operators were responsible for coordinating all these moves manually with varying results. After implementation of the automation solution, we have seen a general improvement in feed switch duration and much lower variability in transition execution, indicating that the automation solution allows the unit to complete the feed switch and engage the steady-state multivariable controller more quickly.

Results

Manual Transitions

Typically during a manual transition a board operator makes about ~300 moves in the 8 hours that a feed switch typically takes to complete. Figure 3 shows the average distribution of these moves over a seven hour feed switch executed manually.

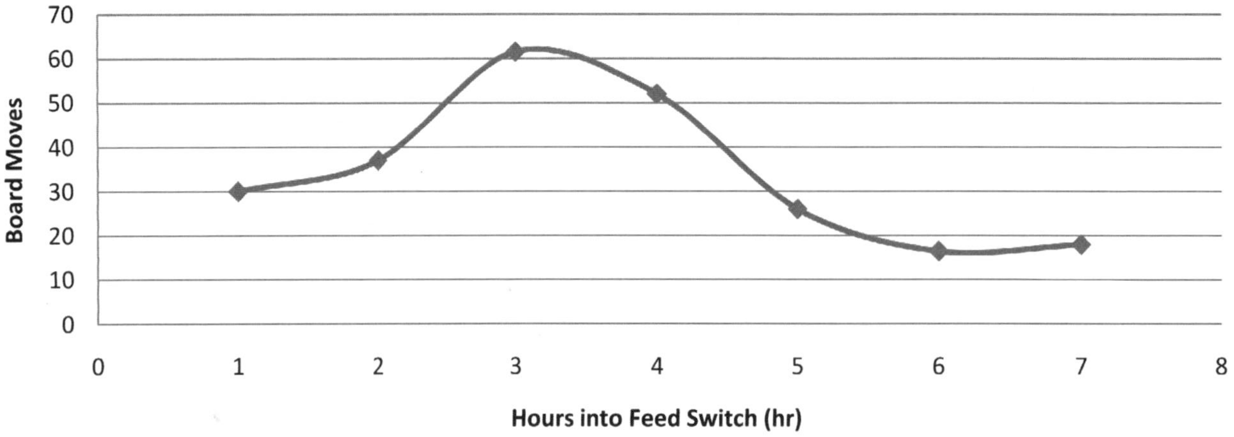

Figure 3- Number of Board Moves Per Hour for a Typical Feed Switch

Note that a single console operator is responsible for multiple plants in addition to executing the feed switch. Figure 3 shows that the console operator makes a board move every minute at the height of the transition. This means that he or she is fully concentrating on making the proper moves for the feed switch; naturally, this diminishes his/her capacity to monitor other aspects of the plant. Also, the operator for a four hour period between hours 2-5 is required to fully concentrate on making the switch with no downtime for any sort of absence. This creates a situation where a stressed operator is fully focusing on a single aspect of his job, potentially leaving other parts of the plant unmonitored. Figure 4 shows feed and reactor average temperature curve for a feed switch done in manual by an operator.

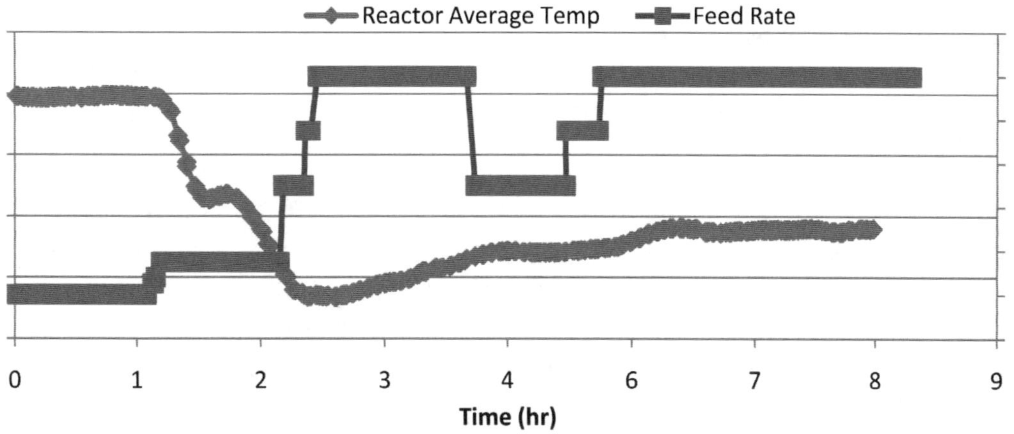

Figure 4 – Operator Manual Feed Switch Curve

We can see that the switch by the operator has inconsistent move sizes and timing, resulting in a jerky feed switch that is an hour longer than an ideal one resulting in a slight loss in throughput. In this case we can postulate that the operator may have been distracted between some of the delays in moves, or may have just left the control board temporarily. Chevron saw the potential benefit in throughput, safety, and feed switch consistency here and decided to try and automate this task and help the operator do his or her job better.

Software Assisted Operator Feed Switches

A instruction logic based graphical application was used to create a program that interacts with the operator during a feed switch and automates repetitive tasks such as feed set point ramping while monitoring unit limits. The actions and timing the program uses are based on the same sequences and tasks the "best operator" would undergo when doing a feed switch. The program executes the feed switch using these parameters at the fastest possible safe rate, while ensuring key timing events are executed in the proper sequence. For example, the program will predict when a new feed introduced into the plant will reach the product rundown manifold and require the product tank routing to be switched. This ensures that in a seamless "flying" feed switch product integrities are not compromised while throughput is maximized. Figure 5 shows an example of the feed and reactor average temperature curve of an automated feed switch. It should be noted that when comparing figure 4 and 5, the time it takes to complete the switch appears to be similar. However, the feed and reactor average temperature final targets are twice as far from the original values in figure 5, meaning that the switch is completed at twice the manual rate.

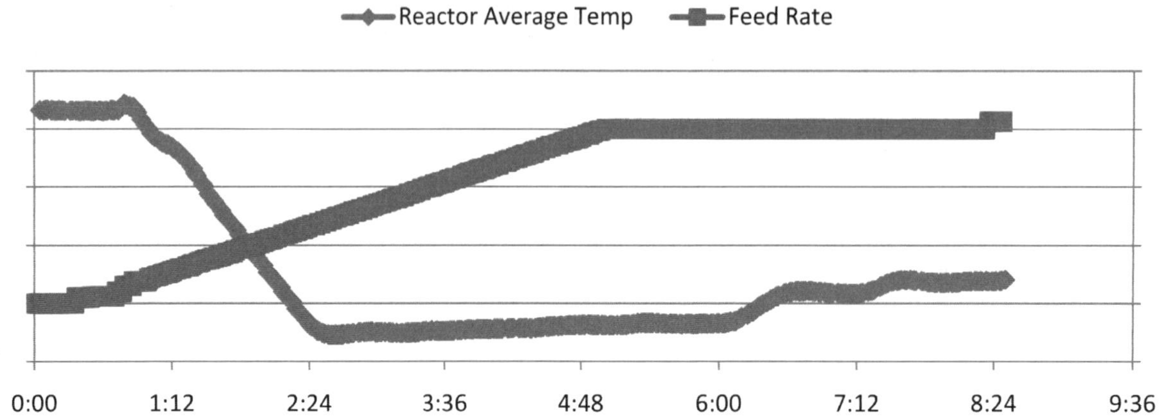

Figure 5- Feed and Reactor Average Temperature Curve of Automated Feed Switch

In comparison to the ~300 manual moves an operator would make during a manual feed switch, the automation solution instead performs these moves in greater resolution, making ~3000 moves during a feed switch, resulting in a smoother transition that takes less time as shown in Figure 5. Figure 6 shows the operator load at the board using the automation program compared to doing the feed switch manually.

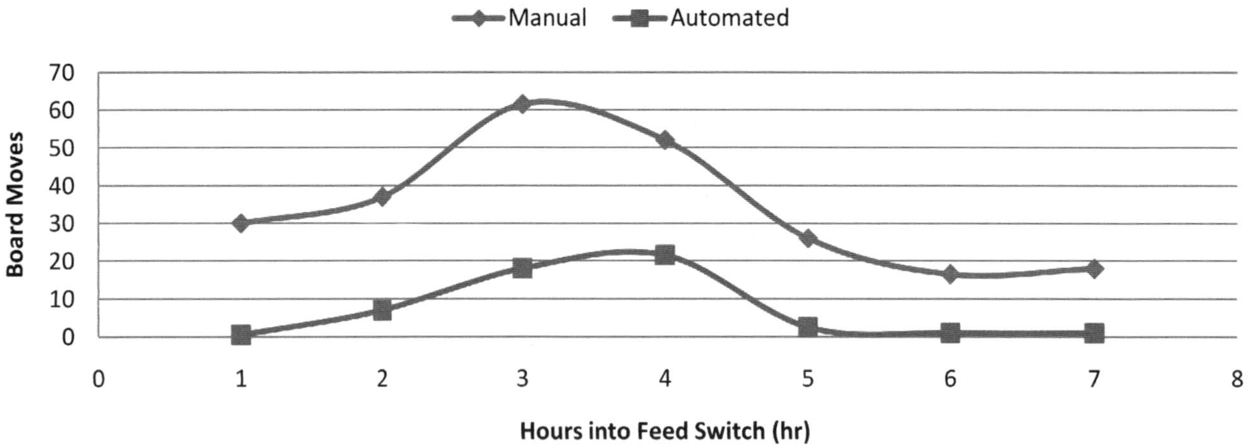

Figure 6- Console Operator Moves per Hour Comparison

From Figure 6 we can see that the load on the operator is reduced by approximately 60%. The operator now can focus on monitoring and optimizing other aspects of his/her plant rather than devoting his/her attention completely to the feed switch.

Benefits

Using automation to assist in routine transitions produces a multitude of benefits. Some of these benefits include: decreased operational variability between operating personnel, faster transitions, increased throughput, smoother transitions, increased computer monitoring of critical process variables, and increased accuracy in executing the manufacturing plan. For this study we will look at the two more straightforward benefit indicators, increased throughput and faster transitions. Because every transition is slightly different, we will quantify transition duration as transition rate rather than an absolute time, and increased throughput as a percentage of total expected throughput for the transition. Table 1 below shows two months of data that were studied and compared.

	% Improvement
Feed	36%
Reactor Average Temperature	60%
Product Yield Loss	42%
Increased Feed Throughput during Transition	18%

Table 1 – Benefits from Automated Transition versus Manual Transition

From process knowledge we know that feed rate and reactor temperature as related; higher feed rates will require a higher reactor catalyst average temperature. Therefore, depending on the type of switch we want to ensure that both feed rate and reactor temperature are at final target as quickly as possible so that little off-spec product is made. However, as we can see in figure 5, reactor average temperature will ramp linearly while feed can be ramped linearly. This means the two ramps may finish in different times. The switch is not complete till the slower of the two is at target and the unit back in a steady state operation. It was found that operators were doing an average job ramping reactor average temperature to targets. Aggressive logic was implemented into the program to ensure that reactor average temperature is ramped aggressively and reaches final temperature in the shortest possible time. We can see from Table 1 that the automated program decreases the transition time when compared to the average manual transition. This results in increased throughput to the unit. A secondary benefit shown in Table 1 is a decreased loss of yield in the distillation section. When a unit is transitioned to a new feed, at new temperature targets the distillation section heat and material balances will move away from optimal operation. This transitory state results in a large decrease in yield and product quality throughout the transition. Decreasing the transition time also decreases this transitory state and reduces product yield loss.

It should be noted that, unlike typical large scale automation initiatives done almost completely in programming code, this automation application is almost completely graphical and directly interfaces with the operator. This results in decreased time between revisions and increased operator acceptance. Chevron saw an average of one revision per month, leading to quickly increasing operator acceptance and usage. Figure 7 shows operator usage of the program over the first year.

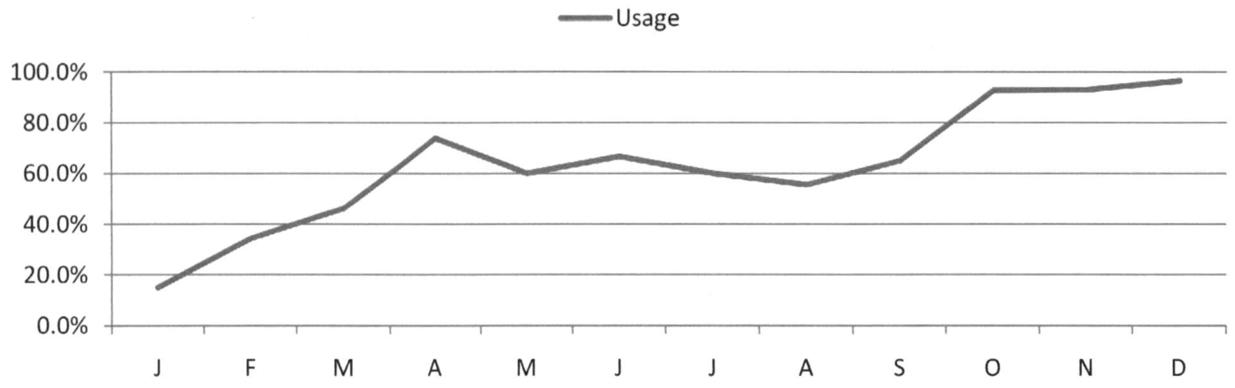

Figure 7- Operator Use of the Program in First Year

In Figure 7, usage is defined instances of an operator using the program over the total opportunities present for using the tool at the plant. We can see that within four months operators were using the program in more than 70% of opportunities. At this point most of the major bugs had been fixed and the tool was doing the task successfully to completion without issues. Minor hurdles keeping the program from reaching higher acceptance at this point were limited issues with graphical nuances, textual clarity, and minor program bugs. In the next eight months as these minor issues were fixed operators began to universally use the program for all transitions.

Conclusion

With the success Chevron saw with this automation application, additional projects were undertaken to expand this technology. We now use this technology in both large installations (crude unit, sulfur recovery unit, complete start up and shutdown of lube Hydroprocessing unit) as well as small installations (molecular sieve dryer, vessel dump and charge). However, like many other advanced control applications it takes a large sample size of nine months or more to get a definitive answer on the success of the projects. We hope that transition automation improves the safety and profitability of refinery units and bridge the gap between various modes of stead state advanced control and optimization technologies. Selection of proper units with sufficient base level control, instrumentation, and enough transitions to produce a large economic benefit will be the challenge going forward, but we hope to learn to leverage this new technology more effectively and expand its use.

References

1. Sequeira, Avilino Jr.(1994), *Lubricant Base Oil and Wax Processing.*
2. Wauquier, Jean-Pierre and Favennec, Jean-Pierre (2001), *Petroleum Refining: Refinery Operation and Management.*

(85b) Practical PID Control: Use Filters for Better Disturbance Rejection and Setpoint Tracking

Shabroz Gill, Doug Nicholson, Sebastien Osta and **Patrick Thorpe**, IPCOSAptitude, Cambridge, United Kingdom

Many of the practical problems in achieving high performance feedback control for challenging applications have been addressed by modifications to the basic PID control law. These modifications include set-point filters that allow both load rejection and set-point following objectives to be satisfied simultaneously and the use of low pass filters to improve stability, particularly when derivative action is employed.

The drawback of course is that these modifications introduce more tuning parameters that have to be adjusted in order to achieve the desired performance. Non linear optimization techniques allow the control engineer to tune this extended set of parameters taking into account practical considerations such as model estimation uncertainty, interactive control loops, actuator wear etc.

This paper will consider the performance of various optimally tuned feedback control strategies to a simulated refining process with interactive control variables. A multivariable controller will be used as a benchmark reference. The impact of model error and process constraints will also be considered. We will demonstrate how control performance can be improved by the correct use of the tuning parameters that are available in most commercial distributed control systems, allowing significant benefits to be achieved for a relatively modest investment.

See more of this Session: Advanced Control and Applications II

See more of this Group/Topical: Topical 7: 14th Topical on Refinery Processing

(85c) Inferential Monitoring & Optimization of Crude Units Via Hybrid Models

Yoel Sanchez and **Vladimir Mahalec**, Chemical Engineering, McMaster University, Hamilton, ON, Canada

Refinery crude units often process variety of feedstocks. If properties of the feed entering the crude unit are not precisely known, it is difficult to predict how changes in the tower operating conditions will affect properties of the products. Similarly, if on-line analyzers are not available, it is not possible to implement very accurate control of the product properties.

This work introduces a new class of models (hybrid models) which can be used to: (i) identify feed distillation curve, (ii) predict product properties from tower operating conditions, and (iii) predict product properties from feed properties and material and energy balances. Hybrid models consist of first principles equations (e.g. balances) and of Partial Least Squares (PLS) models for prediction of product or feed properties. A model of a crude unit has several hundred equations and is practically linear; this model predicts target properties within 1% of the rigorous tray to tray model which consists of 10k-20k nonlinear equations.

In industrial practice there is a very limited number of tray temperature measurements; the models we have developed enable us to achieve the accuracy of prediction that is the same as if there are tray temperature measurements available on every tray. In addition, inclusion of material and energy balances in the hybrid model equations enables optimization of tower operating conditions based on the hybrid model.

We present results of a) optimization of tower operating conditions based on the hybrid model and based on the equation-oriented AspenPlus model; b) identification of the crude feed distillation curve based on the limited number of tray temepratures and the hybrid model variables and compare it to predictions one would obtain if there were tray temperatures available on every tray; c) prediction of product properties from the available tray temperatures and hybrid model variables.

Accuracy of the hybrid model predictions and almost linear nature of its equations makes it suitable for applications in on-line monitoring, or as inferential predictor for model-based controllers, or for optimization of operating conditions, or for crude unit modeling in the refienry plannign and scheduling tools.

See more of this Session: Advanced Control and Applications II

See more of this Group/Topical: Topical 7: 14th Topical on Refinery Processing

Post-Optimality Analysis of Optimal Refinery Production

Abstract

The aim of refinery optimization is to find the best operation conditions that can maximize production of oil fractions and minimize operating costs with respect to operational and environmental constraints. Uncertainty or variations in some parameters (e.g., demand and supply of each processing unit or product prices) can seriously affect the optimization results and lead to inefficient operation. Many techniques of stochastic programming were proposed to handle the effect of variations in process parameters in order to determine a robust operation conditions; however, applications of post-optimality analysis has received less attention especially in refinery industry. In this work, post-optimality analysis is used to study the effect of such variations on the optimal solution of refinery process that is simplified and formulated as a LP model.

The general objective of this study is to try to bridge the gap between the theory and practice of the post-optimality analysis in process engineering. This work attempts to use a modified method of post-optimality analysis that jointly use sensitivity relations and stability region calculations to provide the decision maker in the refinery with valuable and easy-to-use information that helps in handling the effect of variation in process parameters. The results of this study can help the decision maker to identify sensitive parameters that need accurate estimate or intensive monitoring; and compute their stability limits, i.e., allowable variation ranges, within which the operation remains optimum. Independent and simultaneous variations in the objective function coefficients (e.g., products' price) and in the RHS of the constraints (e.g., product specifications, demand and supply) are considered in this study.

1. Introduction

Petroleum refining is one of the most important industries, comprising many different and complicated processes with various possible configurations. Optimization of refinery production levels is an essential task to maximize company profit margins and to remain in the competitive market. Many refinery models and optimization techniques were proposed to handle this task; however, it was recognized that optimization results can suffer from the existence of uncertainty in model parameters due to inaccurate estimates of some parameters. In addition, variations in input data, such as supply and cost of raw materials, can easily affect the process profitability. Many studies were conducted to understand the effect of such uncertainty or variation on the optimized model. There are two general approaches for dealing with the existence of uncertainty in optimization problem: incorporating the uncertainty directly into the optimization problem formulation, such as stochastic programming, and analysis of the effect of uncertainty on the optimal solution, post-optimality analysis.

Many techniques of stochastic and fuzzy programming were proposed to handle the effect of variation in process parameters in order to determine a robust operation conditions[1]. In the other hand, applications of post-optimality analysis have received less attention especially in refinery industry. Post-optimality analysis can help the decision maker to determine how much actual values of parameters may differ from the estimates used in the model before the optimal results become irrelevant. Generally, post-optimality analysis can provide the decision-maker with valuable information about the process and answer many crucial questions such as:

- How the optimal value and solution are affected by individual or simultaneous change(s) in process parameters?
- What are the perturbation limits that do not change the optimality of the solution?
- What are the sensitive constraints that need further monitoring?
- What are the sensitive parameters that need accurate estimation?
- How can we modify the process to make it more profitable or stable?

In addition, many ``what if" type questions can be answered by performing post-optimality analysis. Despite its value, applications of the post-optimality analysis (e.g., in production planning) have received less attention compared with stochastic programming. This may be due to the challenge of studying the effect of simultaneous variations in the model parameters. In addition, current state-of-art post-optimality analysis methods for different linear optimization problems (e.g., LP and MILP) are rarely used due to the high computational complexity.

In this project, we proposed to use post-optimality analysis, mainly stability analysis, in order to determine how variations in model parameters can affect the optimal production levels of the refinery. A LP model of refinery is used as a case study. A stability analysis technique, modified tolerance approach, is applied to determine sensitive parameters that need accurate estimate or intensive monitoring; and compute their stability limits, i.e., allowable variation ranges, in order to maintain efficient plant operation.

2. Refinery optimization model

In this project, the production levels of main products are optimized in order to determine the most efficient distributions for the intermediate products in order to maximize the net profit. However, the model dose not consider the optimization of individual units or equipments but the interaction between the main units of a refinery. Six main refinery units were considered in modeling the problem. They are fractionation, fluidized catalytic cracking, reforming, isomerization, desulphurization and blending units, as in Figure 1.

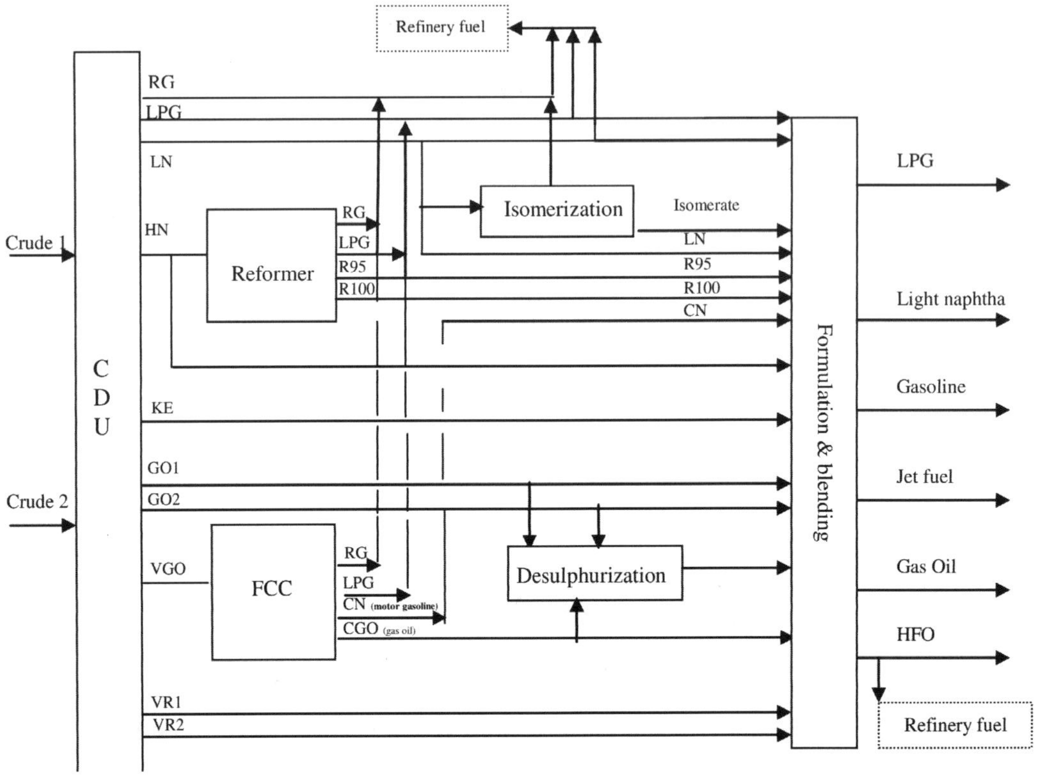

Figure 1: A simplified sketch of a typical oil refinery

Considering the above process units, the objective of optimization problem is to find the optimum production levels (quantities of raw materials, intermediate product, and final products) for each unit such that they lead to the maximum profit with satisfying process and quality constrains.

The refinery optimization model is formulated as a LP problem as follows:

$$max \quad \sum_{j \in M} p_j y_j - \sum_{i \in N} c_i y_i - \sum_{l \in L} \sum_{k \in K} q_l x_{l,k} \tag{1}$$

$s.t.$

$$\sum_{k \in K} a_k x_{l,k} = 0 \quad (for\ each\ component) \tag{2}$$

$$\sum_{k \in K} a_k x_{l,k} = D_j \qquad for \quad j \in M \tag{3}$$

$$\sum_{i \in N} y_i \le S_i \tag{4}$$

$$\sum_{j \in M} y_j \geq D_j \qquad (5)$$

$$\sum_{k \in K} \frac{v_k}{\rho_k} a_k x_{l,k} \leq v_{j,max} y_j \qquad (6)$$

$$for\ j \in M$$

$$\sum_{k \in K} \frac{v_k}{\rho_k} a_k x_{l,k} \geq v_{j,min} y_j \qquad (7)$$

$$\sum_{k \in K} x_{l,k} \leq O_l \qquad (8)$$

The economic objective function (1) attempts to minimize the costs of raw materials and process operations and maximize the revenues from the products. The constraints of this optimization problem can be classified into four different types. Balances of components (2&3) for each product component around each major unit; demand and supply limits (4&5); products quality constraints where refinery products have certain specifications need to be satisfied such as gasoline octane number, butane content, and also maximum and minimum vapor pressure as in (6 &7); and equipments or unit capacity constraints like the capacity of the CDU (8).

3. Post-optimality analysis

After the optimal solution has been computed for a given model, it is important to know how the solution behaves under different variations in problem parameters. Sensitivity analysis and stability analysis are used to evaluate the effects of variations on the optimal solution or basis of the LP problem.

3.1 Sensitivity analysis

Consider a LP problem in form:

$$Max\ \{\ P = \mathbf{c}^{\mathbf{T}}\mathbf{x} : \mathbf{Ax} \leq \mathbf{b},\ \ \mathbf{x} \geq \mathbf{0},\ \ \mathbf{x} \in \mathbf{R}^{\mathbf{n}} \} \qquad (8)$$

sensitivity analysis is usually associated with the determination of the values of the Lagrange multipliers, $\boldsymbol{\lambda}$, that describe the change in the optimal solution with respect to the variations in RHS coefficients. The sensitivity relations are important and useful for the decision maker, but the major challenge is to determine when they are valid. For example, the Lagrange multiplier, λ_i, presents the increase in the optimal value for a maximization problem when the associated RHS coefficient, i.e., b_i, is increased by one unit; however, we do not know by how much the coefficient can be increased under simultaneous variations in vector $\boldsymbol{b}$ before the optimal basis changes and the value of the Lagrange multiplier becomes invalid. This shows the importance of computing stability limits for each coefficient under simultaneous variations and within which the optimal basis remains unchanged.

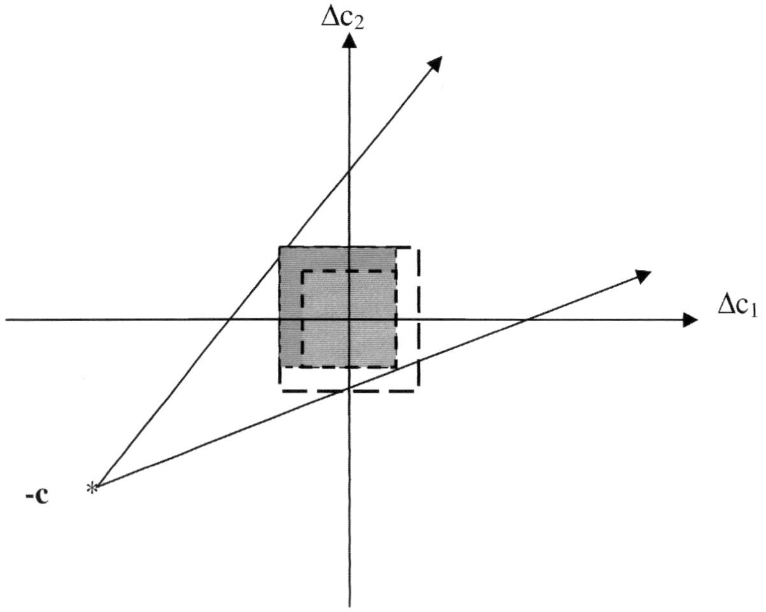

Figure 2: Obtained stability limits inside the stability cone.

3.2 Stability analysis

During the last few decades, many stability approaches have been proposed, for variation in parameters of LP[2]. To date there is no single approach that dominates. In contrast to other approaches, the tolerance approach[2,3] leads to easy-to-use results and considers simultaneous and independent variation in the problem parameters. It basically depends on optimality conditions and uses the concept of basic and non-basic variables to modify matrix $\boldsymbol{A}$ at each iteration. In this study, the modified tolerance approach proposed by Al-Shammari[4] is used to determine the stability ranges. The proposed method provides a new perspective on the problem and has two steps for computing the stability region or limits. First, it defines the entire stability region as a cone and studies the relation between the sensitivity information, Lagrange multipliers, and model parameters. Second, it determines maximum stability limits presented by the maximum rectangular parallelepiped or hyperbox that can be built inside the cone. This hyperbox offers flexible and easy-to-use allowable variation limits as shown later on for variation in objective coefficients, i.e. prices or raw materials and products.

3.2.1 Variations in vector $\boldsymbol{c}$

To demonstrate the approach, consider problem (8) that has a unique optimal solution. To define the entire stability region or stability cone for variations in the coefficients of the objective function, duality information or Lagrange multipliers are used:

$$\nabla_b \boldsymbol{P}^* = \boldsymbol{\lambda} = \boldsymbol{c}^T \boldsymbol{A}_A^{-1} \tag{9}$$

where $\boldsymbol{A}_A$ is the matrix of active constraints. By introducing the perturbations vector, $\Delta\boldsymbol{c}^T$:

$$\boldsymbol{\lambda}' = (\boldsymbol{c}^T + \Delta\boldsymbol{c}^T)\boldsymbol{A}_A^{-1} \tag{10}$$

by using the non-negativity condition on the optimal solution, there is no change in optimal solution if $\boldsymbol{\lambda}' \geq \boldsymbol{0}$. By substituting and rearranging:

$$-\Delta\boldsymbol{c}^T \boldsymbol{A}_A^{-1} \leq \boldsymbol{\lambda} \tag{11}$$

This inequality relation represents the stability region. This stability region can be defined as a stability cone because it satisfies the definition of a cone. In other words, the optimal solution and basis (not objective value) remain optimal under any scalar positive multiplication in the objective function. The solution remains optimal for any variations that satisfy equation (11). The stability cone is shown in Figure 2 for a maximization problem with two variables. Next step is defining the largest possible stability ranges starting with computing ordinary (individual) stability limits for Δc_i as follows:

$$\left(\max_h \quad \Delta c_h = \frac{\lambda_g}{A^{-1}_{Agh}} : \Delta c_h < 0\right) < \Delta c_h^{ord} < \left(\min_h \quad \Delta c_h = \frac{\lambda_g}{A^{-1}_{Agh}} : \Delta c_h > 0\right) \tag{12}$$

where h and g are indices of objective coefficients and constraints, respectively. In Figure 2, the ordinary stability limits of the coefficient c_i are the intersections between the cone's constraints and Δc_i axis. The main challenge in LP stability analysis is the presentation of this cone to the decision maker in a simple and useful way, especially for simultaneous variations. The most useful approach is to construct the largest possible hyperbox inside the cone. Extension of stability analysis for simultaneous variations is discussed in details in Al-Shammari[4].

3.2.2 Variations in vector ***b***

For variations in the RHS coefficients, a similar stability analysis is employed for the dual problem :

$$min \ \{ \boldsymbol{\lambda}^{\mathrm{T}}\boldsymbol{b} : \boldsymbol{\lambda}^{\mathrm{T}}\boldsymbol{A} \geq \boldsymbol{c}^{T} \ \& \ \boldsymbol{\lambda} \geq \boldsymbol{0}\} \tag{13}$$

to determine the variation limits before the optimal basis changes. In this analysis the optimal solution and slack variables are used in the same manner as the Lagrange multipliers were used in the variations analysis of vector ***c***.

4. Results and discussion

A refinery with capacity of 210,000 barrel per day and two different feedstocks is considered in this study. The simplified LP problem has 52 continuous decision variables and 53 constraints in addition to nonnegative constraints[5-7]. The problem was solved using Matlab and the maximum profit was found to be 3.4082×10^7 dollars per month without considering the costs of maintenance, storage, transporting and also fixed costs. Optimal solutions of key process variables, production levels, are shown in Table 1; in addition to the results obtained form stability analysis of the objective coefficients. Table 1 shows the allowable range of price changes for each feedstock or product within which the obtained optimal production levels remain optimum. In other words, the optimal solution would change if the price of feedstock 1 increased by more than 14.5 $/ton and the process would consume more from feedstock 2 because it became relatively cheaper. Such information can help the decision maker to understand the interaction between production levels and other factors (e.g., prices); and determine when the process need to be re-optimized.

Obviously, some parameters have infinite limits in one direction of changes because they are not purchased and any increase in their prices would not affect the solution. Nevertheless, this is not the real case for feedstock 2. There is a constraint on the minimum consumption of feedstock 2 and it is active for the current solution.

In contrast, individual stability limits for individual variations is grater than those for simultaneous variations since the obtained stability range for each parameter decreases with increasing the number of uncertain parameters. Presented limits of simultaneous variation were obtained using same weighting factor (k_i=1) for all parameters in Table 1; however, different weighting factors can be used based on the relative important or frequent change of each parameters. Analysis and results obtained for variations in RHS coefficients (e.g. supply and demand of materials and capacity constraints) will be discussed and presented in the extended paper.

Table 1:Stability limits of individual and simultaneous price variations of some variables.

Selected variables	x_i^* 1000 ton per month	c_i \$/ton	Individual stability limits, \$/ton		Simultaneous stability limits, \$/ton	
			lower	upper	lower	upper
Feedstock 1	442.2	530	- Inf	14.5	-Inf	8.3
Feedstock 2	350.0	500	-27.5	Inf	-11.7	Inf
Reformer feedstock imported	6.138	604	-113.6	119.4	-8.3	33.4
Cracker feedstock imported	0.0	581	-108.3	Inf	-52.2	Inf
Imported gasoline 95	0.0	715	-35.0	Inf	-17.5	Inf
Exported gasoline 95	128.26	680	-7.3	35.0	-1.8	17.5
Imported jet fuel	0.0	657	-32.0	Inf	-16.0	Inf
Exported jet fuel	115.10	625	-42.8	32.0	-8.3	16.0
Imported gas oil	0.0	620	-30.0	Inf	-28.2	Inf
Exported gas oil	227.82	590	-39.4	3.7	-8.3	1.8
Imported HFO	0.0	452	-37.0	Inf	-27.9	Inf
Exported HFO	210.96	415	-6.4	37.0	-1.8	11.7

Conclusion

In this work, post-optimality analysis was applied to a simplified refinery model, formulated as LP problem, in order to investigate the effect of uncertainty or variation in model parameters, mainly RHS and objective coefficients. Modified tolerance approach was used to compute the allowable variation limits, for individual and simultaneous variations, within which the operation levels remain optimum. The main objective of the obtained results is to supply the decision maker in the refinery with useful and easy-to-use information that can help to understand the interaction between production levels and other factors (e.g., prices) and to enable effective use of sensitivity information such as Lagrange multipliers.

Acknowledgment

The authors would like to acknowledge King Fahd University of Petroleum & Minerals for supporting this project (FT090005).

Nomenclature

a	*Molar fraction for each component*
c	*Cost of raw materials or imported streams in $/ton*
i	*Stream index*
j	*Stream index*
k	*Index of feed (or product) of each unit*
l	*Unit index*
q	*Operation cost of each stream entering each unit $/ton*
p	*Price of each product $/ton*
v, v_{max}	*Vapor pressure*
x, y	*Streams flow rate, 1000 ton per month*
D	*Demand of products, 1000 ton*
K	*Number of streams entering all units*
L	*Number of units*
M	*Number of final products*
N	*Number of raw materials and imported streams*
S	*Supply of raw materials*

References

1. Sahinidis, Nikolaos. Optimization under uncertainty: state-of-the-art and opportunities. Comp. Chem. Eng., 2004, 28, 971-983.

2. Filippi, Carlo., A fresh view on the tolerance approach to sensitivity analysis in linear programming., Eur J Oper Res., 2005,167,1-19.

3. Wendell, Richard. Sensitivity analysis revisited and extended., Decision Sci., 1992, 23, 1127-1142.

4. Al-Shammari, Abdallah., Post-optimality analysis of linear operations optimization problems., PhD thesis, University of Alberta., 2008.

5. Wauquier j., Trambouze p., and Favennec, J., Petroleum refining: refinery operation and management, Éditions Technip, Paris, 2001.

6. Parkash, Surinder., "Refining processes handbook", Elsevier, 2003.

7. Web page of U.S Energy Information Administration.

(88a) Catalytic Formation of Carbonyl Sulfide During Treatment of Coal-Derived Fuel Gas with Pd/γ-Al_2O_3 Sorbents

Erik C. Rupp, Evan J. Granite and Dennis C. Stanko, National Energy Technology Laboratory, Pittsburgh, PA

This group has developed a Pd/γ-Al_2O_3 sorbent capable of removing trace contaminants at moderate to high temperatures (200 – 450 °C) from coal-derived fuel gas. In the development of the sorbent, the catalytic formation of carbonyl sulfide (COS) was noted, which would require modification to sulfur removal processes. The formation of COS over a Pd catalyst is also a major concern for Pd alloy based hydrogen separation membranes proposed for use with fuel gas. The conditions required for catalytic formation have been explored, with up to 80 ppmv COS formed under gas-solid contact times of under 0.1 s. A reaction mechanism is proposed and multiple kinetic models are explored based on experimental results. The potential for hydrolysis of COS based on the γ-Al_2O_3 support is considered.

See more of this Session: Clean Coal Technology
See more of this Group/Topical: Energy and Transport Processes

(88b) Sustainability Assessment of Polygeneration Systems

Abhishek Jayswal[1], Dr. Helen Lou[2], Preeti Gangadharan[1], Kailang Zeng[1] and Anand Zambar[1], (1)Dan F. Smith Department of Chemical Engineering, Lamar University, Beaumont, TX, (2)Chemical Engineering, Lamar University, Beaumont, TX

Production of chemicals, electricity and fuels from natural resources via polygeneration systems is one of the emerging clean technologies for reducing greenhouse emissions. The polygeneration systems unite chemical, energy and fuel sectors and provide excellent efficiencies, in terms of materials and energy conversion by receiving multiple natural resources as feeds and producing many useful products through various interrelated processes. The complex nature of polygeneration systems presents a challenge for analysis of their sustainability footprints. This study presents a systematic approach for sustainability assessment of such processes by merging exergy efficiency analysis and inherently safe design approach into conventional economic, environmental and social aspects of sustainability. This work compares several different Polygeneration systems, which uses coal and natural gas as feeds, on sustainability criteria.

See more of this Session: Clean Coal Technology
See more of this Group/Topical: Energy and Transport Processes

(88c) Analysis of Freshwater Usage In Advanced Gasification Based Power Systems

Michael E. Walker and Javad Abbasian, Chemical and Biological Engineering, Illinois Institute of Technology, Chicago, IL

The thermoelectric power industry in the U.S. uses a large amount of freshwater. The large water demand is increasingly a problem, especially for new power plant development, as the availability of freshwater diminishes in the United States. Gasification based systems are next generation coal conversion processes that achieve higher process efficiencies than conventional power systems. Furthermore, gasification based systems use significantly less water than PC-Boiler systems, particularly when carbon dioxide capture strategies are applied.

A wide array of work has been focused on reducing water usage in existing PC plants. This paper addresses the applicability of developed water conservation methods to gasification based systems, e.g. Integrated Gasification Combined Cycle (IGCC). This work will provide insight into the challenges and limitations of water reduction within the IGCC system through careful analysis of detailed process models. In addition, water savings from novel process configurations will be discussed.

See more of this Session: Clean Coal Technology
See more of this Group/Topical: Energy and Transport Processes

(100a) Effects of Process Parameters on Carbon Deposition During Iron Catalyzed Fischer-Tropsch Synthesis Studied with a Tapered Element Oscillating Microbalance

Ard C. J. Koeken[1], Matthijs Ruitenbeek[2] and Krijn P. de Jong[1], (1)Chemistry - Inorganic Chemistry and Catalysis, Utrecht University, Utrecht, Netherlands, (2)Hydrocarbons R&D, DOW Benelux BV, Terneuzen, Netherlands

Effects of Process Parameters on Carbon Deposition during Iron Catalyzed Fischer-Tropsch Synthesis studied with a Tapered Element Oscillating Microbalance

Ard C. J. Koeken, * *Matthijs Ruitenbeek,*# *Krijn P. de Jong**

* *Inorganic Chemistry and Catalysis - Debye Institute for Nanomaterials Science - Utrecht University - Sorbonnelaan 16 - 3584 CA - Utrecht - The Netherlands*

DOW Benelux BV, Hydrocarbons R&D, P.O. Box 48, NL-4530 AA Terneuzen - The Netherlands

Production of lower olefins from renewable feedstock or coal is of great interest to the chemical industry.[1] The Fischer-Tropsch synthesis (FTS) can be the key process to directly convert synthesis gas, a mixture of carbon monoxide and hydrogen gas produced from renewable feedstock or coal, into lower olefins. Iron catalysts have promising characteristics such as high olefin selectivity, high water-gas shift activity, and low cost.[2] Transformation of the catalytic phase, carbon deposition, sintering, and poisoning are the main causes for the deactivation of Fe catalysts during FTS.[2] In particular, at the high reaction temperatures required for a high lower olefin selectivity, carbon deposition is an issue as it may result in catalyst particle disintegration.[2] As the deposition of carbon on the catalytic surface increases the mass of the solid catalyst, measuring the mass *in situ* is a means to investigate the kinetics of carbon deposition. The Tapered Element Oscillating Microbalance (TEOM) allows for such a measurement at varying temperature and pressure under plug-flow conditions.[3]

Here, the first TEOM study of carbon deposition during FT is presented for an a-Al_2O_3 supported Fe catalyst. The figure shows the influence of CO pressure on the carbon build-up and the weight-time yield (WTY). A clear increase in rate of carbon deposition with CO pressure is found.

THE GRAPHICS (Figure A and Figure B) ON THIS PAGE WERE NOT AVAILABLE ON THE E-MEDIA VERSION OF THIS DOCUMENT

Figure- A) Carbon build-up.

Figure- B) Product weight-time yields.

Conditions: 350 °C, 10 bar H_2, 50 mL/min H_2, 0.02 g catalyst (10wt% Fe).

In the full presentation we discuss extensively the effect of process parameters (e.g. P, T, H_2/CO ratio, GHSV) on carbon deposition. It is demonstrated that the TEOM is a powerful tool to investigate carbon deposition along with activity and selectivity under FTS conditions that are industrially relevant.

[1] R. Diercks *et al*, Chem. Eng. Tech. 31 (2008) 631.

[2] M. E. Dry, Appl. Catal A 138 (1996) 319-344

[3] D. Chen *et al*, in Advances in Catalysis, Vol 51, Vol. 51, Elsevier Academic Press Inc, San Diego, 2007, pp. 351.

E-mail: a.c.j.koeken@uu.nl

www: http://www.anorg.chem.uu.nl/home/index.html

See more of this Session: Kinetics, Design and Modeling of Fischer-Tropsch Catalysts and Reactions I
See more of this Group/Topical: Topical 6: 11th Topical Conference on Gas Utilization

(100b) Fischer-Tropsch Synthesis: Metal-Support Interfacial Contact Governs Oxygenates Selectivity Over CeO_2 Supported Pt-Co Catalysts

Muthu Kumaran Gnanamani, Mauro C. Ribeiro, Weinping Ma, Gary Jacobs, Uschi M. Graham, Wilson D. Shafer and Burtron H. Davis, Center for Applied Energy Research, University of Kentucky, Lexington, KY

Abstract

The reaction of carbon monoxide and hydrogen to form hydrocarbons using either an iron or cobalt catalyst usually produces a broad range of products which consists of mainly paraffins and olefins with a carbon number of up to and greater than 100, and low levels of oxygenates including alcohols, aldehydes, ketones and acids. Iron catalysts usually produce more olefins and oxygenates compared to supported cobalt catalysts at elevated temperatures and pressures [1,2]. For Fischer-tropsch (FT) synthesis, cobalt has been reported to produce mainly n-alkanes over a wide range of molecular weights. The cobalt active phase is generally deposited over an oxide support (e.g., Al_2O_3, SiO_2, or TiO_2), which provides good mechanical strength (i.e., attrition resistance) and thermal stability under reaction conditions. Factors such as particle size and shape, metal loading and dispersion, nature of the support and strength of metal-support interactions have been widely studied over supported cobalt-based catalysts [3,4]. Many interesting studies have been devoted to ascertain the influence of oxide support on the FT activity, selectivity, and stability of cobalt catalysts.

Information regarding the influence of CeO_2 during FT synthesis is scarce and only limited data are available in the open literature considering the use of ceria as either a promoter or a support. Recently, Spadaro et al. [5] obtained a higher selectivity to oxygenates using CeO_2 supported cobalt catalysts at very low CO conversions (~1.6); however, the authors did not provide an explanation regarding their mechanism of formation. In general, the amount of oxygenates produced under typical cobalt FT synthesis conditions is very small, and they are therefore often neglected. The objective of the present study is to examine the FT activity and selectivity of Pt promoted cobalt catalysts supported over high surface area ceria at moderate temperatures and pressures and compare oxygenates selectivity to a standard Pt promoted Co/γ-Al_2O_3 catalyst. In addition, different cobalt loadings were used to achieve different cobalt particle sizes, and thus, different degrees of interaction with the support. In this way, the aim was to assess the plausibility of a metal-support bifunctional pathway in the formation of higher alcohols. From this analysis, a plausible mechanism (Scheme 1) is proposed for the formation of oxygenates over Pt promoted Co-Ceria catalysts.

THE GRAPHIC (Scheme 1) ON THIS PAGE WAS NOT AVAILABLE

ON THE E-MEDIA VERSION OF THIS DOCUMENT

Scheme 1 A plausible pathway of alcohol formation on Pt promoted Co-CeO_2 catalysts.

Acknowledgements

The work carried out at the CAER was supported in part by funding from the Commonwealth of Kentucky.

References

1. G.P. van der Laan, A.A.C.M. Beenackers, Catal. Rev.-Sci. Eng. 41 (1999) 255.
2. M.E. Dry, Stud. Surf. Sci. Catal. 152, chapter 7, (2004) 533.
3. E. Iglesia, Appl. Catal. A: Gen. 161 (1997) 59.
4. G. Jacobs, T.K. Das, Y. Zhang, J. Li, G. Racoillet, Burtron H. Davis, Appl. Catal. A: Gen, 233 (2002) 263.
5. L. Spadaro, F. Arena, M.L. Granadeos, M. Ojeda, J.L.G. Fierro, F. Frusteri, J. Catal. 234 (2005) 451-462.

See more of this Session: Kinetics, Design and Modeling of Fischer-Tropsch Catalysts and Reactions I
See more of this Group/Topical: Topical 6: 11th Topical Conference on Gas Utilization

(100c) Synthesis of Higher Alcohols From Syngas Over a K Promoted Cu-Co-Zn Catalyst In Supercritical n-Hexane

Rui Xu[1], Sihe Zhang[2], Ed Durham[2] and Christopher B. Roberts[3], (1)Chemical Engineering, Auburn University, Auburn, AL, (2)Department of Chemical Engineering, Auburn University, Auburn, AL, (3) Department of Chemical Engineering, Auburn University, Auburn University, AL

Synthesis of higher alcohols, isobutanol in particular, from coal, biomass or natural gas via synthesis gas (syngas) has been attracting researchers' interests due to the potential for producing clean and high octane value fuels. Higher alcohols could be used directly as fuels, as fuels additives for octane or cetane enhancement, and as feedstocks in the chemical industry. Due to the exothermic nature of the reactions and the kinetic restrictions, this procedure suffers from poor productivity and selectivity towards higher alcohols. For the last two decades, there has been intensive research centered on shifting the synthesis away from methanol and toward higher alcohols, including; i. the use of modified catalysts, ii. utilizing dual-bed reactors, or iii. slurry phase reactors. In this study, a supercritical medium (i.e. hexane) was employed to investigate the supercritical solvent effect on the formation of higher alcohols. A Cu-Co based catalyst was prepared by a coprecipitation method and evaluated in a fixed bed reactor. The results of catalytic investigations with and without the supercritical hexane medium demonstrated that supercritical hexane has a significant effect on the syngas conversion and the productivity towards higher alcohols.

See more of this Session: Kinetics, Design and Modeling of Fischer-Tropsch Catalysts and Reactions I
See more of this Group/Topical: Topical 6: 11th Topical Conference on Gas Utilization

(100d) Kinetic Monte Carlo Simulations of the Fischer-Tropsch Reaction

Mohammed Minhaj Ghouri[1], Ionel Mugurel CiobÓc•[2], A. P. J. Jansen[1] and Rutger van Santen[1], (1) Department of Chemical Engineering and Chemistry, Eindhoven University of Technology, Eindhoven, Netherlands, (2)Sasol Technology (PTY) Ltd., Eindhoven, Netherlands

Kinetic Monte Carlo (kMC)[1] simulations are the most relevant tool to accurately study the kinetics of the surface catalytic processes such as the Fischer-Tropsch (FT) sysnthesis. Quantum mechanically calculated activation barriers[2] and rate constants of the elementary processes such as adsorption, diffusion, desorption etc. are related to their actual occurance and therefore to the overall kinetics by the kMC methods. In this work, kMC simulations are used to perform a detailed analysis of the effect of barriers and the thermodynamics of the intermediates in the methanation reaction. We will present models on methanation representataive of different surfaces of transition metals as Ru or Co. We present also the trends in the methanation yield rate and the surface compositions as a function of temperatures and CO activation energies. We then proceed to two different models which incorporate the C-C coupling in the formation of ethylene and ethane respectively. We observe that the reaction rates are very sensitive wth respect to activation energies of CO and rate of oxygen removal. References: 1. Tonek Jansen, Kinetic Monte Carlo, pgs 183-197, Computational Methods in Catalysis and Materials Science. (Eds) Rutger A. van Santen and Phillipe Sautet, WILEY-VCH Verlag GmbH & Co. KGaA, Weinheim, 2009. 2. I. M. Ciobica, G. J. Kramer, Q. Ge, M. Neurock, and R. A. van Santen Mechanisms for Chain Growth in Fischer-Tropsch Synthesis over Ru(0001), Journal of Catalysis 212, 136-144 (2002).

See more of this Session: Kinetics, Design and Modeling of Fischer-Tropsch Catalysts and Reactions I
See more of this Group/Topical: Topical 6: 11th Topical Conference on Gas Utilization

(100e) Preconditioning Effects Upon Co-Containing Pt-Promoted and Unpromoted Fischer-Tropsch Catalysts

Donald Cronauer, Jeffrey W. Elam, A. Jeremy Kropf and Joseph A. Libera, Argonne National Laboratory, Argonne, IL

As part of a research program to create, prepare, and evaluate Fischer-Tropsch (FT) catalysts, active catalysts formed by atomic-layer deposition [ALD] of major components were demonstrated. ALD-deposited Co on a silica support was more effective than a similar catalyst deposited upon a support of ALD-deposited Al_2O_3 on silica. The activities of Co on Al_2O_3 catalysts prepared by the Center for Applied Energy Research (CAER) were compared with those of these catalysts. The addition of Pt, by either ALD or incipient wetness techniques, as a promoter of Co-containing catalyst has been shown to primarily affect the catalyst pre-conditioning step. CO conversion results were consistent with previously reported [1] TPR-XANES/EXAFS experiments observing the nature of the chemical transformations occurring during the activation of cobalt-based FT catalysts in H_2. Specifically, there exists a 2-step reduction process involving Co_3O_4 to CoO to Co^0 transformations. The extent of catalyst preconditioning was strongly affected by the reduction temperature and the level of Pt deposition. A continuous-flow catalytic-bed unit was used for this FT study. The gaseous feed consisted of either a 1:1 or 2:1 molar blend of H_2:CO, and the experiments were carried out at temperatures ranging from about 260°C to 300°C, pressures averaging 13.1 MPa (190 psia), a gas feed rate of about 15 mL/min, and a catalyst charge of 0.1 g.

1. G. Jacobs, Y. Ji, B.H. Davis, D.C. Cronauer, A.J. Kropf, C.L. Marshall, *Applied Catalysis A, General* 333 (2007), 177-191

See more of this Session: Kinetics, Design and Modeling of Fischer-Tropsch Catalysts and Reactions I
See more of this Group/Topical: Topical 6: 11th Topical Conference on Gas Utilization

(100f) Development of a Kinetic Model for Fischer-Tropsch Synthesis Over a Ru Promoted Co/Al_2O_3 Catalyst In a Slurry Reactor

Tejas Bhatelia[1], Wenping Ma[2], Gary Jacobs[2], Burtron H. Davis[2] and **Dragomir B. Bukur**[1], (1) Department of Chemical Engineering, Texas A&M University at Qatar, Doha, Qatar, (2)Center for Applied Energy Research, University of Kentucky, Lexington, KY

Abstract:

Depleting oil reserves, environmental pressure, as well as abundant reserves of coal, natural gas and biomass, have all contributed to a revived interest in Fischer-Tropsch (F-T) technology for producing ultra-clean, virtually sulfur-free, transportation fuels and chemicals. F-T technology involves conversion of synthesis gas (i.e., a mixture of H_2 and CO) to hydrocarbons, which can be upgraded by processes such as hydrocracking to produce liquid transportation fuels and chemicals - especially diesel, jet fuels, lubricants and waxes. F-T synthesis is a very complex reaction that produces a large number of products. The reaction is catalyzed by certain metals or metal carbides of the transition metal elements Co, Ru, Ni, and Ru (Dry, 1996). Due to the low intrinsic water-gas-shift activity of Co and the fact that it is easier to separate from the products than Fe based catalysts, cobalt based catalysts are used in commercial reactors for gas-to-liquids (GTL) conversion.

F-T synthesis is a very complex reaction that converts synthesis gas in a product spectrum consisting of a complex multi-component mixture of linear and branched hydrocarbons and oxygenated products. Main products are linear paraffin and α-olefin. The overall reaction stoichiometry may be approximated as

$CO + (1 + m/2n)\, H_2 \rightarrow 1/n\, C_nH_m + H_2O$ - FT (SEQ Equation * ARABIC 1)

$CO + H_2O \leftrightarrow CO_2 + H_2$ - WGS (SEQ Equation * ARABIC 2)

Cobalt is not very active for the water gas shift (WGS) reaction thus, in contrast to most iron-based Fischer-Tropsch catalysts; only a small fraction of the water produced is subsequently converted to carbon dioxide.

A number of kinetic studies have been performed by many researchers, for various cobalt based catalysts. They proposed expressions that are either in the form of empirical power laws or mechanistic equation of LangmuirñHinshelwoodñHougenñWatson (LHHW) type. However, most of these studies used experimental data that were conducted over a narrow range of process conditions in integral fixed bed reactors. Morever, LHHW type of rate laws proposed by researchers were simplifiied by neglecting some of the adsorbed species (Sarup and Wojciechowski, 1989). In this study, we have attempted to overcome these shortcomings and propose a detailed kinetic model for hydrocarbon formation rate over a wide range of operating conditions by using a stirred tank slurry reactor (STSR), which more closely mimics the kinetics of the commercialized slurry process.

The kinetics of the F-T synthesis over 0.27% Ru 25% Co/Al_2O_3 catalyst was studied using the STSR. Experiments were conducted at reactor pressures of 1.41 MPa and 2.4 MPa, temperatures of 205°C and 220°C, H_2/CO feed ratios of 1.4 and 2.1 and gas space velocities ranging from 2 to 15 NL/g-cat∑h. LHHW type rate equations were derived on the basis of a detailed set of possible reactions originating from carbide and enolic pathways for hydrocarbon formation. In some of the previous work on Co catalysts it was assumed that water molecules do not occupy a significant fraction of the active sites (Sarup and Wojciechowski, 1989). No such a priori assumptions were made with regard to the adsorption coefficients of any species in this work. Experimental rates were modified to account for catalyst aging. Derived rate equations were fitted to the corrected experimental rate using Levenberg-Marqurdt method to obtain model parameters.

Statistical and physical tests were performed to discriminate between the rival models. Models yielding unrealistic values of adsorption coefficients i.e. non- positive adsorption coefficients were excluded for further model discriminations. Remaining models were then tested for statistical significance. R-square values were calculated in order to determine what proportions of the variances the model accounts. F-tests were also performed to measure variances between the experimental and modeling rates. It was found that hydrogen-assisted dissociative adsorption of CO followed by hydrogenation of dissociatively adsorbed CO is the likely path for formation of the monomer (methylene) and is the likely rate controlling step in F-T synthesis. Rates obtained from the best kinetic model were able to provide a satisfactory fit to the experimental data.

References:

DRY, M. E. (1996) Practical and theoretical aspects of the catalytic Fischer-Tropsch process. *Applied Catalysis A: General,* 138, 319-344.

SARUP, B. & WOJCIECHOWSKI, B. W. (1989) Studies of the Fischer-Tropsch synthesis on a cobalt catalyst II. Kinetics of carbon monoxide conversion to methane and to higher hydrocarbons. *The Canadian Journal of Chemical Engineering,* 67, 62-74.

Acknowledgement:

The authors would like to acknowledge the financial support of this project by Qatar National Research

Funding under grant (NPRP 08-173-2-050).

* *Corresponding Author: Dragomir Bukur, dragomir.bukur@qatar.tamu.edu,Tel:+974-444230134.*

See more of this Session: Kinetics, Design and Modeling of Fischer-Tropsch Catalysts and Reactions I
See more of this Group/Topical: Topical 6: 11th Topical Conference on Gas Utilization

(106a) Effect of Si/Al Ratio on CoZr/MFI Zeolite Composite Catalysts for Direct Isoparaffins Synthesis From Syngas

Xing Huang[1], Bo Hou[2], Jungang Wang[3], **Debao Li**[2], Litao Jia[2], Jiangang Chen[4] and **Yuhan Sun**[5], (1) State Key Laboratory of Coal Conversion, Institute of Coal Chemistry, The Chinese Academy of Sciences, Taiyuan, China, (2)State Key Laboratory of Coal Conversion, Institute of Coal Chemistry, The Chinese Academy of Science, Taiyuan, China, (3)State Key Laboratory of Coal Conversion,, Institute of Coal Chemistry, The Chinese Academy of Science, Taiyuan, China, (4)State key Laboratory of Coal Conversion, Institute of Coal Chemistry, Chinese Academy of Sciences, Taiyuan, China, (5)State Key Laboratory of Coal Conversion, Shanghai Advanced Research Institute, Chinese Academy of Sciences, Shanghai, China

The effect of acidity on direct isoparaffins synthesis in CO hydrogenation over CoZr/MFI (HZSM-5 and Silicalite-1) zeolite composite catalysts was investigated using zeolites with different Si/A1 ratios by a direct hydrothermal synthesis method. The composite catalysts were characterized by XRD, SEM, H_2-TPR, NH_3-TPD and XPS, and were comparatively investigated for Fischer-Tropsch synthesis in a single fixed-bed reactor under the conditions of 2.0 MPa, 533K, H_2/CO=2.0, 1000h^{-1}. The effect of the Si/Al ratio on the selectivity of isomers was correlated with the acidity and crystallization of the MFI zeolites, and the Co-support interaction, which strongly affected the Co reducibility, was also explored. The composite catalyst with a Si/Al ratio of 40 exhibited significantly higher selectivity of isoparaffins (23.72 wt% of iso-C_{5-11}) and lower methane selectivity (14.35 wt%) than other composite catalysts, while the CoZr/Silicalite-1 catalyst showed the largest amount of heavy hydrocarbons (11.92 wt% of C_{18+}) and the lowest selectivity to isoparaffins (only 2.00 wt% of iso-C_{5-11}) due to no Al source was introduced in the hydrothermal synthesis. The results indicate that zeolite composite catalyst with low Si/A1 ratio can be used to efficiently realize high selectivity of isoparaffins and low selectivity to methane from syngas.

See more of this Session: Kinetics, Design and Modeling of Fischer-Tropsch Catalysts and Reactions II
See more of this Group/Topical: Topical 6: 11th Topical Conference on Gas Utilization

(106b) Influence of Preparation Methods on Catalytic Activity of Bi-Modal Mesoporous Cobalt-Based Catalysts for Fischer-Tropsch Synthesis

Jungang Wang[1], **Debao Li**[2], Bo Hou[2], Litao Jia[2] and **Yuhan Sun**[3], (1)State Key Laboratory of Coal Conversion,, Institute of Coal Chemistry, The Chinese Academy of Science, Taiyuan, China, (2)State Key Laboratory of Coal Conversion, Institute of Coal Chemistry, The Chinese Academy of Science, Taiyuan, China, (3)State Key Laboratory of Coal Conversion, Shanghai Advanced Research Institute, Chinese Academy of Sciences, Shanghai, China

A series of bimodal mesoporous cobalt-based catalysts for Fischer-Tropsch synthesis were prepared using three kinds of methods: incipient-wetness impregnation (IWI), excessive impregnation (EI) and vapor-induced internal hydrolysis (VIH). The catalysts were characterized by N_2 physisorption, XRD, HRTEM, H_2-TPR and Raman spectroscopy. The N_2 physisorption results showed the catalysts prepared by IWI and VIH well retained the double mesoporous character, and the cobalt particles were mainly located in the pore proved by HRTEM. The FTS results showed the reducibility and dispersibility of catalysts prepared by IWI and EI were very high. Especially the catalyst prepared by IWI method with smallest Co_3O_4 crystallite size and highest dispersibility showed the higher FTS activity, lower methane selectivity and higher C_5^+ selectivity.

Key words: bi-modal; preparation method; cobalt-based catalyst; Fischer-Tropsch synthesis

See more of this Session: Kinetics, Design and Modeling of Fischer-Tropsch Catalysts and Reactions II
See more of this Group/Topical: Topical 6: 11th Topical Conference on Gas Utilization

(106c) Kinetics of Methane Formation During Fischer-Tropsch Synthesis Over Co/SiO_2 and Co/ Al_2O_3 Catalysts

Wenping Ma, Gary Jacobs, Tapan K. Das and Burtron H. Davis, Center for Applied Energy Research, University of Kentucky, Lexington, KY

Kinetics of methane formation during Fischer-Tropsch synthesis over Co/SiO_2 and Co/Al_2O_3 catalysts

Wenping Ma, Gary Jacobs, Tapan K. Das, and Burtron H. Davis*Center for Applied Energy Research, University of Kentucky, 2540 Research Park Drive, Lexington, Kentucky, 40511, USA.Abstract Wate produced during Fischer-Tropsch synthesis (FTS) was found to decrease CH_4 selectivity significantly. This was assumed to take place by way of water inhibiting the secondary hydrogenation of primary olefins [1-3]. This implies that the power law kinetic model which was previously used to describe the rate of methane formation over cobalt- [4] and iron-based catalysts [5] can be improved to include a parameter to define the impact of water. In the present contribution, a new empirical model incorporating a water effect term () in the power law model - , where $k = k_0 \, exp\,(-E_a/RT)$, is the CH_4 formation rate constant, *a* and *b* are the reaction orders for CO and H_2, respectively, and *m* is a water effect constant - was used to fit the methane kinetic data obtained over 25%Co/HP14 g-Al_2O_3 15%Co/SiO_2 catalysts. Therefore, the new methane kinetic model has the same form as the CO consumption model developed by CAER [6-9]. Kinetic experiments over the two catalysts were conducted under a range of conditions: 205-220 ^oC, 280 psig, H_2/CO = 1.0-2.5 and 3.3-22 Nl/g-cat/h a 1-L continuously stirred tank reactor (CSTR). The results indicate that water exhibits a negative ef on CH_4 formation during FTS for both types of catalyst, suggesting that the impact of water on CH_4 formation during FTS is kinetic in nature, consistent with interpretations reported in the literature. T both CO conversion and CH_4 selectivity could be accurately described by the CAER kinetic model [over the range of conditions used in the work. Activation energies of methane formation obtained or two supported Co catalysts were found to be close (~135.5 kJ/mol), in agreement with the range of values reported in the open (i.e., 98-145 kJ/mol). The kinetics equations obtained at 220 ^oC over 2 Co/HP14 g-Al_2O_3 and 15%Co/SiO_2 catalysts are described as follows:25%Co/HP14 g-Al_2O_3 : mol/ g-cat/h;15%Co/SiO_2: , mol/g-cat/h The values of reaction orders *a* and *b* calculation for CH_4 formation, as well as the CH_4 formation activation energies, are different from those of CO consumptic kinetic results reported in [8] and [9]. This may reflect that CH_4 formation on the catalysts is involved in not only normal FTS polymerization pathways, but also another pathway, for example methanation, which may take place on different active sites.**Acknowledgements**This work was supported by NASA contract, #NNX07AB93A and the commonwealth of Kentucky.

References`[1] H. Schulz, M. Claeys, S. Harms, Stud. Surf. Sci. Catal. 107 (1997) 193.
[2] [4] E. Iglesia, S.C. Reyes, R.J. Madon, S.L. Soled, Adv. Catal. 39 (1993) 221.
[3] Ø. Borg, S. Eri, E.A. Blekkan, S. Storsæter, H. Wigum, E. Rytter, A. Holmen, J. Catal. 248 (2007) 89.
[4] C.H. Bartholomew, W. **Lee,** Stud. Surf. Sci. Catal. 130 (2000) 1151.
[5] R.A. Dictor, A.T. Bell, J. Catal. 97 (1986) 121.
[6] J.L. Li, X.D. Zhan, G. Jacobs, B. H. Davis, Appl. Catal. 208 (2002) 203.

[7] T.K. Das, W.A. Conner, J. L. Li, G. Jacobs, M.E. Dry, B.H. Davis, Energy & Fuels 19 (2005) 1430.
>[8] W. Ma, G. Jacobs, M.K. Gnanamani, U.M. Graham, in Advances in Fischer Tropsch Synthesis, Catalysts, and Catalysis, eds. Burt H. Davis and Mario L. Occelli. Taylor & Francis Group, 2009, p.31.

[9] W. Ma, G. Jacobs, D. Sparks, M.K. Gnanamani, R. Pendyala, B.H. Davis, C.H. Yen, J.L.S. Klettlinger, T.M. Tomsik, Fuel, accepted, in press.
[10] W.H. Lee, C.H. Bartholomew, J. Catal. 120 (1989) 256.
[11] K.B. Kester, J.L. Falconer, J. Catal. 89 (1984) 380.

* Corresponding author: Burtron H. Davis, davis@caer.uky.edu, Tel: 859-257-0251.
See more of this Session: Kinetics, Design and Modeling of Fischer-Tropsch Catalysts and Reactions II
See more of this Group/Topical: Topical 6: 11th Topical Conference on Gas Utilization

(106d) An Investigation Into the Role Played by the Initial Iron Oxide Phase In Determining Catalyst Performance

P. Julius Pretorius, Saskatchewan Research Council, Saskatoon, SK, Canada and Frank C. Hawthorne, Dept of Geological Sciences, University of Manitoba, Winnipeg, MB, Canada

Iron-based Fisher-Tropsch (FeFT) catalysts have a number of characteristics that make them attractive in the Biomass to Liquids (BTL) arena. Feedstock logistics dictate that BTL plants will be smaller than GTL and CTL plants. Furthermore, syngas obtained from biomass gasification typically have a H_2/CO ratio of 1 – 1.5. Co-based FT catalysts require a $H_2/CO \geq 2$ to prevent deactivation due to coke formation, thus requiring a separate WGS plant. This may add cost and diminish the overall thermodynamic efficiency of the process.[1] Low Temperature FeFT (LT FeFT) catalysts are prepared by precipitation from a copper-containing iron nitrate solution, prepared by dissolving metallic copper and iron with concentrated nitric acid. The resulting precipitate is washed and promoters are added. We have recently found that the mode of preparation exerts a significant influence on catalyst activity.[2] In this paper, the relationship between observed synthesis results and different iron phases observed in fresh and spent catalyst, as determined by Mössbauer spectroscopy and XRD, is examined.

(1) Bartholomew, C. H.; Hecker, W. C. *Short Course on Fischer-Tropsch Catalysis: Fundamentals and Practice*, Brigham Young University, 2009.

(2) Pretorius, P. J. *Preparing low temperature iron Fischer-Tropsch catalysts: Strategies that work and those that do not.*, In *240th ACS National Meeting* Boston, MA, 2010.

See more of this Session: Kinetics, Design and Modeling of Fischer-Tropsch Catalysts and Reactions II
See more of this Group/Topical: Topical 6: 11th Topical Conference on Gas Utilization

(106e) The Influence of Co Nanoparticles Distribution In Fischer-Tropsch Co-Supported Catalysts on Selectivity

Galina Pankina[1], P.A. Chernavskii[2], Andrei Y. Khodakov[3] and V.V. Lunin[2], (1)Chemical Department, M.V. Lomonosov MSU, Moscow, Russia, (2)Chemical, M.V. Lomonosov MSU, Moscow, Russia, (3) Unitex de Catalyse et de Chimie du Solide, Universite des Sciences et Technologies de Lille, Lille, France

It' s known that realistic FT catalysts are characterized by metall particles size distribution. Moreover, as the average particles size and the distribution during the catalytic process are changing. The distributions parameters of Co particles in Co/Al_2O_3 catalysts (dispersion and average value) were modified by alteration the calcination temperature of (200^0C – 500^0C) by Pt promotion and sucrose addition to catalyst at study of preparation and by changing the water vapors pressure [[i]]. For characterization of catalysts the magnetometric method ***in situ*** was employed. The distributions calculated from dates of magnetic method are almost the same as received by from TEM's dates. In this study are showed that increase of temperature calcination result in increase of average Co particles size and the shape of distribution. So the distributions parameters maximum is achieving in 400^0C [[ii]]. The addition to Co/SiO2 catalyst of Pt and the promotion Co-Ru/SiO2 by sucrose leads to decrease of average Me particles size down to superparamagnetic sizes and the dispersion of distribution [[iii]].

So, it was showed that the width of distribution is principal factor, which influence on FT selectivity. The smaller the width of dispersion, the more C_{5+} - selectivity (1). For Co nanoparticles with diameter less than 6 nm we see the increase of methane catalysts function, but with diameter more than 6-7 nm the influence of Me particles size on selectivity is absent.

References

[i] A.V. Mugtasimov, P.A. Chernavskii., G.V. Pankina, V, V. Lunin. //Russ. J. Phys. Chem., 2008, V.82, ¹3,

pp.495-498

[ii] Chernavskii P.A., Chu Wei, Khodakov A.Y., Pankina G.V., Peskov N.V.// J. Phys. Chem., 2008, V.

82, ¹6, pp. 1083-1088.

[iii] Chernavskii P.A., Khodakov A.Y., Pankina G.V., Girardon J.-S., Quinet E. // Appl. Catal., 2006, V. 306, pp.108-119.

See more of this Session: Kinetics, Design and Modeling of Fischer-Tropsch Catalysts and Reactions II
See more of this Group/Topical: Topical 6: 11th Topical Conference on Gas Utilization

(111a) Simulation In Early Stage Process Development

Joe Schroer, Clearwaterbay Technology, Inc., Pomona, CA

Simulation tools are frequently used in process development. When properly applied, they can provide powerful results, such as the evaluation of many flowsheet alternatives in a short amount of time, and numerical calculation and representation of the complex chemical system. When used for process development, there are challenges. One concern is the limited availability of the properties of new chemical systems and mixture properties for previously not evaluated systems. Key questions that help to focus the efforts are "What is the simulation objective?", and "What is the process development objective?"

See more of this Session: Best Practices Utilizing Simulation Tools, Pre and Post-Project
See more of this Group/Topical: Topical 7: 14th Topical on Refinery Processing

(111b) Understanding Design and Operational Issues Using Dynamic Simulations

Dipanjan Bhattacharya[1], Manish Jain[1] and Ramachandra P. Tekumalla[2], (1)Bechtel Oil and Gas, Houston, TX, (2)Advanced Simulation & Analysis, Bechtel Corporation, Houston, TX

Simulation has traditionally been used for steady state process design. In reality, many process design controlling scenarios are dictated by transient conditions. Dynamic simulation effectively predicts these transient situations during process design and helps in reducing future operational and start-up risks. This article goes through multiple case studies where dynamic simulation has been effectively used to design equipments as well as predicting methods to minimize operational problems. The article also shows cases where dynamic simulation validates design calculations conducted by traditional methods.

See more of this Session: Best Practices Utilizing Simulation Tools, Pre and Post-Project
See more of this Group/Topical: Topical 7: 14th Topical on Refinery Processing

(111c) Business Justification for the Investment In Dynamic Simulators

Martin Berutti, MYNAH Technologies, Chesterfield, MO

Almost everyone who has used a dynamic simulator for testing automation system configuration, validating process design, or training operators claim that they received great value from the simulation tools. But how do users build a solid economic justification for this investment? This paper will show how to justify the investment in dynamic simulation by increasing the Net Present Value (NPV) and Internal Rate of Return (IRR) of the total automation system capital investment. Case studies are shown to demonstrate the operational effectiveness and project efficiency that influence these calculations, plus examples how the simulation tools were applied.

See more of this Session: Best Practices Utilizing Simulation Tools, Pre and Post-Project
See more of this Group/Topical: Topical 7: 14th Topical on Refinery Processing

Operational Inputs into the Mechanical Integrity Program

F. Russ Davis, CSP
Aon Energy Risk Engineering
6455 South Shore Blvd. Suite 400, League City, TX 77573
russ.davis@aon.com

Prepared for Presentation at
American Institute of Chemical Engineers
2011 Spring Meeting
7th Global Congress on Process Safety
Chicago, Illinois
March 13-16, 2011

UNPUBLISHED

Operational Inputs into the Mechanical Integrity Program

F. Russ Davis, CSP
Aon Energy Risk Engineering
6455 South Shore Blvd. Suite 400, League City, TX 77573
russ.davis@aon.com

Keywords: mechanical integrity (MI), corrosive, temperature, pressure, pH

Abstract

A predictive mechanical integrity (MI) program is based on the following:

Damage mechanisms that affect process equipment

Rate at which degradation progresses

Ability of the inspection department to identify and correct equipment damage before equipment failure

Operational temperature, pressure, flow and composition are process variables that greatly affect process equipment life cycles.

A risk-based inspection program or any proactive inspection program should establish mechanical integrity (MI) plans for all equipment included in the program. An equipment plan should be based on anticipated damage, suitable inspection techniques capable of assuring the reliability of the process equipment in order to reduce the risk of operating the equipment to the lowest practicable level. Equipment plans are created based on the chemicals in the operational environment and the damage that may be anticipated from those chemicals.

Any time that the operational environment changes, the damage that affects the process equipment may also change. This paper will explore the process parameters and their affect on process equipment and the mechanical integrity (MI) program. How changes made to operational parameters such as temperature, flow, pH, pressure and process stream consistency will be reviewed and discussed. This paper will result in recommendations on how to incorporate process changes and their affect into the mechanical integrity (MI) program.

1. Introduction

The primary purpose of a mechanical integrity (MI) program is to prevent critical equipment failures leading to a loss of containment of highly hazardous chemicals. Critical equipment should be defined as equipment whose failure could reasonably and directly lead to the occurrence of a hazardous event. Equipment failures have resulted in very serious fires and explosions with injury, loss of life and significant impact to corporation assets. Recent accident investigations such as the 2005 BP Texas City and the 2009 Silver Eagle Refinery explosions have listed failure of the mechanical integrity (MI) program as root cause or as contributing causes.

Note: Do not add page numbers. Do not refer to page numbers when referencing different portions of the paper

The mechanical integrity (MI) program should be predicated on a risk basis. Equipment should be inspected, tested, and maintained based on the risk associated with operating the equipment to personnel, the environment and to the fiscal well being of the company. Mechanical integrity risk is defined as the product of the consequence of a failure and the likelihood that the failure will occur. Mechanical integrity programs impact the likelihood of failure by recognizing a damage mechanism, evaluating the effects of the damage mechanism on the process equipment and inspecting and testing the equipment on a frequency that will predict the life cycle of the equipment. Equipment is then repaired or replaced prior to hazardous failure. The damage mechanism part of the equation is the operational environment. The operational environment includes, but is not limited to:

- Chemicals in the process stream
- Temperature of operations
- Pressure of operations
- Flow/erosivity of the process stream
- pH of the process stream

Changes to the operational environment can have an impact on the ability of the mechanical integrity department to predict the life-cycle of the equipment. All changes to operational parameters should be conveyed to the mechanical integrity (MI) department. The mechanical integrity (MI) department can then assess the impact to process equipment and adjust equipment inspection and testing plans if needed. Operations departments should have a system in place to recognize process changes and communicate the changes to the mechanical integrity (MI) department.

2. Risk-Based Mechanical Integrity (RBMI) Program Overview

In a risk-based mechanical integrity (RBMI) program each equipment item receives a risk evaluation to determine the criticality of the equipment within the inspection and testing program. RBMI programs are primarily directed at fixed equipment but the concept can be used to determine the relative risk ranking for any process equipment or system. To determine the relative risk of an equipment failure the consequence of the failure is compared with the likelihood of the failure. Likelihood of failure (LOF) is determined by analyzing all damage mechanisms that can adversely affect the equipment item dependent on the material of construction of the item. Damage mechanisms are most closely associated with the operating parameters in which the equipment resides. A simplified example is carbon steel equipment operating in a process stream with water and oxygen making equipment susceptible to corrosion.

Temperature and flow rates can have a significant effect on the rate of corrosion and the anticipated life of the equipment. Higher temperatures can speed corrosion rates in some environments. Higher operating temperatures contribute to their own damage mechanisms such as creep, thermal shock and high temperature hydrogen attack.

A risk-based analysis is performed using data supplied by the operations department. Anticipated damage mechanisms and assumed corrosion rates are developed based on that operational data. Inspection plans are then developed for high risk equipment. If there are significant changes in operational parameters from those used to perform the initial risk assessment the RBMI department may have to re-evaluate the risk analysis and reformulate inspection and testing plans.

3. Operational Parameter Changes can affect the Mechanical Integrity (MI) Program

3.1 Erosion/Erosion-corrosion

Erosion is the removal of surface material as a result of impingement by solids, liquids, or vapors. All typical metals and refractory are susceptible to some degree to erosion.

Erosion-corrosion is the damage that occurs when corrosion contributes to the erosion process by removing protective films or scale and exposing the metal surface to further corrosion. The erosion process then impinges on the metal surface and removes the metal or refractory.

Erosion-corrosion is the more common of the two mechanisms. The erosion element can be increased by the corrosiveness of the operating environment. Increased velocity of the process stream will increase the rate of metal loss due to the mechanism. Increasing the corrosivity of the process stream will also increase the metal loss. Increases in temperature can also be a contributor to a higher than anticipated corrosion rate.

The mechanical integrity (MI) department should understand what equipment is susceptible to erosion-corrosion in the process. Operations personnel should be part of the process analysis so the correct process data can be used to evaluate risk of equipment failure. Operations should provide worst-case process information. Operational conditions such as temperature and flow rates used to determine the likelihood of equipment failure should be based on the worst possible operating conditions. When temperatures, flow rates and pH levels change the mechanical integrity (MI) department should be informed. Systems for making sure information flows from operations to the mechanical integrity (MI) department will be discussed in later paragraphs.

3.2 Corrosion Fatigue

Corrosion fatigue is cracking that develops in a corrosive environment when cyclic loading is present. Cracking can often initiate from pit locations caused by corrosion. The cyclic loading then propagates the cracks. Deaerators and boilers are examples of equipment that is susceptible to this damage mechanism. Inspection plans are typically established based on plant history or industry average. If operations is experiencing a larger than normal cyclic operations (i.e., shutdowns and startups), the mechanical integrity (MI) department should reassess the frequency between inspections. Corrosion fatigue cracking and crack propagation is directly tied to the number of cycles experienced by the equipment.

3.3 Chloride Stress Corrosion Cracking (Cl SCC)

Chloride Stress Corrosion Cracking (Cl SCC) is cracking caused by chlorides in the operating environment, primarily affecting 300 series stainless steel materials and effecting other alloys to a lesser degree. Cl SCC occurs under the combined action of tensile stress, temperature and aqueous chloride. Tensile stresses can be either induced due to operating environment or from residual manufacturing stresses. An increase in temperature increases the likelihood of cracking. Cracking usually occurs at a temperature of 140°F or higher but cracking has been found in environments where temperatures were lower than 140°F.

If operating temperatures are lower than 100°F, Cl SCC may not have been a major factor when the mechanical integrity (MI) department assessed the damage mechanism associated with this process equipment. Equipment inspection plans will be specified that may not include an emphasis of testing for cracking. If operating temperatures are raised equipment assessment should be re-evaluated. The same can be said for chloride concentration. When chloride concentrations are raised cracking can occur more aggressively.

Operational changes can have a severe impact on process equipment where chlorides are present. Changes in feedstock pH or percentage of chlorides and especially in temperature can make equipment susceptible to Cl SCC. A raise in process temperature can lead to chloride cracking or accelerate the Cl SCC damage mechanism. The water used for cleaning or as a part of the process can also have an impact on the amount of chlorides that can affect the equipment. Using water with high chloride content can cause Cl SCC in process equipment over time. Idling of process equipment containing chlorides, without clearing, can allow chloride concentration to build up leading to Cl SCC.

3.4 Caustic Corrosion

Caustic corrosion can be localized or general corrosion depending upon the concentration and strength. The caustic can be a constituent of the process stream or used for cleaning process equipment. Caustic corrosion primarily affects carbon steels, low alloy steels and 300 series stainless steels. Caustic corrosion can cause an accelerated loss of material leading to failures and loss of containment.

Process equipment preventative maintenance and inspection plans are typically established based on operating conditions. The use of caustic for cleaning equipment should be analyzed by the mechanical integrity (MI) department. Inspection plans can be established that recognize that there can be caustic corrosion present.

Problems arise when operations either adds caustic to streams where it was not previously found or starts using a cleaning substance containing caustic. The mechanical integrity (MI) department should be notified so that inspection plans can be modified.

3.5 Caustic Stress Corrosion Cracking

Caustic stress corrosion cracking is caused by embrittlement of susceptible materials in the presence of stress and elevated temperature. Increasing caustic concentrations and temperature increase the likelihood of cracking, but cracking has been found in concentrations as low as 50 ppm. Stresses can be either residual from manufacturing or induced from the operating pressures or installation. Carbon steel, low alloy steels and 300 series stainless steels are especially susceptible to caustic stress corrosion cracking.

Operating temperature is a critical factor used by the mechanical integrity (MI) department when equipment inspection plans are developed. It is understood that higher temperatures accelerate caustic cracking. When operating temperatures are raised or when temperature excursions occur the mechanical integrity (MI) department should re-evaluate frequency of inspection and testing for caustic cracking. When caustic concentrations are changed the mechanical integrity (MI) department should re-evaluate equipment inspection plans. As can be seen in the service graph from API 571[1] below, the susceptibility of carbon steel to caustic cracking is dependent on caustic concentration and temperature.

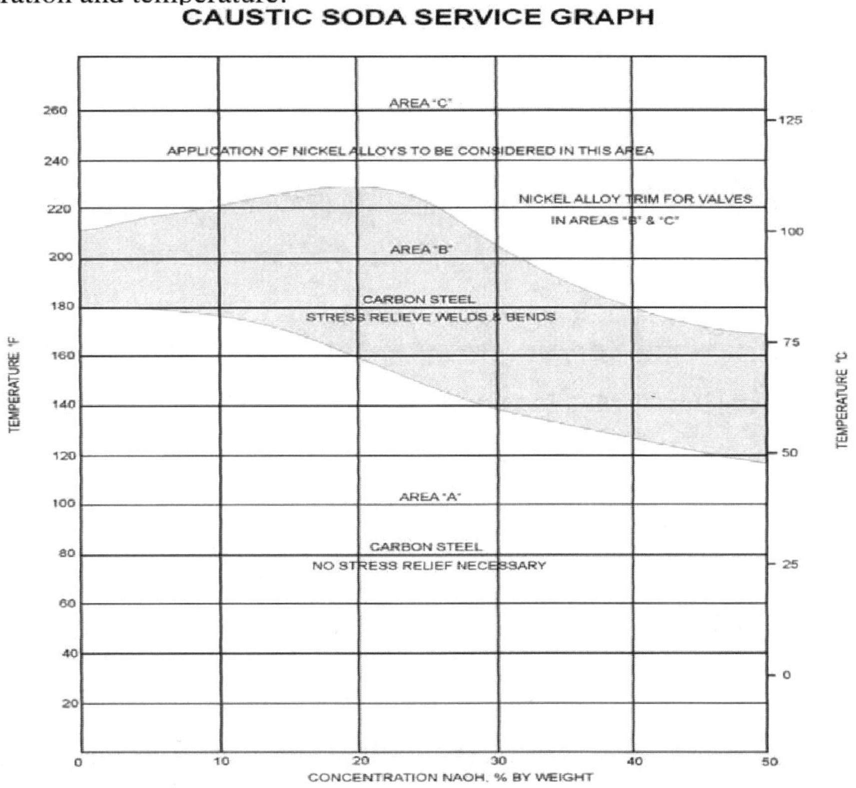

3.6 Microbial Corrosion (MIC)

Microbial corrosion, often referred to as microbiologically influenced corrosion (MIC), is a form of corrosion caused by living organisms. Microbial corrosion is usually found in aqueous environments especially prevalent where water flow is stagnant or very low. Microbial corrosion can attack many materials including carbon steel, stainless steel, aluminum and copper and some nickel based alloys. Microbial corrosion can be found in heat exchangers, storage tanks and piping where the flow is low or there are times of stagnant flow or when the water is not properly treated.

In many plant organizations operations is in control of water treating. The mechanical integrity (MI) department should be aware that microbial corrosion is a possible damage mechanism in systems where water is present and should be made aware of all changes to the process. If equipment is going to be left idle for a period of time the mechanical integrity (MI) department should be part of the discussion on how best to leave equipment idle. Operations should be the first to know when there are problems with water treating systems. If the treatment system is lost or if the treatment chemical is changed the mechanical integrity (MI) department should be notified. The mechanical integrity (MI) department can then re-assess equipment testing and inspection plans where needed.

3.7 Hydrogen Embrittlement

Hydrogen embrittlement is a loss of ductility in high strength steels due to penetration by atomic hydrogen. Hydrogen embrittlement can lead to brittle cracking and failure of process equipment. Hydrogen embrittlement can affect carbon steel, low alloy steels, some stainless steel and also some nickel based alloys. The three (3) factors that must be present for hydrogen embrittlement to occur are:

1. Stress must be present; either residual stress from manufacture or applied stress from the process
2. Hydrogen must be present at the critical concentration within the material
3. Steel or alloy must be susceptible to hydrogen embrittlement

Services where hydrogen embrittlement can be found include wet H_2S or HF acid services. Temperature is a critical element in hydrogen embrittlement. Hydrogen embrittlement is most likely to occur at room temperature operations. Hydrogen embrittlement is less likely to occur at higher temperatures, above 180°F.

Changes in operating temperature can have an impact on the possibility of hydrogen embrittlement occurrence. If operating temperatures are lowered the mechanical integrity (MI) department should be notified and equipment plans re-assessed. Equipment that was operating above the hydrogen embrittlement temperature may now need to be inspected and inspection plans should be changed.

Note: Do not add page numbers. Do not refer to page numbers when referencing different portions of the paper

3.8 Wet H_2S Damage

Wet H_2S damage includes the following:

- Hydrogen blistering
- Hydrogen induced cracking
- Stress oriented hydrogen induced cracking (SOHIC)
- Sulfide stress corrosion cracking (SSC)

Wet H_2S affects carbon steel and low alloy steels. Process stream pH, H_2S levels, contaminants, and temperature are all critical factors that affect process equipments susceptibility to wet H_2S damage.

Operational changes should always be relayed to the mechanical integrity (MI) department. When the pH of the feed stock changes the damage mechanism affecting process equipment can quickly change as well. Free water content changes can also make equipment susceptible to wet H_2S damage. Operations should work in conjunction with the mechanical integrity (MI) department when free water is noted in equipment where it is not usually found or when the free water content increases. When free water in liquid phase is present the likelihood of damage occurring is much higher when H_2S is present. The operations or process department monitors the daily parameters of the process. This group should recognize and be aware of the possibility of equipment damage occurring when parameters such as amount of free water in the process change.

3.9 Amine Stress Corrosion Cracking

Amine stress corrosion cracking can be found in Amine Scrubber Systems where H_2S and CO_2 are removed. Carbon steel and low alloy steels are susceptible to amine stress corrosion cracking. Stress, amine concentration and temperature are the critical factors for determining if material is susceptible to amine stress corrosion cracking. Stresses can be residual from manufacturing or induced from operations. Increase in temperature increases the severity and likelihood of amine stress corrosion cracking. Lower amine concentrations are less susceptible to amine stress corrosion cracking.

Amine concentration and operational temperature of the amine process are parameters monitored by the operations/process department. Changes made to the operation parameters can alter whether equipment is susceptible to amine stress corrosion cracking. When process parameters change the mechanical integrity (MI) department should be informed because equipment plans may need to be re-assessed.

3.10 Creep Damage

Equipment operated at high temperature can slowly deform over time due to stress. The deformation can eventually lead to equipment rupture. All metals and alloys are susceptible to creep damage. Equipment inspection and testing plans are based on the operating temperature and the time in service. Operating pressure is a critical factor in creep damage as a stress inducer. High pressures induce more stress into the equipment. API 571 notes that an increase of about 25°F, or a 15% increase in stress, can shorten equipment remaining life by as much as 50%.

If operations run a furnace at higher temperatures or at higher pressures the mechanical integrity (MI) department should be consulted. The equipment inspection and test frequency may need to be changed dramatically.

3.11 Thermal Fatigue Cracking

Thermal fatigue cracking can occur due to a thermal shock. Metals will expand and contract when affected by temperature differences. If hot metal comes in contact with a cold process stream rapid contraction can lead to cracking. Process upsets that send cold material to hot equipment should quickly be reported to the mechanical integrity (MI) department. Equipment testing and inspection plans may need to be revised.

4. Communications – Operational Information to Mechanical Integrity

Section 3 discussed some typical damage mechanisms experienced in refineries and chemical plants and how operational parameters can affect these damage mechanisms. By understanding the effects operational changes or upsets can have on process equipment it becomes clear that the mechanical integrity (MI) department should receive and analyze operational data. There are various ways that the mechanical integrity (MI) department should receive this information. The following sections discuss various options.

4.1 Process Trending and Monitoring

The operations department typically has a system in place for monitoring process trends such as Aspen's IP21. The mechanical integrity (MI) department should be trained in the use of process monitoring systems. When trends are developed to monitor critical equipment parameters the mechanical integrity (MI) department can quickly receive information that may have an effect on process equipment. When temperatures and pressures are changed the mechanical integrity (MI) department should be able to spot these changes. Possible damaging effects can be assessed and inspection and test plans can be modified if necessary.

4.2 Operational Changes

In too many facilities the mechanical integrity department is not a required approver of process changes. The mechanical integrity (MI) department should always be required to review process changes that may affect equipment. Higher or lower temperatures, higher or lower pressures, and higher or lower flow rates can significantly affect equipment life. As we have seen in the discussion on various damage mechanisms, each mechanism has critical factors that affect the possibility and severity of the damage that can be experienced by process equipment. When operational parameters change equipment that may not have been susceptible to a specific damage mechanism can now experience severe damage leading to equipment failure and loss of containment. Changes that the operations department may not necessarily see as adverse can affect the life of process equipment. Changes in feed pH can greatly affect the corrosion rate experienced by process equipment. Changes in sulfur or hydrogen content of raw materials can cause cracking in process equipment. Higher flow rates of abrasive processes leads to high erosion/corrosion rates. Operations should ensure that the mechanical integrity (MI) department is a required reviewer of all process changes.

4.3 Operational Upsets

Operational upsets are typically noted in operational reports. These reports should be forwarded to the mechanical integrity (MI) department and used in the evaluation of equipment test and inspection plans. Temperature excursions, pH excursions, and pressure excursions should always be relayed to the mechanical integrity (MI) department. Water breakthrough or build-up in processes where it is not normally found should be noted as an operational deviance and this information should be transmitted to the mechanical integrity (MI) department. Quality testing information may not be recognized as a factor that affects process equipment but in some cases factors affecting quality such as un-reacted chemical breakthrough in the process may allow corrosive materials into equipment where it is not normally found. The mechanical integrity (MI) department should be made aware of these quality issues in order to analyze possible damage effects.

5. Path Forward

Risk-based mechanical integrity systems evaluate the possible damage mechanisms that may adversely affect process equipment. Inspection and testing plans are developed to detect equipment damage and to repair or replace equipment prior to failure and loss of containment. These equipment plans are based on information provided by the operations group. When operational parameters change process equipment damage mechanism will change as well. The mechanical integrity (MI) department should be informed about any changes that will alter the basis for the equipment risk analysis and the inspection and testing plans developed for the process equipment. A system should be instituted that will convey operational information to the mechanical integrity (MI) department. The system should contain administrative controls such as a management of change (MOC) system as well as the on-line systems such as IP 21 that Operation uses to track process trends.

Plant operational designs should consult the mechanical integrity (MI) department for information when new units are designed or modified. Typically if the operational group needs the information so does the mechanical integrity (MI) department. It is important that adequate temperature, pH, flow and pressure instrumentation be located to provide sufficient data. The flow of data from the process to the mechanical integrity (MI) department should be considered in plant design.

Awareness training for operational personnel should be part of every operator's and engineer's required training matrix. Operations personnel need to understand how process changes could affect equipment. Operations personnel and the mechanical integrity (MI) department need to understand where process information comes from and how it is available. The mechanical integrity (MI) department should be trained in the use of the process trending software.

Safer plants, equipment reliability and less downtime are the primary benefits of a proactive Risk-based Mechanical Integrity (RBMI) program. The risk based mechanical integrity (RBMI) program is evergreen. Process information is the critical parameter in the design and maintenance of the RBMI program.

References

[1] API 571, "Damage Mechanisms Affecting Fixed Equipment in the Refining Industry", 1st ed. December 2003.

[2] API 580, "Risk-Based Inspection", 2nd ed. November, 2009.

[3] API 581, "Risk-Based Inspection Base Resource Documents", 1st ed., October 2000.

From Process Simulation to Operator Training - A Success Story of Business Improvement Enabled by Simulation Technologies
Rajat Basu, Yuon Chiu
Honeywell Specialty Materials, Buffalo, NY
Eddy Yap
Honeywell Process Solutions, Houston, TX

Honeywell

Honeywell SM and ACS: A Winning Linkage

Honeywell

Honeywell Specialty Materials

Honeywell Specialty Materials is a $4.14 billion world-leader in supplying high performance chemicals. The Geismar site is a leader in fluorine technology and provides acids and non-ozone depleting refrigerants to a broad customer base.

Honeywell Automation and Control

Honeywell Automation and Control Solutions is a $12.6 billion business applying sensing and control expertise that helps create safer, more comfortable, more secure and more productive environments. Honeywell is the leader in providing value added industrial solutions to the chemical marketplace.

Honeywell - A world leader in fluorine technology that applies leading control technology to create a safer, more comfortable, secure, productive and responsive solution for the chemical marketplace

We are owner-operators as well as solution providers

Lifecycle for Chemical Process Modeling

Honeywell

Operations Simulation

- Off-Line Training (Simulator)
- Operator Certification
- Error Trapping-Locate & Diagnose
- Upset/Start-Up/Shutdown Simulator

Helps Mfg Eng'rs

Dynamic Simulation

- Optimize Transitions
- Reduce Waste
- Test New Control Systems
- Improve Quality (Reduce Variability)
- Improve Loop Tuning
- Reduce Recovery Time
- Multivariate Control
- Inferential Sensors
- Product by Process

Helps Process Eng'rs

Steady State Simulation

- Design new plants/unit operations
- Reduce Operating Costs
- Increase Capacity
- Reduce Cycle Times
- Define Conditions for New Products
- "What-If Scenarios"
- Improve Quality
- Train Optimization
- Reduce Extractables
- Capture Expert knowledge

Increasing Knowledge and Reach

Impact Depends on Knowledge and Reach/Depth

Development of operator training simulator

Honeywell

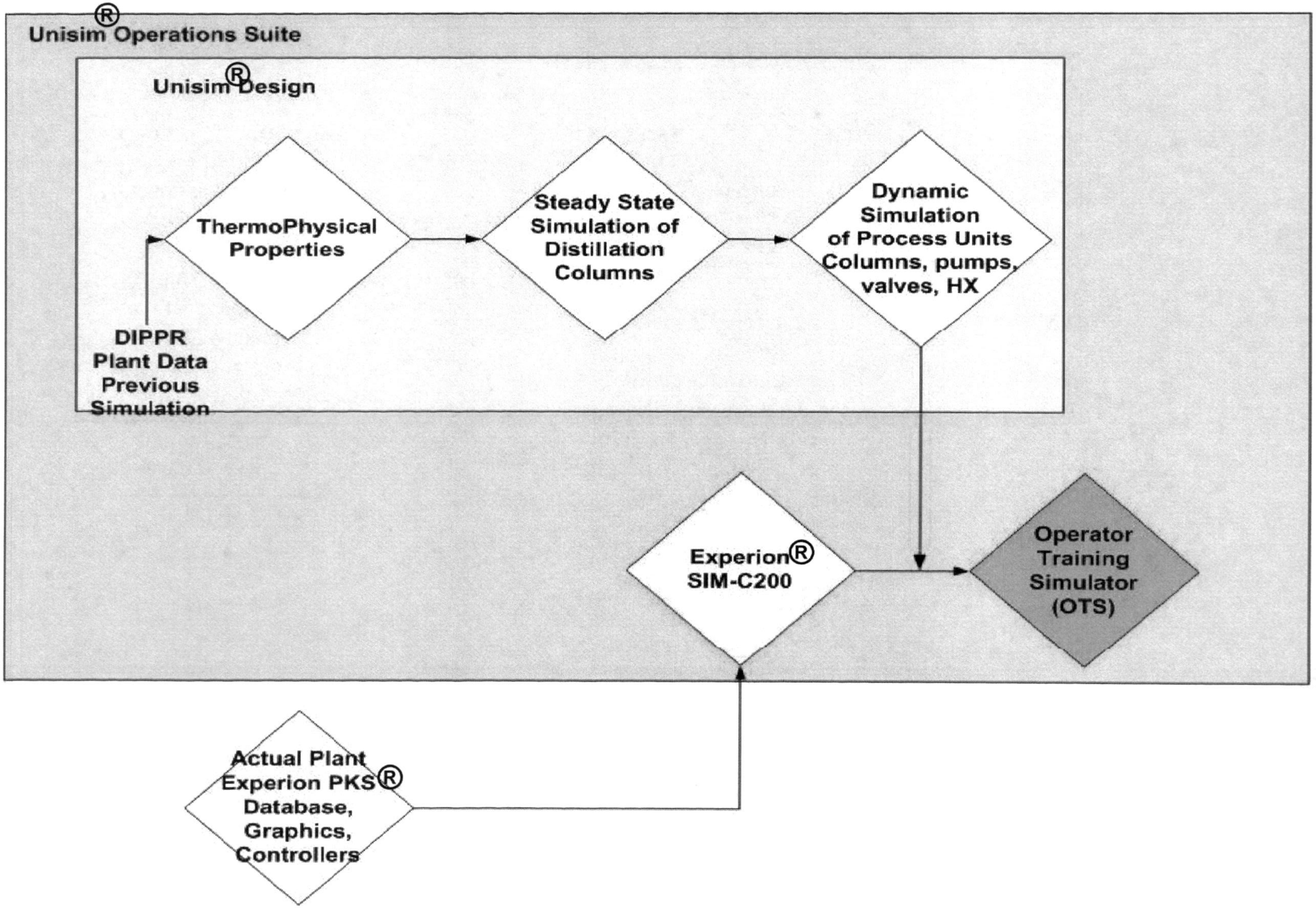

Model Development

- **Thermodynamic Models –**
  - Thermodynamic properties were measured and correlated and later added to process simulator to be used for plant simulation
  - Both equation of state and activity coefficient models were developed to be used in various parts of the plant
- **Steady State Models**
  - Thermo models were incorporated in UniSim® process simulator and plant models were developed
  - Models were validated on-line with plant data from UNIFORMANCE® data historian
- **Dynamic Models**
  - Steady state models were converted to dynamic models in UniSim
  - Dynamic models validated with known plant disturbances

Steady State on-line Model Validation

Honeywell

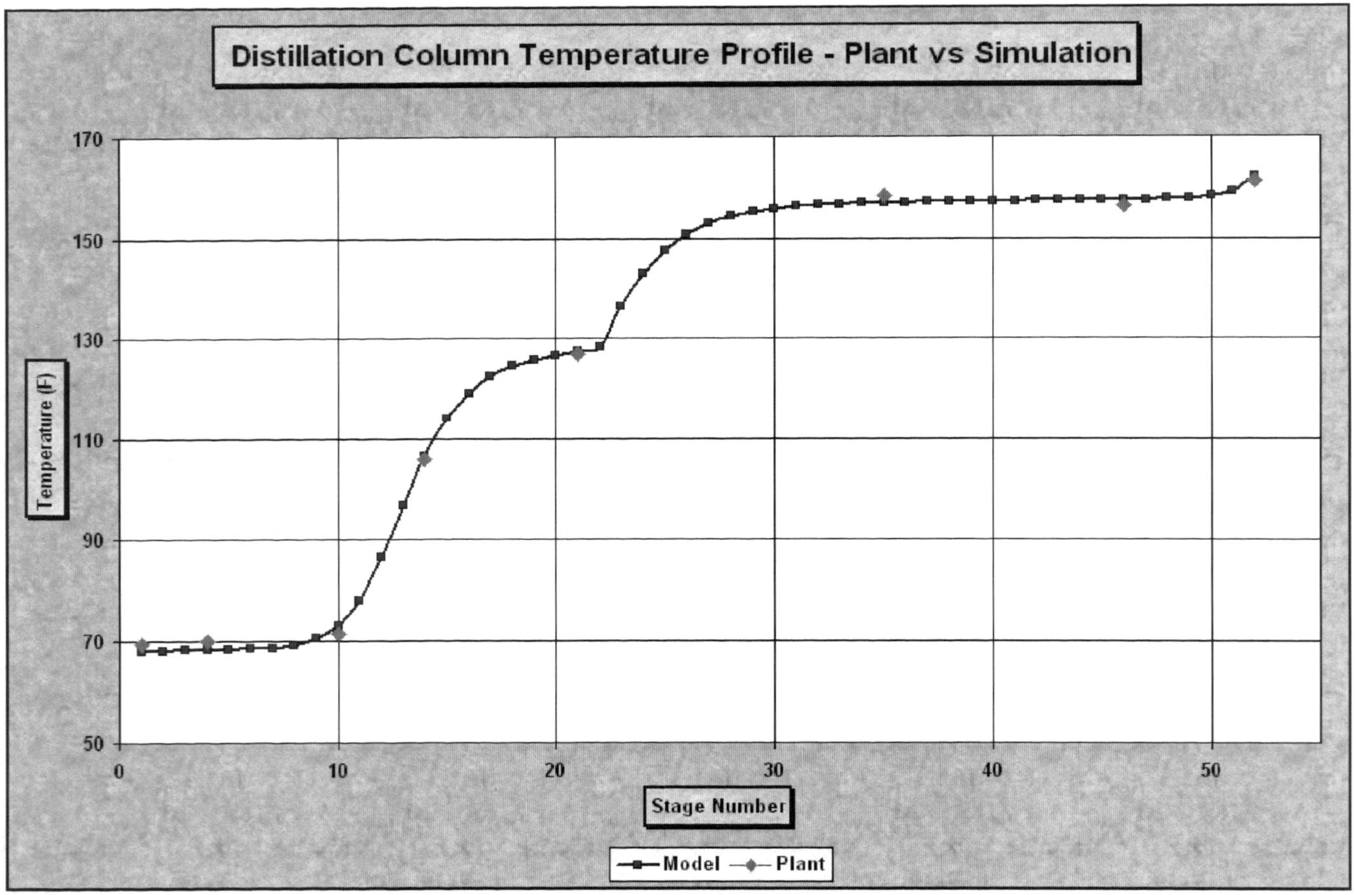

On-line plant data from data historian is validated with model calculation – aids plant operation and troubleshooting

UniSim®-Experion® PKS C200 Architecture

Honeywell

UniSim® – Experion® PKS C200

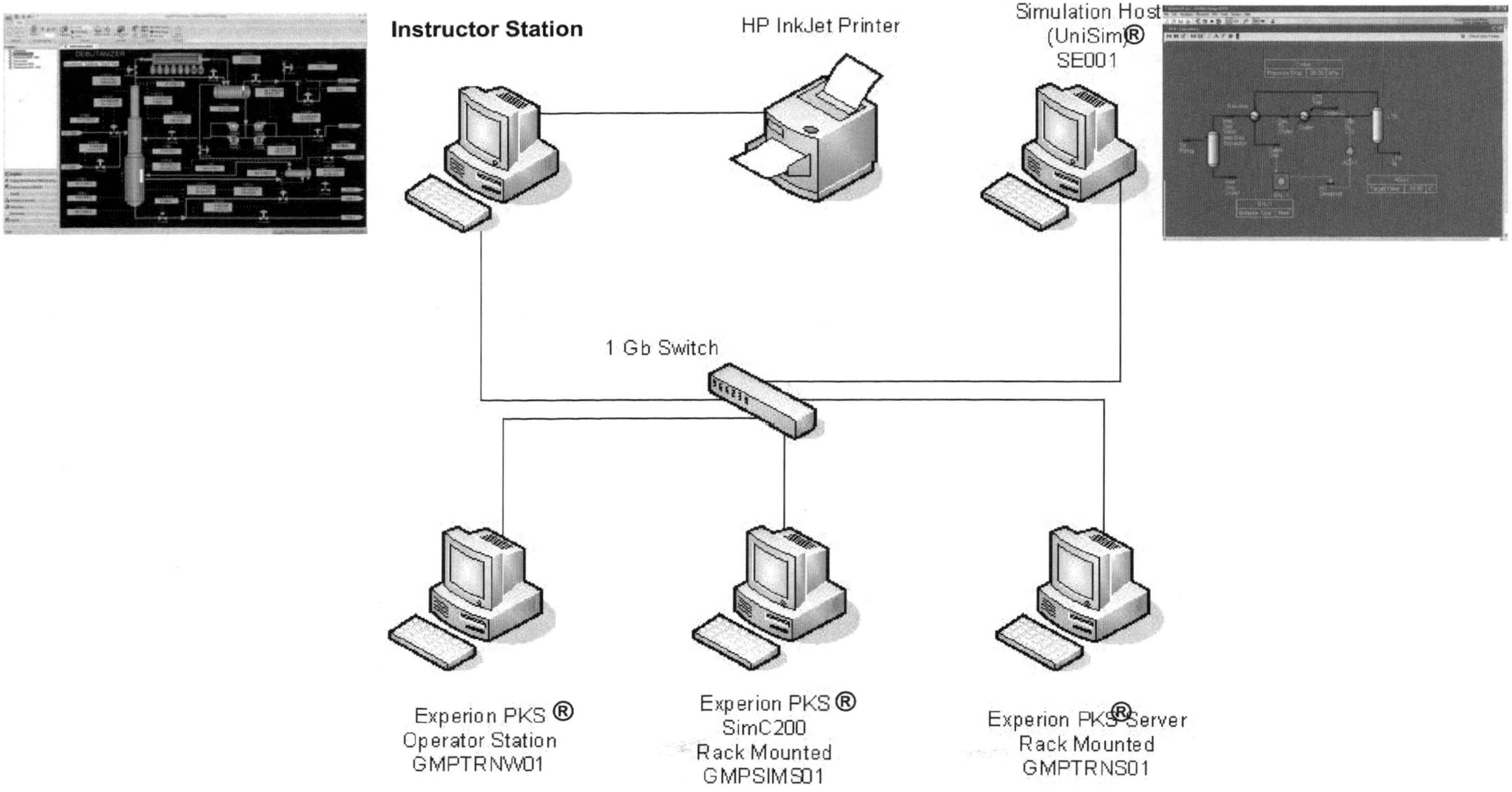

AIChE Spring Meeting - 2011

Maintenance of OTS

Honeywell

- Advantages of Virtualization[1]
  - Lower Cost – Multiple Virtual Machines (VM) can run on a single server
  - Greater flexibility – Users can add, remove or change computers and network components at will.
  - Better manageability – Snapshots provide a mechanism to roll back and forth between various backups. This substantially lowers the risk of upgrades and other changes.
  - Easier migration – Data centers can migrate to new hardware by simply copying VMs to a new server.
  - Easier remote support – Easier to duplicate identical system at remote location or easier remote access to the system.

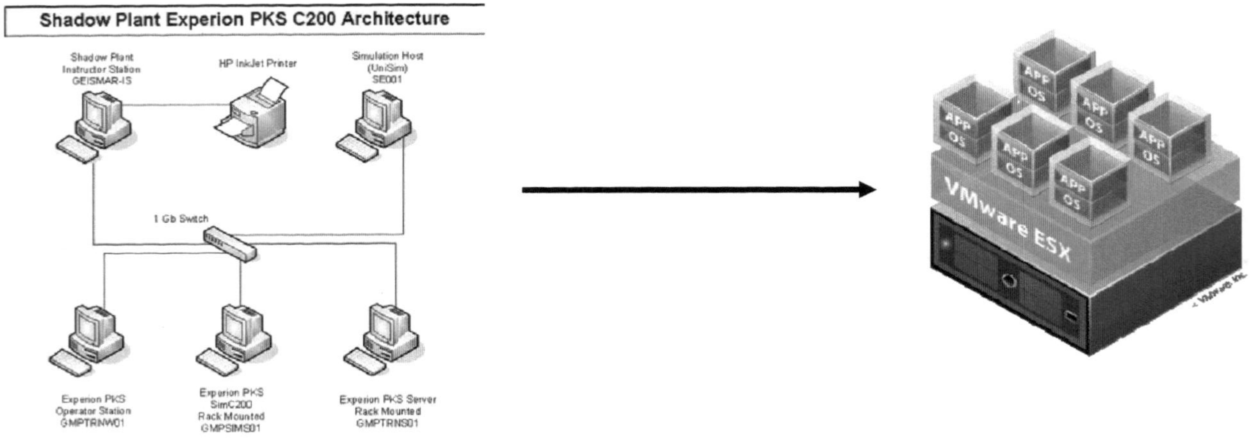

1. Virtualization of Operator Training & Production Systems at Honeywell Specialty Materials by John McIlwain

How Operator training is working

- **Effectiveness and Benefits of OTS has been well documented[2]**
  - **Operator Training System (OTS) applied to Fluorocarbon Plant**
    - **First Principles Models built in UniSim® coupled with Experion® System perfectly mimics actual control environment**
    - **Built-in and Custom Scenarios train operators for Start-Up, Shut-Down, plus Abnormal Situations.**
  - **Results:**
    - **Reduced time to qualify an operator from >18 months to six months.**
    - **Reduced start-up time by 20 hours on average.**
    - **Used OTS in virtual mode to commission new EPKS versions.**

OTS simulates virtual plants and helps operators

2. The Role of Simulator Technology In Operator Training Programs, Jose Alamo, Nov. 2010 Control Engineering

How a Training Simulator Helps Operations

Honeywell

- Highly Effective Training:
  - Looks & behaves just like the real plant (Like a Flight Simulator !)
  - No risk to failure
  - Endlessly repeatable
  - Visual: compare performance with best in class.
  - Motivates: fun to do
  - Builds confidence
- Very Specific & Practical Training:
  - Address start-up, shut-down, and recurring problems
  - Demonstrate impact of instrument & equipment failures.
- Develop better procedures:
  - Test improvements before they go on-line
  - Check impact of anticipated problems: power or service failures, etc.
- Identify and Solve Process Control Issues:
  - Makes control issues obvious
  - Allows us to test new strategies off-line

Operator Training System in Use

Instructors training operator using UniSim®/Experion® training screens

Increasing Operator Confidence in Procedures & Best Practices

Projected Financial Benefits

Honeywell

- Estimated Benefits:
  - Start-Up Speed: Gain 1 day production each product switch; Develop better procedures off-line. Std Work → Improve Std Work
  - Disturbance response improvement - Gain 7-10 hours production each major disturbance
  - Significant $$ benefit payback in about 1-2 years
  - This is in addition to the benefits obtained from steady-state and dynamic simulations used for various plant operations improvement

Conclusion

Honeywell

- This presentation highlights the development, validation and use of plant models from steady state to dynamic and then to operator training module development
- The advantage of steady state and dynamic in one simulation platform (UniSim®) helped ease the conversion from steady-state to dynamic
- In every stage there are advantages to the engineering operations
- Ultimately these models helped develop the Operator Training System for the plant
- Significant $$ benefit attained by improving operations in the plant

Acknowledgements

Honeywell

- Sanjoy Saha – Honeywell ACS
- Laurie Wang – Honeywell ACS
- Fred Gabriel – SM Plant Engineer
- John McIlwain – SM Control Engineer
- Charles Shivers – SM Operator training specialist
- Wendy Cart – SM operator training specialist

(111f) Use of Process Simulators for Control Studies and Operator Training at RasGas

Abdulla Al-Subaey[1], Ferdi Pangan[1], John W. Womack[2] and **Ahmet Pekediz**[3], (1)Ras Gas Company Limited, Doha, Qatar, (2)Markham Dynamics Corporation, Princeton, NJ, (3)Honeywell Process Solutions, London, ON, Canada

Ras Gas Company Limited (Ras Gas) operates LNG production as well as pipeline quality gas production facilities in the state of Qatar. RasGas has made extensive use of Process Simulators (POTS) during pre-project commissioning and later during normal plant operation. The primary use is for Distributed Control System (DCS) training of central control room Operators. The paper explains the training use of Process Simulator made by RasGas, and provides examples of other benefits.

See more of this Session: Best Practices Utilizing Simulation Tools, Pre and Post-Project

See more of this Group/Topical: Topical 7: 14th Topical on Refinery Processing

AUTHOR INDEX

AUTHOR INDEX

AUTHOR INDEX